Achieving sustainable production of poultry meat

Volume 2: Breeding and nutrition

It is widely recognised that agriculture is a significant contributor to global warming and climate change. Agriculture needs to reduce its environmental impact and adapt to current climate change whilst still feeding a growing population, i.e. become more 'climate-smart'. Burleigh Dodds Science Publishing is playing its part in achieving this by bringing together key research on making the production of the world's most important crops and livestock products more sustainable.

Based on extensive research, our publications specifically target the challenge of climate-smart agriculture. In this way we are using 'smart publishing' to help achieve climate-smart agriculture.

Burleigh Dodds Science Publishing is an independent and innovative publisher delivering high quality customer-focused agricultural science content in both print and online formats for the academic and research communities. Our aim is to build a foundation of knowledge on which researchers can build to meet the challenge of climate-smart agriculture.

For more information about Burleigh Dodds Science Publishing simply call us on +44 (0) 1223 839365, email info@bdspublishing.com or alternatively please visit our website at www.bdspublishing.com.

Related titles:

Achieving sustainable production of poultry meat Volume 1: Safety, quality and sustainability
Print (ISBN 978-1-78676-064-7); Online (ISBN 978-1-78676-066-1, 978-1-78676-067-8)

Achieving sustainable production of poultry meat Volume 3: Animal health and welfare
Print (ISBN 978-1-78676-072-2); Online (ISBN 978-1-78676-074-6, 978-1-78676-075-3)

Achieving sustainable production of eggs Volume 1: Safety and quality
Print (ISBN 978-1-78676-076-0); Online (ISBN 978-1-78676-078-4, 978-1-78676-079-1)

Achieving sustainable production of eggs Volume 2: Animal welfare and sustainability
Print (ISBN 978-1-78676-080-7); Online (ISBN 978-1-78676-082-1, 978-1-78676-083-8)

Chapters are available individually from our online bookshop: https://shop.bdspublishing.com

climate•SMART•publishing
IN AGRICULTURAL SCIENCE

BURLEIGH DODDS SERIES IN AGRICULTURAL SCIENCE

NUMBER 14

Achieving sustainable production of poultry meat

Volume 2: Breeding and nutrition

Edited by Professor Todd Applegate, University of Georgia, USA

burleigh dodds
SCIENCE PUBLISHING

Published by Burleigh Dodds Science Publishing Limited
82 High Street, Sawston, Cambridge CB22 3HJ, UK
www.bdspublishing.com

Burleigh Dodds Science Publishing, 1518 Walnut Street, Suite 900, Philadelphia, PA 19102-3406, USA

First published 2017 by Burleigh Dodds Science Publishing Limited

Library of Congress Control Number: 2016956458

British Library Cataloguing in Publication Data
A catalogue record for this book is available from the British Library

ISBN 978-1-78676-068-5 (print)
ISBN 978-1-78676-070-8 (online)
ISBN 978-1-78676-071-5 (online)
ISSN 2059-6936 (print)
ISSN 2059-6944 (online)

Typeset by Deanta Global Publishing Services, Chennai, India
Printed by Lightning Source

Contents

Part 2 Animal nutrition

Series list

Title	Series number
Achieving sustainable cultivation of maize - Vol 1 From improved varieties to local applications Edited by: Dr Dave Watson, CGIAR Maize Research Program Manager, CIMMYT, Mexico	001
Achieving sustainable cultivation of maize - Vol 2 Cultivation techniques, pest and disease control Edited by: Dr Dave Watson, CGIAR Maize Research Program Manager, CIMMYT, Mexico	002
Achieving sustainable cultivation of rice - Vol 1 Breeding for higher yield and quality Edited by: Prof. Takuji Sasaki, Tokyo University of Agriculture, Japan	003
Achieving sustainable cultivation of rice - Vol 2 Cultivation, pest and disease management Edited by: Prof. Takuji Sasaki, Tokyo University of Agriculture, Japan	004
Achieving sustainable cultivation of wheat - Vol 1 Breeding, quality traits, pests and diseases Edited by: Prof. Peter Langridge, The University of Adelaide, Australia	005
Achieving sustainable cultivation of wheat - Vol 2 Cultivation techniques Edited by: Prof. Peter Langridge, The University of Adelaide, Australia	006
Achieving sustainable cultivation of tomatoes Edited by: Dr Autar Mattoo, USDA-ARS, USA & Prof. Avtar Handa, Purdue University, USA	007
Achieving sustainable production of milk - Vol 1 Milk composition, genetics and breeding Edited by: Dr Nico van Belzen, International Dairy Federation (IDF), Belgium	008
Achieving sustainable production of milk - Vol 2 Safety, quality and sustainability Edited by: Dr Nico van Belzen, International Dairy Federation (IDF), Belgium	009
Achieving sustainable production of milk - Vol 3 Dairy herd management and welfare Edited by: Prof. John Webster, University of Bristol, UK	010
Ensuring safety and quality in the production of beef - Vol 1 Safety Edited by: Prof. Gary Acuff, Texas A&M University, USA & Prof.James Dickson, Iowa State University, USA	011
Ensuring safety and quality in the production of beef - Vol 2 Quality Edited by: Prof. Michael Dikeman, Kansas State University, USA	012
Achieving sustainable production of poultry meat - Vol 1 Safety, quality and sustainability Edited by: Prof. Steven C. Ricke, University of Arkansas, USA	013
Achieving sustainable production of poultry meat - Vol 2 Breeding and nutrition Edited by: Prof. Todd Applegate, University of Georgia, USA	014
Achieving sustainable production of poultry meat - Vol 3 Health and welfare Edited by: Prof. Todd Applegate, University of Georgia, USA	015

Introduction

Poultry production faces a range of challenges. These are addressed in the three volumes of *Achieving sustainable production of poultry meat.* The three volumes are:

- Volume 1: Safety, quality and sustainability
- Volume 2: Breeding and nutrition
- Volume 3: Health and welfare

Volume 2 discusses recent developments in breeding and improving poultry nutrition.

Part 1 Genetics and breeding

The first group of chapters looks at developments in genetics and breeding. Production of poultry meat and eggs have long been key targets for improvement in poultry breeding programmes. Functional traits can be defined as the morphological, biochemical, physiological, immunological and behavioral attributes considered essential for the optimal functioning of the individual bird. Historically, functional traits of poultry have received less attention in breeding. However, over the past decade, the changes in rearing practices and increased awareness of poultry welfare has led to broadening of selection programs to include more functional traits in genetic improvement programs.

Chapter 1 discusses the key functional traits of reproductive capacity, skeletal integrity, cardiovascular fitness and disease resistance. It reviews how advances in molecular biology and sequencing of the chicken genome have made it possible to identify quantitative trait loci (QTLs) and gene variants associated with functional traits. These developments have increased the understanding of the genetic basis of these traits, their relationships with production traits, and their potential for incorporation into genetic improvement programs.

Building on Chapter 1, Chapter 2 discusses issues in balancing production and functional traits in breeding. Achievement of commercial genetic potential for growth and yield characteristics does not always result in the fittest individual, as indicated by the negative correlation between growth and reproductive performance. The challenge in breeding is for the animal is able to achieve its genetic potential for all aspects of growth. As Chapter 2 points out, for genetic homeostasis, a population must maintain enough genetic variation to moderately perform with regard to growth, reproduction and immune response. In addition, the population must maintain enough heterozygosity to buffer against environmental challenges.

Traits of economic importance such as growth, yield and feed conversion ratio will continue to be the primary focus of breeding. However, new challenges, many driven by the consumer, will test the ability of the commercial broiler and layer to maintain genetic homeostasis. Examples include the move to cage-free systems, where there is increased opportunity for more bird to bird interactions in establishing the pecking order. This will lead to a shift in resource allocation away from production traits as birds expend energy while protecting territory. The elimination of antibiotics in poultry production will present new challenges with disease organisms. The development of organic and free range markets creates yet another set of environmental shifts that will redirect resources from growth and yield related traits.

As Chapter 2 discusses, the industry has matured from single trait or simple index selection to the present day selection program that monitors more than fifty traits. It is a hallmark of the modern poultry industry to produce a product focused on improved health and well-being. Therefore, half of the fifty traits are directed toward fitness. The remaining traits are focused on economically producing a rapidly growing, high yielding, feed efficient product. Improved monitoring of these traits and how they can be balanced will allow for further refinement of breeding decisions.

Chapter 3 explores new developments in breeding to improve functional and other traits. Marker assisted selection (MAS) is a form of indirect selection that depends on the accuracy of measuring the marker and the genetic correlation between them. Chapter 3 describes the development of microsatellite markers and single nucleotide polymorphic (SNP) markers associated with high throughput automated technology, which has made MAS applicable to whole genome prediction of breeding values. The chapter explores how these techniques have been successfully implemented in commercial poultry breeding programmes. The development of technology for rapid genotyping of large numbers of DNA markers as SNPs cheaply and on large numbers of individual birds has made the application of MAS to commercial poultry breeding programmes both feasible and practical. The benefits are considerable in terms of improved accuracy of estimation of breeding values in combination with phenotypic measurements. Considerable benefits have been realised for traits with low heritability or measured in one sex such as egg production. These advances could also lead to a reduction in the generation interval in males of egg-laying strains, for example, where early selection decisions based on predicted merit before sexual maturity could be made.

The chapter also highlights how future technological developments may permit the use of whole genome sequencing rather than high density SNP chips for prediction of breeding values of selection candidates. As the chapter points out, genomic selection in poultry breeding programmes promises to increase the accuracy of selection, especially for traits that are expensive to measure (such as feed conversion ratio) or measured late in the life of an individual (such as female fertility). Additional benefits could include the incorporation of welfare traits with no loss of gains in production traits because of the greater accuracy of estimating breeding values.

Part 2 Animal nutrition

The second group of chapters reviews key developments in understanding and optimising poultry nutrition. As Chapter 4 points out, feed is one of the most significant costs in animal production, and feed efficiency is therefore a very important genetic trait in livestock production. A clear link between breast muscle mitochondrial function within cells and feed efficiency has been reported in poultry. This chapter provides an overview of the mitochondrial processes which occur in muscle cells, presents the evidence that enhanced mitochondrial functions lead to high feed efficiency, and then considers the role of enhanced nucleotide metabolism and muscle cytoarchitecture in the feed efficiency of broilers. As it points out high feed efficiency is associated with factors such as enhanced mitochondrial energy (ATP) production and enhanced capability for synthesis and metabolism of purine and pyrimidine nucleotides. These insights provide new opportunities for breeding and other interventions to further improve feed efficiency,

Chapter 5 points out that selection for phenotypic feed efficiency has tremendously improved livestock productivity over the past 50 years. However, echoing themes in Part 1, the chapter points out that, associated with this success, there have been a number of undesirable changes in the regulation of energy homeostasis and water balance. Feed and water efficiency encompasses complex mechanisms regulating feed and water intake, energy expenditure, water retention and excretion, and intermediary metabolism related to nutrient and water utilization and partition. Knowledge of these should be used to guide more effective selection.

Chapter 5 focuses first on feed intake regulation, reviewing current understanding of both central feed intake regulation and peripheral and hormonal regulation. The chapter then turns to the issue of the regulation of water homeostasis. The chapter reviews research which may help in developing new strategies to improve both feed and water efficiency. New techniques involving genomics, epigenetics, proteomics, transcriptomics, mobilomics and metabolomics are helping to show the relationships between nutrients, water, genes and performance. These approaches have the potential to lead to more targeted management based on optimisation of nutrient and water intake fine-tuned with an animal's genetic profile. The identified molecular signatures could subsequently be used to improve water and feed efficiency via genetic selection, nutrition and livestock management.

As noted earlier, feed represents a major cost of poultry production. Broilers and layers are highly efficient in converting feed to muscle, but they still excrete significant amounts of unutilised nutrients. There remains therefore considerable room to improve the efficiency of conversion of feed to animal products. A good portion of this inefficiency results from incomplete digestion and/or utilisation of nutrients. Chapter 6 provides an authoritative overview of major advances in poultry feeding to overcome these challenges. Key topics discussed include advances in: understanding of nutrient metabolism and nutrient requirements; quantification of the availability of nutrients in raw materials; formulation of least-cost diets that bring nutrient requirements and nutrient supply together in an effective manner; the contribution of new feed additives; and progress in feed processing. The chapter provides a context for a number of following chapters which discuss specific aspects of nutrition. Advances include improvements in composition and ingredient quality, better feed formulation (covered in Chapter 12), the use of additives such as crystalline and synthetic amino acids (discussed in Chapter 7), feed enzymes (discussed in Chapter 8) and probiotics for better gut function and health (surveyed in Chapter 10), as well as improvements in modelling feed efficiency (reviewed in more detail in Chapter 11).

Dietary amino acids are central to optimizing growth performance, meat yield, and egg production of poultry. Chapter 7 reviews recent research on amino acid digestibility coefficients for feed ingredients, digestible amino acid requirements of poultry based on production efficiency, and the role of supplementation of crystalline and synthetic amino acids on nitrogen balance and ammonia output of poultry. The chapter discusses specific amino acids such as arginine, threonine, and branch chain amino acids. As the chapter points out, significant advances have been made in understanding factors influencing amino acid digestibility coefficients for poultry. However, amino acid requirements for immune function, disease/microbial load, and physiological needs may differ from growth performance and meat yield, providing a target for future research in the move to antibiotic-free poultry production.

Chapter 8 reviews the current status of research on feed enzymes with an emphasis on identifying the key challenges researchers face in terms of enzyme development, mechanisms of action and enzyme efficacy. The key chapter discusses current drawbacks and opportunities in the application of phytase, carbohydrases, protease and their combinations in poultry nutrition. Sections cover the advances and continuing challenges in the application of particular enzymes. The chapter looks first at the use of phytase in poultry diets, reviewing research on efficacy, phosphorus content and environmental impacts. It then discusses non-starch polysaccharides (NSP) and NSP enzymes, including their physiological effects, and prebiotic potential. The chapter also discusses β-mannanase in poultry nutrition, starch digestion and supplemental α-amylase as well as microbial protease supplementation. The chapter shows the potential of enzymes in both improving feed efficiency and making poultry production more sustainable.

Chapter 9 discusses key advances in understanding the role of phytate in phosphorus and calcium nutrition of poultry. Poultry depend upon a continuous supply of phosphates for the formation of bones as well as other physiological functions. In plant seeds, phytic acid (InsP6) is the primary storage form of phosphorus (P), and it is usually present in salt form (phytate). The move to minimize the use of feed phosphates makes it imperative to better understand the interacting factors related to InsP6 breakdown in the digestive tract. As the chapter points out, the potential to utilize InsP6-P is very high in broiler chickens. However, degradation of InsP6 in the gastrointestinal tract is variable and affected by supplements of calcium, P, and other dietary factors. The chapter discusses ways of adjusting feed ingredients and supplements of P, calcium, and phytase to optimize feed formulation.

Optimising gut function and immunity is an important goal for poultry producers to improve bird productivity, health and food safety. Probiotics and prebiotics are attractive approaches to use in the pursuit of optimal gut health, especially with the ongoing need to reduce the use of in-feed antimicrobial growth promoters. Chapter 10 reviews research to address the three main obstacles to the use of probiotics in particular: concerns about effectiveness and reproducibility of action; concerns about lack of knowledge regarding mechanisms of action; and making an informed choice about which product to use from amongst the many that are available. As Chapter 10 points out, recent research has increased our knowledge of the effects of both probiotic and prebiotic treatment, their possible modes of action, and the strengths and limitations of their use. The chapter summarises studies on why some products may give variable outcomes and what may be done to further validate the performance of existing products. It also explores ways of developing a new generation of more reliable and effective probiotics and prebiotics.

Animal nutritionists face various problems when formulating feeds for poultry. Advances in simulation modelling have made it possible to look into all aspects of production when formulating feeds for animals. Chapter 11 looks at the use of simulation models to optimize poultry nutrition. It discusses modelling methods, their strengths and weaknesses. The chapter then reviews ways of predicting responses of poultry to nutrients and predicting food intake. The chapter also describes the methods used to predict potential laying performance of hens and the environmental factors that affect feed intake. Finally, this chapter shows how models can be used to optimize feeding programmes.

Systematic evaluation of each stage of the feed manufacturing has the potential to identify opportunities for improvement in manufacturing efficiency and reduced nutrient variation in finished feed. This will ultimately result in lower cost sustainable poultry production. Chapter 12 examines the role of automation technology in composing and delivering feed, and addresses the issues of batching, mixing and pelleting feed as well as means of assessing feed quality. It discusses such issues as particle size reduction in feed, improving feed pelleting as well as post-pellet liquid application.

Soybean and canola meal are the conventional ingredients used to provide protein in poultry feed. However, they are relatively expensive and must be imported to many poultry producing areas. Developing alternative protein sources for poultry nutrition will reduce the pressure on these key protein sources, as well as promoting the development and sustainability of the poultry industry. Chapter 13 first reviews the supply of conventional protein sources for poultry, and then considers the range of alternative protein sources which might be developed. These alternative sources include grain by-products, oil seed and fruit by-products, grain legumes or pulses, algae and duckweed. The chapter considers how birds respond to diets containing alternative protein sources and current constraints on the use of these sources. Finally, it provides recommendations for improving the nutritive value of these alternative sources of nutrition.

Contaminants in poultry feed may result in deteriorated feed quality, reduced performance and increased incidence of disease in poultry as well as potential safety issues. Chapter 14 focuses on those contaminants considered to pose the most significant risk to poultry and human health: mycotoxins, dioxins, and bacterial pathogens. It discusses both the risks from these contaminants as well as research on current best practice to control feed contamination, including good manufacturing practices (GMPs) and Hazard Analysis and Critical Control Point (HACCP) systems in feed sourcing and manufacturing. It also discusses particular techniques for managing contaminants. As an example it discusses the use of binding agents to control of mycotoxins, limitations in their efficacy and novel techniques such as the use of microorganisms and their enzymes to detoxify specific non-absorbable mycotoxins. The chapter also discusses control of pathogens using natural feed additives such as prebiotics, probiotics as well as phytogenic feed additives. As the chapter points out, genomics, transcriptomics, and proteomics are now transforming our approaches to the detection, prevention, and treatment of biological feed contaminants. These advances are also allowing the development of control measures and treatments that are more specific in terms of their targets.

As pointed out in Chapter 5, enhanced feed efficiency and increased growth rates have led to a reduced ability in poultry to balance energy expenditure and maintain body water balance under variable environmental conditions. Chapter 15 focuses on the effect of combinations of environmental conditions (temperature, ventilation, relative humidity) on the thermal status and performance parameters of broilers, turkeys and laying hens. It discusses body temperature control by endothermic birds, including neuronal and endocrine regulation. It then reviews what we know about different strategies used by birds to cope with changes in temperature, including physiological and cellular responses to changes in the environment. Finally, it highlights the role of epigenetic temperature adaptation during embryogenesis as a tool to improve poultry tolerance to heat.

Part 1

Genetics and breeding

Chapter 1

Genes associated with functional traits in poultry: implications for sustainable genetic improvement

Samuel E. Aggrey, University of Georgia, USA; Fernando González-Cerón, Chapingo Autonomous University, Mexico; and Romdhane Rekaya, University of Georgia, USA

1 Introduction

Poultry production primarily focuses on the supply of saleable products, for example, meat and eggs since they are the primary sources of income for most poultry producers. As such, meat and eggs are classified as production traits, and have been the focus of improvement in poultry breeding organizations. Genetic improvements in production traits over the last six decades have been very successful and have led to a significant increase in productivity (Havenstein et al., 2003). Chick survival is essential to productivity and necessitates the normal functioning of basic biological mechanisms under reasonable environmental conditions. Functional traits of birds have not received much attention compared to production traits due to several reasons, including (1) availability, expense and practicality of direct measurements; (2) accuracy of indirect measurements; (3) low heritability and accuracy of genetic parameters; and (4) negatively correlated with production traits. Functional traits may be defined as the morphological, biochemical, physiological, immunological and behavioural attributes that are considered to be essential

http://dx.doi.org/10.19103/AS.2016.0011.02

for the optimal functioning of the individual and also the response of such individuals to the environment. Functional traits are therefore at the crossroad between responses to production traits and the environment. A suboptimal production environment directly limits the expression of functional traits and indirectly affects production responses. Conversely, genetic improvements in production traits could adversely affect functional traits and limit a bird's ability to thrive optimally in the production environment.

Over the past decade, the decline in some functional traits and increased awareness of poultry welfare has led to the inclusion of functional traits in genetic improvement programmes (Katanbaf and Hardiman, 2010). The pendulum of poultry breeding has swung from exclusively selecting based on production traits towards more balanced breeding plans, which includes selection on both production and functional traits. Table 1 lists functions associated with functional traits and some of the methods for indirect measurement. It does not list every possible functional trait in poultry but rather the ones that can be measured directly or have some form of indirect measurement. In most cases, functional traits associated with cardiovascular fitness can be accurately observed *post-mortem*. Therefore, we generally accept the lack of observation of those functional traits as an indication of proper cardiovascular capacity. In the same way, we assume the lack of observation of clinical symptoms of diseases to be an indication of health, as sub-clinical expressions are difficult to observe.

Advances in molecular biology and sequencing of the chicken genome (International Chicken Polymorphism Consortium, 2004) have made it possible to identify quantitative trait loci (QTLs) and gene variants associated with functional traits, raising the possibility of understanding their genetic basis, their relationships with production traits and their potential for incorporation into genetic improvement programmes. Advances in

Table 1 Function and functional traits in poultry

Biological function	Functional trait	Indirect measurement
Reproductive capacity	Egg number and size	
	Fertility	Sperm count, mobility, natural ability to mate
	Hatchability	Embryonic development
	Sexual maturity	Age at first egg
		Nesting
Skeletal integrity	Bone strength	X-ray lixiscope
	Leg structure and shape	Valgus–varus, angular, twisted
	Osteoporosis	Fractures
Cardiovascular fitness	Ascites	Pulse oximeter
	Pulmonary hypertension	
	Sudden death	
Health	Disease resistance	Antibody response to antigens
		Mortality
		MHC haplotypes

high-throughput technologies have made it possible to generate genome-wide sequence variants, gene and protein expression levels, metabolites and epigenetic changes of several tissues under different experimental conditions. The establishment of an international consortium for the functional annotation of domestic animal genomes (FAANG) to accelerate the genome-to-phenome efforts (Andersson et al., 2015) will delineate the molecular, physiological and environmental mechanisms that underlie traits of economic importance in poultry and other species. A better understanding of the biology of traits and how they genetically interrelate would further enhance the possibility for incorporation of functional traits into breeding goals. In this chapter, we will review how extensive genetic information on traits that control saleable products and functional traits and their genetic relationships could be used to implement sustainable genetic improvement.

2 Reproductive capacity

The egg-laying mechanism is part of the reproductive system of a hen. Prior to the advances in molecular tools in the 1990s, only a few gene variants were identified to be associated with parts of the reproductive system of the hen. Sexual maturity as a trait is difficult to measure, and age at first egg (AFE) has been used as an indirect measure of sexual maturity. Egg production is usually measured as egg number (EN), egg production rate (EPR) or hen day rate (HDR). With the availability of genetic markers (Aggrey and Okimoto, 2003) several QTLs have been identified. Several AFE QTLs have been identified on chromosomes (GGA) 1–7, 16 and 23 (Tuiskula-Haavisto et al., 2004; Podisi et al., 2011; Goto et al., 2011; Liu, 2011; Shen et al., 2012; Zhang et al., 2012; Yuan et al., 2015). Some variants in the growth hormone (GH), GH receptor (GHR) and insulin-like GH genes have been found to be associated with the onset of ovulation and AFE (Kuhnlein et al., 1997; Feng et al., 1997). A single-nucleotide polymorphism (SNP) in the odd Oz/ten-m homolog 2 (ODZ2) (also known as Teneurin 2) was associated with AFE (Liu et al., 2011). Bone morphogenetic protein 15 (BMP15) is an oocyte-secreted growth factor required for ovarian follicle development and ovulation. SNPs detected in the BMP15 gene were also found to be associated with egg production-related traits in chickens (Han et al., 2015). Polymorphisms in the BMP15 and growth differentiation factor 9 (GDF9) were found to be associated with AFE in a maternal chicken line. SNPs in the GH-releasing hormone receptor (GHRHR) gene were also found to be associated with EN in chickens (Liu et al., 2012). Haplotypes from the IGF-binding proteins 1 and 3 and signal transducers and activators of transcription 5B (STAT5B) genes were found to be associated with sexual maturity in chickens (Ou et al., 2009).

Polymorphisms in several other genes have been found to affect different aspects of reproductive capacity (Jiang et al., 2005; Qin et al., 2015a,b). Polymorphisms in BMP15 exon1 and GDF9 exon2 were detected by DNA sequencing and Polymerase chain reaction-restriction fragment length polymorphism (PCR-RFLP). Three SNPs were detected in BMP15 (A111G, C231T and C34T) and GDF9 (G593A, T824C and C896T). C34T leads to the substitution of Leu by Phe, which was predicted to affect protein function. Results of the association analysis indicated that C34T had an effect on total egg production at 300 d of age (EN) and age at first laying (AFE). G593A affected EN, and both C231T and C896T influenced AFE. The TGC1TGC1 diplotype in BMP15 had the highest EN. Collectively, these associations results point to the possibility of improvement of EN through selection for markers in the BMP15 gene in maternal lines of Shaobo hens (Huang et al., 2015).

Follicle-stimulating hormone and its receptor (FSHR) play an important role in follicular development. Kang et al. (2012) reported associations of a major insertion/deletion (INDEL) marker in the FSHR gene with AFE and EN. Other SNPs in the FSHR associated with EN have also been identified (Li et al., 2011). Several QTLs have been identified for either EN at a specified age or HDR. Multiple QTLs for EN have been identified on GGA 1–10 and on GGAZ. Hansen et al. (2005) identified a major EN QTL on GGA1. Using a multigenerational population, Atzmon et al. (2008) also reported a major QTL for EN and EPR on GGA1, GGA3, GGA5 and GGAZ. Xu et al. (2011) identified a major QTL for EN at 300 d of age, and further identified the vasoactive intestinal peptide receptor 1 (VIPR1) as the putative candidate gene for this QTL. A major QTL for EN was identified on GGA3 by Zhou et al. (2010), and the VIP as the putative candidate gene. Other significant QTLs on GGA4–10 and GGAZ for EN have also been found (Tuiskula-Haavisto et al., 2002; Fatemi et al., 2012; Goraga et al., 2012; Shen et al., 2012; Huang et al., 2015; Qin et al., 2015a). Several EPR QTLs have been identified on GGA 9 (Uemoto et al., 2009), GGA11 (Goto et al., 2011), GGA17 (Atzmon et al., 2007), GGA14 and GGA27 (Wolc et al., 2014). A pleiotropic QTL was identified for egg weight on GGA4, 11, 27 and Z (Sasaki et al., 2004). In addition to QTLs, several genetic markers in candidate genes have been found to be associated with egg production.

The ovocalyxin 32 (OCX32) gene that encodes for a matrix protein found in the outer layers of the eggshell and also in the cuticle has been found to be associated with several egg production and egg quality traits (Dunn et al., 2009; Takahashi et al., 2010). Uemoto et al. (2009) identified two novel non-synonymous polymorphisms (c.267T>G and c.494A>C) and one known non-synonymous polymorphism (c.381G>C) in the coding sequences of the chicken OCX32 gene. All three polymorphisms were found to be associated with egg production. Fulton et al. (2012) demonstrated strong associations between SNPs in the OCX32 gene and several egg production traits in commercial poultry lines. A 24 bp indel in the promotor region of the prolactin (PRL) gene has been found to be associated with egg production (Cui et al., 2006). Markers in the GH, GHR and IGF-I genes were found to be associated with persistency of lay (Kuhnlein et al., 1997), eggshell and egg weight (Nagaraja et al., 2000). IGF and IGFR were reported to regulate follicular development and ovarian functions in chickens (McMutry et al., 1997). Most recently, Yuan et al. (2015) reported associations between markers in the general transcription factor IIA, 1, 19/37kDa (GTF2A1) and the claspin (CLSPN) genes to be significantly associated with EN. A SNP in the growth factor receptor-bound protein 14 (GRB14) was found to be associated with EN (Liu et al., 2011). Transcription factor forkhead box L2 (FOXL2) and GDF9 genes have critical roles in the regulation of hen ovarian development. SNPs in the FOXL2 and GDF9 genes were significantly correlated with egg production and egg weight in laying hens (Qin et al., 2015b).

Improvement in egg production in commercial layers has been successful partly due to the inhibition of incubation behaviour or broodiness. The PRL gene is known to be critical for the onset and maintenance of reproductive behaviour in avian species. The prolactin receptor (PRLR) plays an important role in the PRL signal transduction cascade. Genetic markers in the PRL and PRLR genes have been shown to be associated with broodiness in chickens (Jiang et al., 2005). The dopamine D1 receptor (DRD1) has also been shown to regulate reproductive behaviour in birds. SNPs in the DRD1 gene have been shown to be associated with egg production and broodiness in chickens (Xu et al., 2010a). Genetic variants in the dopamine D2 receptor gene have been found to be associated with the duration of broodiness (Xu et al., 2010b). The SLIT/ROBO pathway has been suggested to be involved in the pre-hierarchical follicular development of the hen ovary by an intra-follicular autocrine and/or paracrine action, and is influenced by activin A and inhibin A hormones (Qin et al., 2015).

Brooding is an integral function of reproductive behaviour exhibited in avian species. Whereas broodiness is observed in unimproved breeds and strains, the behaviour is either eliminated or significantly reduced in commercial egg-laying strains (Hutt, 1949). The breed differences in broodiness and the reduction in broodiness associated with increased selection for ENs demonstrate that the trait is hereditary (Hutt, 1949; Nestor et al., 1996). Romanov (2001) reviewed a long-held hypothesis that broodiness is polygenic with a major contribution from the Z chromosome. Romanov concluded that broodiness in chicken is not controlled by a major gene (or genes) on the Z chromosome and speculated that major autosomal genes may be contributing to the expression of this trait. Basheer et al. (2015) reported that loci on chromosomes 1, 8, 13, 18 and 19 and linkage group E22C19W28 contribute towards the variation in broodiness. They further asserted that an over-dominant locus on chromosome 5 coincides with the strongest selection sweep in chickens and together with loci on chromosomes 1 and 8, and genes of the thyrotrophic axis (thyroid hormones) may play a role in the loss of broodiness in the White Leghorn breed. Recently, Shen et al. (2016) reported that the PRL, CGA, PGR, RLN3 and GRP genes initiate and maintain brooding behaviour in chickens. Jiang et al. (2005) reported that a 24-bp insertion in the promoter region of the PRL gene was strongly associated with broodiness in chickens.

Regarding male fertility, Froman et al. (1999) demonstrated that sperm mobility is a primary determinant of fertility in chicken. Froman and Roads (2013) suggested that markers on chicken chromosome Z may influence sperm mobility. A genetic variant in the mitochondrial gene encoding for $tRNA^{arg}$ was found to be associated with sperm motility (Froman and Kirby, 2005). Using a proteomic approach, several genes (phosphoglucose isomerase, phosphofructokinase, aldolase enolase, pyruvate kinase and lactate dehydrogenase) encoding for glycolytic enzymes were found to be differentially expressed between roosters divergently selected for sperm mobility (Froman et al., 2011).

Using microarray technology 54 and 84 genes were differentially expressed between germinal disc regions of F1 maturation stage oocytes from hens exhibiting either high (100%) or low (from 22% to 80%) fertility rate from laying and meat lines, respectively (Elis et al. 2009). The same study concluded that the mechanisms involved in the decrease in fertility rate of laying and meat-type lines were independent. Among the major genes were VWC2, CR407412, TAPA, FGL2 and TRAP6.

Optimal hatchability is needed for a successful poultry operation. Hatchability is the proportion of fertile eggs that develop into viable chicks and is a complex interaction between the developing embryo and incubation properties. With standardized incubation properties, hatchability is more related to the development of the embryo than the incubation process. Thus, genes affecting embryonic development, viability and mortality will affect hatchability.

The molecular mechanisms that underlie the oocyte to embryo transition depend exclusively on maternal RNAs and proteins accumulated during growth of the oocyte (Evsikov et al., 2006). During this transition period, certain genes become essential for fertilization, first cleavage and embryonic genome activation. The BTG anti-proliferation factor 4 (BTG4), chicken c-mos proto-oncogene (CHKMOS), similar to Wee 1A kinase (WEE), similar to zona pellucida A (ZPA), deleted in azoospermia-like, cvh, similar to zygote arrest 1 (ZAR1) and similar to Kruppel-like transcription factor neptune (KTFN) genes are essential for the maturing oocytes. Also, the chicken vasa homolog protein (CVH), ZAR1 and KTFN are essential for early embryonic development (Elis et al., 2008).

It has been demonstrated that mutations in the ovalbumin and riboflavin-binding protein genes increase the mortality of embryos (Winter et al., 1967; Inafuku et al., 1997). SNP markers in the ovomucoid (Huang et al., 2011), proline-rich nuclear receptor coactivator (PNRC)-1 (Chang et al., 2012) and ovalbumin (Huang et al., 2013) genes were found to be associated with hatchability in ducks. An SNP in the lysozyme gene was significantly associated with hatchability in Tsaiya ducks (Huang and Cheng, 2014).

The usefulness of genetic variation in genes associated with traits of interest should be further evaluated within single- and multiple-trait selection setups. First, 'causative' variants harboured in genes associated with reproductive capacity needs to be determined and validated. This is not a trivial task, especially in the presence of multiple variants with small effects and in high linkage disequilibrium. Second, applied breeding programmes include several traits in selection indices. Consequently, it is equally important to assess the relationships between the aforementioned variants and other traits including production and disease responses.

3 Skeletal integrity

Commercial birds need a developed skeletal structure with sufficient strength to carry the muscle load, ensure reasonable walking abilities, and limit fractures and bone breakages during processing. This makes bone quality and skeletal maladies important in poultry breeding. Several QTLs have been identified for several bone quality traits, including bone morphology, tibia dyschondroplasia (TD) and other skeletal deformities (Hu et al., 2013; Zhou et al., 2007; Zhang et al., 2011). In the study by Zhou et al. (2007), they identified fibroblast growth factor 9 (FGF9), Sox-10, bone morphogenetic protein receptor type II, transforming growth factor-β receptor 3, bone proteoglycan 2, osteocrin, angiotensin II, progesterone receptor, tumour necrosis factor-α, calcitonin receptor and type IIb Na-P co-transporter as candidate genes for bone-related traits. Rubin et al. (2007) reported the involvement of WD-repeat containing protein 5 (WDR5), Wnt inhibitory factor 1 (WIF1), syndecan 3 (SDC3) and immunoglobulin-like receptor CHIR-A (CHIR-A) in bone metabolism. Wu et al. (2003) demonstrated *in vitro* the presence of the Fas receptor in avian-derived osteoclasts whose activity is important for bone turnover.

Li et al. (2003) investigated the role of transforming growth factor-β genes (TGF-β) in skeletal traits of birds belonging to an experimental population, founded with broiler sires and Leghorn and Fayoumi dams. They found that in the F_2 population, TGF-β2 birds homozygous for the RFLP polymorphisms had higher BMC and BMD than Leghorn homozygous birds; likewise, TGF-β4 broiler homozygous birds had higher BMD expressed as percentage of body weight at 8 weeks of age than Fayoumi homozygous birds. In a study carried out with chickens, Houston et al. (2004) reported the role of the novel phosphatase PHOSPHO1 in bone and matrix mineralization. In diaphyseal cortical bone, authors localized PHOSPHO1 in the osteoid layer of the periosteum but not in the endosteum and closed osteons. Likewise, PHOSPHO1 was located in hypertrophic chondrocytes and the ossification groove of Ranvier in the growth plate cartilage. Li et al. (2005) studied the association between very low-density apolipoprotein-II (apoVLDL-II) gene, which participates in chicken lipid transportation, and skeletal traits in the F_2 population of birds described in Li et al. (2003). They found that apoVLDL-II broiler homozygous birds for the RFLP polymorphisms had greater tibia length (TBL) than Fayoumi homozygous birds. However, apoVLDL-II broiler homozygous birds had lower BMD and TBL percentages expressed relative to body weight at eight weeks of age, than Fayoumi homozygous birds. Vitamin D receptor, insulin (INS), insulin-like growth factor 1 (IGF1) and osteopontin (SPP1) genes were also found to be associated with bone strength traits in a broiler x layer F_2 population (Bennett et al., 2006).

Several researchers have investigated the relationship of multiple genes, such as GHR, OPN, MMP9, MMP13, type X collagen, visfatin, RB1 and Hox genes, with bone length. Langhorst and Fechheimer (1985) studied the recessive shankless mutation (shl) located in the proximal region of chromosome 2 and whose origin is an X-ray-induced pericentric inversion. They found that in addition to several bone malformations, mutation increases TBL. Burnside et al. (1992) studied the GHR gene in chicken having the sex-linked dwarf mutation that causes reduction in the long bone growth. Results indicated the dwarf phenotype results from deficient production of a functional GHR. Reich et al. (2005) investigated the role of mechanical stimuli, by loading 10% extra body weight, on the development of long bones in broiler chickens during the first week of age. Mechanically stimulated chicks showed overexpression of osteopontin (OPN), and matrix metalloproteinases-9 (MMP9) and -13 (MMP13) genes in birds with shorter and narrower long bones phenotype. The retinoblastoma 1 (RB1), G-protein-coupled purinergic receptor (P2RY5), fibronectin type III domain containing 3A (FNDC3A), motilin receptor (MLNR) and calcium-binding protein 39-like (CAB39L) genes were found to be associated with bone morphology (Zhang et al., 2011). It was found that the expression of Hoxd11 and Hoxd12 genes in the zeugopod region of the hindlimb in chicken embryos disappeared before the cartilage formation and, in general, the expression pattern of Hox genes characterizes the chicken hindlimb (Kamiyama et al., 2012).

Loveridge et al. (1993) studied the expression of the c-myc proto-oncogene and the transforming growth factor-β3 (TGF-β3) gene through immune-localization of their proteins in chondrocytes of the tibial growth plate from broiler chickens at three weeks of age. While chondrocytes in transition from the proliferative to the differentiation phase from TD-affected chickens had reduced expression of both genes with respect to healthy chickens, proliferative chondrocytes had normal expression of the c-myc proto-oncogene. Some studies using gene expression and dot-blotting hybridization have shown that transforming growth factors-β (TGF-β), type X collagen, Indian hedgehog (IHH), type II collagen, aggrecan, type X collagen, glyceraldehyde 3 phosphate dehydrogenase (GAPDH), transglutaminase 2 (TGMse), runt-related transcription factor (Runx2), matrix MMP-2 and -13, and vascular endothelial growth factor (VEGF) and parathyroid hormone-related protein (PTHrP) are involved in growth plate cartilage formation (Warman et al., 1993; Law et al., 1996; Velleman, 2000; Webster et al., 2003; Rath et al., 2000, 2004, 2005, 2007). Also, the expression of 3 iodothyronine deiodinases (types 1 (DIO1), 2 (DIO2) and 3 (DIO3)), that regulate the availability of 3,3′,5-triiodo-L-thyronine (T_3) (a thyroid hormone participating in chondrocyte differentiation and maturation), VEGF, PTHrP and IHH correlated with TD (Shen et al., 2004).

Several diet-induced models have been used to study the involvement of genes in TD. MMPs-2, -3, -9 and -13; collagen type X (Col X); pro-alpha-1 collagen type I (Col I α1); collagen type IX (Col IX); NADH dehydrogenase (NADH DH); cytochrome C oxidase subunit III (COX III); enolase 1-alpha (ENO1); carbonic anhydrase II (CA2); and heat shock protein 90 (Hsp90) were shown to be upregulated in TD-affected growth plates. Matrilin 3 (MATN3) and chondromodulin-I (ChM-I) were downregulated in the same growth plates. Peptidylprolyl isomerase B, alpha-enolase, G protein, collagen type II precursor and origin recognition complex subunit 1 proteins were found to be downregulated in TD-affected growth plate tissue (Hasky-Negev et al., 2008; Dan et al., 2009; Tian et al., 2009; Velada et al., 2011). Tong et al. (2003) studied the role of the gelatinase B/MMP-9 in avian growth plate angiogenesis and endochondral ossification. Using tibia growth plates from 14-day-old embryos for *in situ* hybridization, they revealed that gelatinase B was expressed by cells surrounding the blood vessels penetrating the growth plate, endothelial cells in the vessels, adjacent chondrocytes to the vascular invasion and cells in the chondro-osseous junction.

Stewart et al. (2006) isolated chondrocytes from three-week-old male chickens and separated them according to their maturational phenotype to study the role of hypoxia-inducible factors (HIF), which participates in the activation of hypoxia-responsive genes, in chondrocyte differentiation, blood vessel formation and endochondral ossification. Results from semi-quantitative RT-PCR indicated an increased expression of VEGF and HIF-2α, during chondrocyte differentiation. Teixeira et al. (2011) investigated the growth plate mineralization mechanism by comparing gene expression profiles of mineralizing and non-mineralizing cultures of chick limb-bud mesenchymal stem cells. Microarray data indicated that expression of genes such as secreted phosphoprotein 24 (SPP2), Na-Pi II co-transporter, bone morphogenic protein 1 (BMP1), extracellular matrix protein dentin matrix protein 1 (DMP1), matrix extracellular phosphoglycoprotein (MEPE), calbindin 29, matrilin, sonic hedgehog and low-density lipoprotein receptor-related protein 6 (LRP6) increased with mineralization. On the contrary, the expression of genes such as fibroblast growth factors 3 (FGF3) and 8 (FGF8), a disintegrin and MMP with thrombospondin motifs 1 (ADAMTS-1), cathepsin O, protocadherin and aggrecanase decreased with mineralization.

Most traits are genetically determined, and understanding the genes underlying skeletal traits is as important as the underlying additive genetic variation associated with these traits. Genetic variability in genes in breeding populations offers the potential to include such traits in breeding strategies for their improvement.

4 Cardiovascular fitness

Ascites is an increase in the amount of lymph (pale yellow and clear fluid) normally found in the peritoneal spaces (abdominal cavity). The lymph could contain clumps or strands of fibrin when there is increased protein in the lymph or oedema fluid (Julian, 1985). Meat-type chickens may have up to 3 ml of fluid in the pericardial sac at six to eight weeks of age, but fluid levels over 4 ml is likely considered to be abnormal (Julian, 1993). According to the same author, ascites is not a disease but may result from four physiological changes that cause an increased production or decreased removal of peritoneal lymph – (1) obstruction of lymph drainage, (2) decreased plasma oncotic pressure, (3) increased vascular permeability and (4) increased hydraulics in the blood pressure system. Ascites can also result from increased blood flow or increased resistant flow in the lungs, valvular insufficiency, and RVF following right ventricular hypertrophy and dilation from pulmonary hypertension (PH) at high altitudes. In low altitudes, primary or spontaneous PH resulting from insufficient capacity of the pulmonary capillaries could also lead to ascites. From an integrative analysis of transcriptomic and metabolomic profiling of ascites syndrome, Shi et al. (2014) showed that two biological pathways of tryptophan biosynthesis and metabolism, and glycerol–phospholipid metabolism may contribute to the induction of ascites in broilers. In a genome-wide transcriptome study, CYP1B1, ALDH7A1, MYLK, CAMK4, BMP7 and INOS were found to be upregulated in the pulmonary artery of ascites broilers (Yang et al., 2016). Liu et al. (2016) reported that 20 miRNAs that correlated with 18 target genes appeared to be involved in pulmonary artery remodelling mainly in three broad physiological processes: the hypoxia-sensing response (HIF1α,NHE1, STAT5 and STAT3), endothelial permeability dysfunction (CD44, TRAF2, CDK2AP1, LZTFL1, JAZF1, PEBP1, LRP1B, RPS14 and THBS2) and inflammation (MEOX2, STAT5, STAT3, IRF8, MAP3K8, IL-1β and TNRFSF1B).

The pathogenesis of sudden death syndrome (SDS) in meat-type chickens is poorly understood, but cardiac arrhythmia has been implicated (Grashorn, 1994). Death in broiler chickens as a result of SDS was shown to be associated with the catastrophic cardiac arrhythmia, ventricular fibrillation (Olkowski and Classen, 1997; Olkowski et al., 2008). Olowski et al. (2008) demonstrated that the combination of stress and changes in the cardiomyocytes and His–Purkinje system are the key requisite features in the pathogenesis of SDS. Reiner et al. (1995) had suggested that weaker calcium regulation might lead to hyperactivation of skeletal muscles, followed by elevated lactic acid concentration and cardiovascular failure. Hassanpour et al. (2009) asserted that impaired NO synthesis and reduction in iNOS gene expression in the heart ventricles may be involved in the aetiology of PH. Additionally, heat shock proteins were thought to be involved in PHS (Hassanpour et al., 2013). Using genome-wide association, a approximate 2 Mbp region on GGA 9 was shown to be associated with ascites phenotype in broilers (Krishnamoorthy et al., 2014). However, SNPs identified in AngII type 1 receptor 1 (AGTR1) and urotensin 2 (UTS2) harboured in this region showed no association with ascites phenotypes (Dey et al., 2016). Therefore, it is imperative that thorough investigative research should be conducted to identify quantitative trait nucleotides that affect cardiovascular and other functional traits to allow for inclusion into breeding programmes for improvement.

5 Health

The criteria to determine health could be complex; however, the absence of a disease state can be taken as a proxy for health. Even though chicks are vaccinated at hatching against several diseases, there are many more diseases that are yet to be controlled. In this chapter, we will only consider Marek's disease (MD) and salmonellosis.

Some gene variants have been found to be associated with resistance or susceptibility to some diseases in poultry. MD is of great importance in poultry, especially layers and is caused by the MD virus (MDV). The MDV is a herpes virus that causes lymphoproliferative disease. Several QTL regions in the chicken genome have been found to exert genetic control over MD (Vallejo et al., 1998; Yonash et al., 1999; Cheng et al., 2007; Heifetz et al., 2007, 2009). Some genes in the chicken major histocompatibility complex (MHC) play a major role in host resistance to autoimmune, viral, bacterial and parasitic diseases (Bacon, 1987; Lamont, 1989). The MHC haplotype B21 has been found to confer genetic resistance to MD (Lamont, 1998). Similarly, MHC genetic recombinants B-F/B-L region have been found to confer resistance to MD (Briles et al., 1983; Hepkema et al., 1993; Schat et al., 1994). According to Lamont (1998), the MHC has pleiotropic effects on genetic control of immune-responsiveness through its role as a restriction element and also through specific MHC-linked immune response genes. It has been reported that genetic complementation of MHC and Ir-glutamic acid-tyrosine (Ir-GAT) gene affects susceptibility to MD (Steadham et al., 1987). Lui et al. (2001) reported that polymorphisms in the GH gene interact to affect the number of tissues with tumours in chickens with the

MHC B2/B15 genotype. This validated a previous report that RFLPs in the GH gene affected MDV resistance and susceptibility (Kuhnlein et al., 1997). Furthermore, chicken stem lymphocyte antigen 6 complex, locus E (LY6E) also known as SCA2, was found to be associated with MD-related traits (Lui et al., 2001a,b, 2003). Several *cis* and *trans* genes have been shown to respond to MDV infection in allele-specific expressions manner (MacEachern et al., 2012; Perumbakkam et al., 2013; Cheng et al., 2015). It has also been shown that several genes (ACTN1, AP1M1, BRAF, CHCHD2,CNDP2, COG5, CYCS, FANCA, GDI2, GPS1, MDH1, METAP2, NDC80, OPTN, RB1, RIPK1, RRP1B, SEC 24B, SLC25A13, ST3GAL5, TCOF1, TLR4 and TMEM164) are directly regulated by Meq, a bZIP transcription factor, that exhibit allele-specific expression in response to MDV infection (Subramaniam et al., 2013). Additionally, mRNA expression differences in IL10, TNFRSF8, IGFBP7 and GZMA were associated with early stages of the MD pathological process (Lian et al., 2012). Recently, Kaya et al. (2016) demonstrated that seven genes (ITGB2, SGPL1, COMMD5, MOCS2, CCBL2, ATAD1 and CHTF18) with alternatively spliced isomers provide putative mechanisms that underlie phenotype variation in chickens with genetic differences to MD.

Salmonella, the cause of human salmonellosis, is a common food-borne pathogen that is of great public concern. The common causes of human salmonellosis are *Salmonella enteritidis* and *S. typhimurium* (Hafez, 1999). For a full review of salmonellosis, see Calenge et al. (2010).

There are several chromosomal regions and candidate genes that enhance the natural resistance of chicken to *Salmonella*. Several QTLs have been identified that confer resistance to *Salmonella* infection in poultry (Tilquin et al., 2005; Calenge et al., 2009, 2011; Beaumont et al., 2010). Based on results from mice, the chicken homologs of NRAMP1 (SCL11A1) and TNC were evaluated for their involvement in *Salmonella* resistance (Malo and Skamene, 1994; Malo et al., 1994; Caron et al., 2002). These two genes conferred early differential resistance to *S. typhimurium* infection (Hu et al., 1997). The association of SC11A1 with *Salmonella* resistance has been confirmed by several investigators (Caron et al., 2002; Girard-Santosuosso et al., 2002; Liu et al., 2003; Beaumont et al., 2003). Lamont et al. (2002) reported that MHC class I, NRAMP1, PSAP and IAP1 genes showed association with *Salmonella* colonization. Other genes such as TGFβ3, TGFβ4, IgL, MD-2, INOS and TRAIL have been found to have some association with genetic resistance against *S. enteritidis* (Malek and Lamont, 2003; Tohidi et al., 2012; 2013). From an advanced intercross, Hasenstein and Lamont (2007) reported that SNPs in adjacent genes of the chicken β-defensin cluster (GAL11, GAL12 and GAL13) were associated with bacterial load in caecal content suggesting the role of gallinacins in defence of poultry against enteric pathogens. Using a global gene expression approach, Hsin-I et al. (2008) reported significant expression changes in several interleukins (IL1β, IL6, IL12B and IL8) and other genes such as TLR7, CCL4, CXC chemokine K60, TRAF7, GAL9 and IRAK2 between *Salmonella* infected and non-infected birds. Swaggerty et al. (2014) confirmed that CXCLi2 and CCLi2 gene expression levels were associated with *Salmonella* resistance. An SNP in the myeloid differentiation primary response gene 88 (MYD88) was found to be associated with susceptibility to *S.* Pullorum infection (Lui et al., 2015). There are myriad of gene variants that affect other diseases that are not reviewed herein. It will be important to include genetic markers that confer resistance to several diseases rather than selecting on markers that confer resistance to a few diseases. For breeding purposes, variants that are associated with the general immunity and health are of special importance as they confer better resistance and produce more robust birds.

6 Breeding to improve functional traits in poultry

Selection for growth has been labelled as the primary cause for most of the ailments in modern commercial poultry used for meat. Comparison of the modern broiler with its counterpart from 60 years ago shows very noticeable phenotypic differences (Havenstein et al., 2003). It was also reported that genetic selection for improved broiler performance has resulted in a decrease in adaptive immunity and an increase in the cell-mediated and inflammatory responses (Cheema et al., 2003). Other studies have reported increases in CVD (Julian, 1998), skeletal problems and difficulty in walking (Kestin et al., 1992; Julian, 1998; Shim et al., 2012). It should be pointed out that, most of the reports on the undesirable aspects of improved growth have come from incidence studies. To date, there are very few reports on the direct genetic interrelationship of growth and functional traits (Kapell et al., 2012; Kalmar et al., 2013; Rekaya et al., 2013; González-Cerón et al., 2015a, b). Even though it is expected that genetic selection for any particular trait may have both positive and negative correlated responses on other traits, the practical changes in functional traits, 60 years ago was inconceivable. This is not a holistic exoneration of the past. The secondary theorem of fitness (Robertson, 1966) points to negative consequences on functional traits as a result of selection for production traits. This begs the question, why were negative consequences of selection for growth not predicted 60 years ago? For example, comparing a 2001 broiler to an aggregate broiler genotype of 1957, Cheema et al. (2003) postulated that genetic selection for performance has resulted in a decrease in adaptive immune response and an increase in cell-mediated and inflammatory responses. However, in turkey, it was found that long-term selection for performance did not have adverse effect on immune function examined prior to sexual maturity (Cheema et al., 2007). From a meta-study, van der Most et al. (2011) concluded that selection for growth compromises immune function, but selection for immune function did not consistently affect growth suggesting that it may be possible to breed animals for increased immune function without compromising growth. However, a more realistic scenario will be a balanced selection between fitness and production traits.

The knowledge in the art of breeding, biology of traits, data collection and computation capacity were limited 60 years ago and coupled with a lack of accurate definition of the phenotype and limited statistical methods for estimating breeding values. Most of the genetic improvement occurred through mass selection mainly for growth, and at best, a few other traits via the selection index. Growth is moderately heritable, and can be measured accurately. Thus, genetic improvement in growth is much easier and faster than for functional and welfare traits. Per capita yearly consumption of broiler meat in the United States in 1960 was 10 kg (2–3 chickens per year) compared to the projected consumption of 41 kg in 2017 (National Chicken Council, 2012). Therefore, with gross undersupply of poultry meat in the 1960s, a trait with moderate heritability such as growth would obviously be the target for genetic improvement. Whereas some of the unexpected consequences could have been predicted, some of them were unexpected (Hockings, 2014). Additionally, most of the functional and welfare traits are scored in a categorical manner with limited statistical and computational tools to link them directly to production traits (Rekaya et al., 2013). It must be stated that not all maladies, or so-called negative effects of selection, are actually due to the genetic improvement of production traits. Husbandry practices including high stocking density and other factors may have also contributed.

7 Genetic determination of a trait and genetic variability

From quantitative genetics theory, traits that are controlled by a large number of genes governed by additive allele actions have moderate to high heritability (Falconer and Mackay, 1996). However, functional and welfare traits have limited within-breed (intra-sub-population) genetic variance and low heritability suggesting that, in part, these traits are determined largely by non-additive gene actions (dominance and epistasis), which makes improvement through direct selection less efficient. There have been several single-trait (Muir et al., 2014) genetic improvements in experimental populations that have shown some genetic progress. The presence of QTLs for functional and welfare traits have raised the stakes for improvement in these traits. Genetic determination of a trait is not synonymous with its genetic variability in a specific population. Thus, the journey from QTLs and associated polymorphisms to genetic improvement is an arduous one. First, the type of gene action (additive or non-additive) that underlies a trait should be determined. Secondly, the trait should be easily measured in a pedigreed population. Thirdly, the cost associated with the measurement of the trait should be reasonable and manageable. Fourthly, the pedigreed population should be large enough to obtain genetic parameters with acceptable statistical accuracy.

The poultry breeding industry now collects direct and indirect data on production, functional and welfare traits. Extensive data sets, modern computational capacity, increased knowledge in the biology of traits, reduction in functional capacity and welfare, and public pressure has resulted in the incorporation of functional, welfare and health traits into breeding poultry breeding programmes. Expected genetic responses in functional traits in a multi-trait selection context will be slow especially when they are lowly heritable and in negative genetic relationship with production traits. Heritability of a trait could change with the age of the animal as shown by several classical quantitative and molecular approaches. In fact, recent data show that even for a trait such as growth, different QTLs control the trait at different ages (Ankra-Badu et al., 2010a; Gao et al., 2010). This implies that taking advantage of genetic variability of traits at different ages can be attractive for multi-stage improvement programmes.

It should be pointed out that in poultry breeding, the selection and commercial populations are different. Several complimentary pure lines that 'nick' well to balance production, welfare and functional traits are undertaken in the creation of parent lines. Ankra-Badu et al. (2010b) showed that there are several sex-limited, sex-antagonistic and epistatic QTLs that affect growth alone. Therefore, a detailed genetic architecture, biology of traits and their genetic interrelationships along the growth period will be necessary to develop novel breeding strategies leading to commercial birds that are productive and free from welfare and functional concerns. Unfortunately, this is much easier said than done.

If the early enthusiasm for QTL mapping and marker-assisted improvement was of any learning value, excitement about new technologies has to be moderated by the complexity of practical genetic improvement programmes. Microarray and high-throughput science (genome and exome sequencing, transcriptomics, proteomics and metabolomics) and genomic selection are the tools of the day. These aforementioned tools coupled with extensive collection of phenotypic data could unearth new traits, subtle traits and easy-to-collect parameters that correlate genetically well with saleable, difficult-to-measure and expensive-to-measure traits. This will significantly also improve genotype-to-phenotype linkages and increase our knowledge in the biology of traits, their genetic determinants, and their genetic and environmental interrelationships. Combining this new information with existing population and quantitative genetics theory could move us towards producing commercial poultry with significantly improved function.

8 Breeding for sustainability

Sustainable breeding can be defined as genetic improvement today that will maintain the future animal's function, welfare and reproductive capacity without a significant carbon footprint or negative effect on the environment. This will require a complete paradigm shift from genetic improvement as we know it today. The degree of the paradigm shift will depend on the impact of current genetic improvement strategies on carbon footprint and other environmental parameters today versus the impact by incorporating these new aforementioned parameters in the breeding programme. The carbon footprints, environmental impact such as nitrogen and phosphorus in the excreta per bird and the amount of ammonia produced per excreta with certain composition should be known. However, it should be pointed out that, at least for egg-type poultry, the environmental footprint has reduced significantly between 1960 and 2010 due to efficiencies of background systems, changes in feed composition and improved bird performance (Pelleteir et al., 2014). Genetic improvements in production, functional and welfare traits, and permissible environmental impact have to be considered simultaneously. This will obviously add significant cost to breeding and could substantially limit productivity. Cost–benefit analysis is required and who bears the cost of sustainable breeding should be adequately debated. There are no straightforward answers or a magic bullet to sustainable breeding. At any given time in history, breeders will breed chickens under the legally allowable production environment (stock density, cage size, animal welfare, environmental impact, etc.), and what consumers' desire, and are willing to pay for.

9 Where to look for further information

Cabello, F., Hormaeche, C., Mastroeni, P. and Bonina, L. (1993), *The Biology of Salmonella,* Springer Verlag.

Cheng, H., Niikura, M., Kim, T., Mao, W., MacLea, K. S., Hunt, H., Dodgson, J., Burnside, J., Morgan, R., Ouyang, M., Lamont, S., Dekkers, J., Fulton, J., Soller, M. and Muir, W. (2008), Using integrative genomics to elucidate genetic resistance to Marek's disease in chickens. In: Pinard, M.-H., Gay, C., Pastoret, P.-P. and Dodet, B. (eds), *Animal Genomics for Animal Health and Developmental Biology.* Karger.

Lee, L. F. Powell, P. C., Rennie, M., Ross, L. J. and Payne, L. N. (1981), 'Nature of genetic resistance to Marek's disease in chickens', *J. Natl Cancer Inst.*, 66(4), 789–96.

Muir, W. H. and Aggrey, S. E. (2003), *Poultry Genetics, Breeding and Biotechnology*, CABI.

Vaino, O. and Imhof, B. A. (2012), *Immunology and Developmental Biology of Chicken,* Springer Verlag.

Whitehead, C. C. (1992), *Bone Biology and Skeletal Disorders in Poultry,* Carfax.

10 Acknowledgements

We thank Marie Carlie Milfort and Ashley Shelton Ling for their editorial assistance.

11 References

Aggrey, S. E. and Okimoto, R. (2003), Genetic markers: Prospects and applications in genetic analysis. In: Muir, W. H. and Aggrey, S. E. (ed.), *Poultry Genetics, Breeding and Biotechnology*, CABI.

Andersson, L., Archibald, A. L., Bottema, C. D., Brauning, R. et al. (2015), 'Coordinated international action to accelerate genome-to-phenome with FAANG, the Functional Annotation of Animal Genomes project The FAANG Consortium', *Genome Biol.*, 16, 57, doi:10.1186/s13059-015-0622-4.

Ankra-Badu, G. A., Le Bihan-Duval, E., Mignon-Grasteau, S., Pitel, F., Beaumont, C., Duclos, M. J., Simon, J., Carré, W., Porter, T. E., Vignal, A., Cogburn, L. A. and Aggrey, S. E. (2010a), 'Mapping QTL for growth and shank traits in chickens divergently selected for high or low body weight', *Anim. Genet.*, 41(4), 400–5.

Ankra-Badu, G. A., Shriner, D., Le Bihan-Duval, E., Mignon-Grasteau, S., Pitel, F., Beaumont, C., Duclos, M. J., Simon, J., Porter, T. E., Vignal, A., Cogburn, L. A., Allison, D. B., Yi, N. and Aggrey, S. E. (2010b), 'Mapping main, epistatic and sex-specific QTL for body composition in a chicken population divergently selected for low or high growth rate', *BMC Genomics*, 11, 107, doi:10.1186/1471-2164-11-107.

Atzmon, G., Blum, S., Feldman, M., Lavi, U. and Hillel, J. (2007), 'Detection of agriculturally important QTLs in chickens and analysis of the factors affecting genotyping strategy', Cytogenet. *Genome Res.*, 117(1–4), 327–37.

Atzmon, G., Blum, S., Feldman, M., Cahaner, A., Lavi, U. and Hillel, J. (2008), 'QTLs detected in a multigenerational resource chicken population', *J. Hered.*, 99(5), 528–38.

Bacon, L. D. (1987), 'Influence of the MHC on disease resistance and productivity', *Poult. Sci.* 66(5), 802–11.

Basheer, A., Haley, C. S., Law, A., Windsor, D., Morrice, D., Talbot, R., Wilson, P. W., Shart, P. J. and Dunn, I. C. (2015), 'Genetic loci inherited from hens lacking maternal behaviour both inhibit and paradoxically promote this behaviour', *Genet. Sel. Evol.*, 47, 100, doi:10.1186/s12711-015-0180-y.

Beaumont, C., Protais, J., Pitel, F., Levesque, G., Malos, D., Lantier, F., Plisson-Petit, F., Colin, P., Protais, M., Le Roy, P., Elsen, J. M., Milan, D., Lanteir, I., Neau, A., Salvat, G. and Vignal, A. (2003), 'Effect of two candidate genes on the *Salmonella* carrier state in fowl', *Poult. Sci.*, 82(5), 721–6.

Beaumont, C., Chapus, H., Sellier, N., Calenge, F., Zongo, P., Velge, P. and Protais, J. (2010), 'Selection for increased resistance to Salmonella carrier-state', *Worlds Poult. Sci. J.*, 66(2), 251–9.

Bennett, A. K., Hester, P. Y. and Spurlock, D. E. M. (2006), 'Polymorphisms in vitamin D receptor, osteopontin, insulin-like growth factor 1 and insulin, and their associations with bone, egg and growth traits in a layer-broiler cross in chickens', *Anim. Genet.*, 37(3), 283–6.

Briles, W. E., Briles, R. W., Taffs, R. E. and Stone, H. A. (1983), 'Resistance to a malignant lymphoma in chickens is mapped to subregion of major histocompatibility (B) complex', *Science*, 219(4587), 977–9.

Burnside, J., Liou, S. S., Zhong, C. and Cogburn, L. A. (1992), 'Abnormal growth hormone receptor gene expression in the sex-linked dwarf chicken', *Gen. Comp. Endocrinol.*, 88(1), 20–8.

Calenge, F., Lecerf, F., Demars, J., Feve, K., Vignoles, F., Pitel, F., Vignal, A., Velge, P., Sellier, N. and Beaumont, C. (2009), 'QTL for resistance to Salmonella carrier state confirmed in both experimental and commercial chickens', *Anim. Genet.*, 40(5), 590–7.

Calenge, F., Kaiser, P., Vignal, A. and Beaumont, C. (2010), 'Genetic control of resistance to salmonellosis and to *Salmonella* carrier-state in fowl: a review', *Genet. Sel. Evol.*, 42, 11, doi:10.1186/1297-9686-42-11.

Calenge, F., Vignal, A., Demars, J., Feve, K., Menanteau, P., Velge, P. and Beaumont, C. (2011), 'New QTL for resistance to Salmonella carrier-state identified on fowl microchromosomes', *Mol. Genet. Genomics*, 285(3), 237–43.

Caron, J., Loredo-Osti, J. C., Laroche, L., Skamene, E., Morgan, K. and Malo, D. (2002), 'Identification of genetic loci controlling bacterial clearance in experimental Salmonella enterididis infection: An unexpected role of NRAMP1 (Slc11a1) in the persistence of infection in mice', *Genes Immun.*, 3(4), 196–204.

Chang, M. T., Cheng, Y. S. and Huang, M. C. (2012), 'A novel SNP of the PNRC1 gene and its association with reproductive traits in Tsaiya ducks', *Theriogenology*, 78(1), 140–6.

Cheema, M. A., Qureshi, M. A. and Havenstein, G. B. (2003), 'A comparison of the immune response of a 2001 commercial broiler with a 1957 randombred broiler strain when fed representative 1957 and 2001 broiler diets', *Poult. Sci.*, 82(10), 1519–29.

Cheema, M. A., Qureshi, M. A., Havenstein, G. B., Ferket, P. R. and Nestor, K. E. (2007), 'A comparison of the immune response of 2003 commercial turkeys and a 1966 randombred strain when fed representative 2003 and 1966 turkey diets' *Poult. Sci.* 86(2), 241–8.

Cheng, H. H., Zhang, Y. and Muir, W. M. (2007), 'Evidence for widespread epistatic interactions influencing Marek's disease virus viremia levels in chicken', *Cytogenet. Genome Res.*, 117(1–4), 313–18.

Cheng, H. H., Perumbakkam, S., Black-Pyrkosz, A., Dunn, J. R., Legarra, A. and Muir, W. M. (2015), 'Fine mapping of QTL and genomic prediction using allele-specific expression SNPs demonstrate that the complex trait of genetic resistance to Marek's disease is predominantly determined by transcriptional regulation'. *BMC Genomics*, 16, 816, doi:10.1186/s12864-015-2016-0.

Chiang, H. I., Swaggerty, C. L., Kogut, M. H., Dowd, S. E., Li, X., Pevzner, I. Y. and Zhou, H. (2008), 'Gene expression profiling using chicken heterophils with Salmonella enteritidis stimulation using a chiken 44 K Agilent microarray' *BMC Genomics*, 9, 526, doi:10.1186/1471-2164-9-526.

Cui, J. X., Du, H. L., Liang, Y., Deng, X. M., Li, N. and Zhang, X. Q. (2006), 'Association of polymorphism in the promoter region of the chicken prolactin with egg production', *Poult. Sci.*, 85(1), 26–31.

Dan, H., Simsa-Maziel, S., Hisdai, A., Sela-Donenfeld, D. and Monsonego-Ornan, E. (2009), 'Expression of matrix metalloproteinases during impairment and recovery of the avian growth plate', *J. Anim. Sci.*, 87(11), 3544–55.

Dey, S., Krishna, S., Anthony, N. B. and Roads, D. D. (2016), 'Further investigation of a quantitative trait loci for ascites on chromosome 9 in broiler chickens', *Poult. Sci.*, doi:10.3382/ps/pew380.

Dunn, I. C., Joseph, N. T., Bain, M., Wilson, P. W., Milona, P., Nys, Y., Gautron, J., Schmutz, M., Preisinger, R. and Waddington, D. (2009), 'Polymorphisms in eggshell matrix genes are associated with eggshell quality measurements in pedigree Rhode Island Red hens', *Anim. Genet.*, 40(1), 110–14.

Elis, S., Batellier, F., Couty, I., Balzergue, S., Martin-Magniette, M. L., Monget, P., Blesbois, E. and Govoroun, M. S. (2008), 'Search for the genes involved in oocyte maturation and early embryo development in the hen', *BMC Genomics*, 9, 110, doi:10.1186/1471-2164-9-110.

Elis, S., Blesbois, E., Couty, I., Balzergue, S., Martin-Magniette, M. L., Batellier, F. and Govoroun, M. S. (2009), 'Identification of germinal disk region derived genes potentially involved in hen fertility', *Mol. Reprod. Dev.*, 76(11), 1043–55.

Evsikov, A. V., Graber, J. H., Brockman, J. M., Hampl, A., Holbrook, A. E., Singh, P., Eppig, J. J., Solter, D. and Knowles, B. B. (2006), 'Cracking the egg: molecular dynamics and evolutionary aspects of the transition from the fully grown oocyte to embryo', *Genes Dev.*, 20(19), 2713–27.

Falconer, D. S. and Mackay, T. F. C. (1996), Introduction to Quantitative Genetics, 4th ed. Longman Group Ltd.

Fatemi, S. A., Mehrabani-Yeganeh, H., Nejati-Javaremi, A. and Niknafs, S. H. (2012), 'Association of neuropeptide Y and gonadotrophin-releasing hormone receptor gene SNPs with breeding value for growth and egg production traits in Mazandaran native chickens', *Genet. Mol. Res.*, 11(3), 2539–47.

Feng, X. P., Kuhnlein, U., Aggrey, S. E., Gavora, J. S. and Zadworny, D. (1997). 'Trait association of genetic markers in the growth hormone and growth hormone receptor gene in a White Leghorn strain', *Poult. Sci.*, 76(12), 1770–5.

Froman, D. P., Feltmann, A. J., Roads, M. L. and Kirby, J. D. (1999), 'Sperm mobility: A primary determinant of fertility in the domestic fowl (*Gallus domesticus*)', *Biol. Reprod.*, 61(2), 400–5.
Froman, D. P. and Kirby, J. D. (2005), 'Sperm mobility: Phenotype in roosters (*Gallus domesticus*) determined by mitochondrial function', *Biol. Reprod.*, 72(3), 562–7.
Froman, D. P., Feltmann, A. J., Pendarvis, K., Coksey, A. M., Burgess, S. C. and Roads, D. D. (2011), 'A proteome-based model for sperm mobility phenotype', *J. Anim. Sci.*, 89(5), 1330–7.
Froman, D. P. and Roads, D. D. (2013), 'A system biology definition for chicken semen quality', *J. Anim. Sci.*, 91(2), 523–9.
Fulton, J. E., Soller, M., Lund, A. R., Arango, J. and Lipkin, E. (2012), 'Variation in the ovocalyxin-32 gene in commercial egg-laying chickens and its relationship with egg production and egg quality traits', *Anim. Genet.*, 43(Suppl. 1), 102–13.
Gao, Y., Du, Z. Q., Feng, C. G., Deng, X. M., Li, N., Da, Y. and Hu, X. X. (2010), 'Identification of quantitative trait loci for shank length and growth at different development stages in chicken', *Anim. Genet.*, 41(1), 101–4.
Girard-Santosuosso, O., Lantier, F., Lantier, I., Bumstead, N., Elsen, J. M. and Beaumont, C. (2002), 'Heritability of susceptibility to *Salmonella enteritidis* infection in fowls and test of the role of the chromosome carrying the NRAMP1 gene', *Genet. Sel. Evol.*, 34(2), 211–19.
González-Cerón, F., Rekaya, R. and Aggrey, S. E. (2015a), 'Genetic analysis of bone quality traits and growth in a random mating broiler population', *Poult. Sci.*, 94(5), 883–9.
González-Cerón, F., Rekaya, R. and Aggrey, S. E. (2015b), 'Genetic relationship between leg problems and bone quality traits in a random mating broiler population', *Poult. Sci.*, 94(8), 1787–90.
Goraga, Z. S., Nassar, M. K. and Brockmann, G. A. (2012), 'Quantitative trait loci segregating in crosses between New Hampshire and White Leghorn chicken lines: I. egg production traits', *Anim. Genet.*, 43(2), 183–9.
Goto, T., Ishikawa, A., Onitsuka, S., Goto, N., Fujikawa, Y., Umino, T., Nishibori, M. and Tsudzuki, M. (2011), 'Mapping quantitative trait loci for egg production traits in an F2 intercross of Oh-Shamo and White Leghorn chickens', *Anim. Genet.*, 42(6), 634–41.
Grashorn, M. (1994), 'Investigation of the aetiology and pathology of sudden death syndrome in meat-type chickens', *Archiv fur Geflugelkunde*, 58(5), 243–4.
Hafez, H. M. (1999), 'Poultry meat and food safety: Pre- and post- harvest approaches to reduce foodborne pathogens', *Worlds Poult. Sci. J.*, 55(3), 269–80.
Han, H., Lei, Q., Zhou, Y., Gao, J., Liu, W., Li, F., Zhang, Q., Lu, Y. and Cao, D. (2015), 'Association between BMP15 gene polymorphism and reproductive traits and its tissue expression characteristics in chickens', *PLoS One*, 10(11), e0143298, doi:10.1371/journal.pone.0143298.
Hansen, C., Yi, N., Zhang, Y. M., Xu, S., Gavora, J. and Cheng, H. H. (2005), 'Identification of QTL for production traits in chickens', *Anim. Biotechnol.*, 16(1), 67–79.
Hasenstein, J. R. and Lamont, S. J. (2007), 'Chicken gallinacin gene cluster associated with Salmonella response in advanced intercrossed line', *Avian Dis.*, 51(2), 561–7.
Hasky-Negev, M., Simsa, S., Tong, A., Genina, O. and Monsonego-Ornan, E. (2008), 'Expression of matrix metalloproteinases during vascularization and ossification of normal and impaired avian growth plate', *J. Anim. Sci.*, 86(6), 1306–15.
Hassanpour, H., Yazdani, A., Soreshjani, K. H. and Asgharzadeth, S. (2009), 'Evaluation of endothelial and inducible nitric oxide synthase genes expression in the heart of broiler chickens with experimental pulmonary hypertension', *Br. Poult. Sci.*, 50(6), 725–32.
Hassanpour, H., Afzali, A. and Bahadoran, S. H. (2013), 'Diminished gene expression of cardiac heat shock proteins in pulmonary hypertensive chickens', *Br. Poult. Sci.*, 54(5), 581–6.
Havenstein, G. B., Ferket, P. R. and Qureshi, M. A. (2003), 'Growth livability, and feed conversion of 1957 verus 2001 broilers when fed representative 1957 and 2001 broiler diets', *Poult. Sci.*, 82(10), 1500–8.
Heifetz, E. M., Fulton, J. E., O'Sullivan, N. P., Arthur, J. A., Wang, J., Dekkers, J. C. M. and Soller, M. (2007), 'Mapping quantitative trait loci affecting susceptibility Marek's disease virus in a backcross population in layer chickens', *Genetics*, 177(4), 2417–31.

Heifetz, E. M., Fulton, J. E., O'Sullivan, N. P., Arthur, J. A., Cheng, H., Wang, J., Soller, M. and Dekkers, J. C. M. (2009), 'Mapping QTL affecting resistance to Marek's disease in an F6 advanced intercross population of commercial layer chickens', *BMC Genomics*, 10, 20, doi:10.1186/1471–2164-10-20.

Hepkema, B. G., Blankert, L. J., Albers, G. A. A., Tilanus, M. G. J., Egberts, M. E., van der Zijpp, A. J. and Hansen, E. J. (1993), 'Mapping of susceptibility to Marek's disease within the major histocompatibility (B) complex refined typing of White Leghorn chickens', *Anim. Genet.*, 24(4), 283–7.

Hockings, P. M. (2014), 'Unexpected consequences of genetic selection in broilers and turkeys: problems and solutions', *Br. Poult. Sci.*, 55(1), 1–12.

Houston, B., Stewart, A. J. and Farquharson, C. (2004), 'PHOSPHO1: a novel phosphatase specifically expressed at sites of mineralisation in bone and cartilage', *Bone*, 34(4), 629–37.

Hu, J., Bumstead, N., Barrow, P., Sebastiani, G., Olien, L., Morgan, K. and Malo, D. (1997), 'Resistance to Salmonellosis in the chicken is linked to NRAMP1 and TNC', *Genome Res.*, 7(7), 693–704.

Hu, Z. L., Park, C. A., Wu, X. L. and Reecy, J. M. (2013), 'Animal QTLdb: an improved database tool for livestock animal QTL/association data dissemination in the post-genome era', *Nucleic Acids Res.*, 41(Database issue), D871–9.

Huang, H. L., Cheng, Y. S., Huang, C. W., Huang, M. C. and Hsu, W. H. (2011), 'A novel genetic marker the ovomucoid gene associated with hatchability in Tsaiya ducks (*Anas platyrhynchos*)', *Anim. Genet.*, 42(4), 421–7.

Huang, H. L., Huang, L. T. and Cheng, Y. S. (2013), 'A novel SNP marker of ovalbumin gene in association with duck hatchability', *Theriogenology*, 79(8), 1218–23.

Huang, H. L. and Cheng, Y. S. (2014), 'A novel minisequencing single-nucleotide polymorphism marker of the lysozyme gene detects high hatchability of Tsaiya ducks (*Anas platyrhynchos*)', *Theriogenology*, 82(8), 1113–20.

Huang, H. Y., Liang, Z., Li, S. F., Li, C. M., Zhao, Z. H. and Wang, Q. B. (2015), 'Polymorphism identification in BMP15 and GDF9 genes and their association with egg production in chickens', *Br. Poult. Sci.*, 56(3), 277–83.

Hutt, F. B. (1949), *Genetics of the Fowl*, McGraw-Hill Book Company, Inc.

Inafuku, K., Maeda, Y., Ishihara, K., Okamoto, S. and Hashiguchi, T. (1997), 'A new mutation of ovalbumin in the chicken', *Jpn. Poult. Sci.*, 34(2), 87–93.

International Chicken Polymorphism Map Consortium. (2004), 'A genetic variation map with 2.8 million single nucleotide polymorphisms', *Nature*, 432(7018), 717–22.

Jiang, R. S., Xu, G. Y., Zhang, X. Q. and Yang, N. (2005), 'Association of polymorphisms for prolactin and prolactin receptor genes with broody traits in chickens', *Poult. Sci.*, 84(6), 839–45.

Julian, R. J. (1985), The peritoneum, retroperitoneum and mesentery. In: Jubb, K. V. F., Kennedy, P. C. and Palmer, N. C. (ed.), *Pathology of Domestic Animals*, 3rd ed., Academic Press.

Julian, R. J. (1993), 'Ascites in poultry', *Avian Pathol.*, 22(3), 419–54.

Julian, R. J. (1998), 'Rapid growth problems: ascites and skeletal deformities in broilers', *Poult. Sci.*, 77(12), 1773–80.

Kalmar, I. D., Vanrompay, D. and Janssens, G. P. J. (2013), 'Broiler ascites syndrome: Collateral damage from efficient feed to meat conversion', *Vet. J.*, 197(2), 169–74.

Kamiyama, N., Seki, R., Yokoyama, H. and Tamura, K. (2012), 'Heterochronically early decline of Hox expression prior to cartilage formation in the avian hindlimb zeugopod', *Develop. Growth Differ.*, 54(6), 619–32.

Kang, L., Zhang, N., Zhang, Y., Yan, H., Tang, H., Yang, C., Wang, H. and Jiang, Y. (2012), 'Molecular characterization and identification of a novel polymorphism of 200 bp indel associated with age at first egg of the follicle-stimulating hormone receptor (FSHR) gene', *Mol. Biol Rep.*, 39(3), 2967–73.

Kapell, D. N. R. G., Hill, W. G., Neeteson, A. M., McAdam, J., Koerhuis, A. N. M. and Avendaño, S. (2012), 'Twenty-five years of selection for improved leg health in purebred broiler lines and underlying genetic parameters', *Poult. Sci.*, 91(12), 3032–43.

Katanbaf, M. N. and Hardimann, J. W. (2010), 'Primary broiler breeding striking a balance between economic and well-being traits', *Poult. Sci.*, 89(4), 822–4.

Kaya, M., Preeyanon, L., Dodgson, J. B. and Cheng, H. H. (2016), 'Validation of alternative transcript splicing in chicken lines that differ in genetic resistance to Marek's disease', *Anim. Biotechnol.*, 27(4), 238–44.

Kestin, S. C., Knowles, T. G., Tinch, A. E. and Gregory, N. (1992), 'Prevalence of leg weakness in broiler chickens and its relationship with genotype', *Vet. Rec.*, 131(9), 190–4.

Krishnamoorthy, S., Smith, C. D., Al-Rubaye, A. A., Erf, G. F., Wideman, R. F., Anthony, N. B. and D. Rhoads, D. (2014), 'A quantitative trait locus for ascites on chromosome 9 in broiler chicken lines', *Poult. Sci.* 93(2), 307–17.

Kuhnlein, U., Ni, L., Weigend, S., Gavora, J. S., Fairfull, W. and Zadworny, D. (1997), 'DNA polymorphisms in the chicken growth hormone gene: response to selection for disease resistance and association with egg production', *Anim. Genet.*, 28(2), 116–23.

Lamont, S. J. (1989), 'The chicken major histocompatibility complex in disease resistance and poultry breeding', *J. Dairy Sci.*, 72(5), 1328–33.

Lamont, S. J. (1998), 'The chicken major histocompatibility complex and disease', *Rev. Sci. Tech.*, 17(1), 128–42.

Lamont, S. J., Kaiser, M. G. and Liu, W. (2002), 'Candidate genes for resistance to Salmonella enteritidis colonization in chickens as detected in a novel genetic cross', *Vet. Immunol. Immunopathol.*, 87(3–4), 423–8.

Langhorst, L. J. and Fechheimer, N. S. (1985), 'Shankless, a new mutation on chromosome 2 in the chicken', *J. Hered.*, 76(3), 182–6.

Law, A. S., Burt, D. W., Alexander, I. and Thorp, B. H. (1996), 'Expression of the gene for transforming growth factor-beta in avian dyschondroplasia', *Res. Vet. Sci.*, 61(2), 120–4.

Li, H., Deeb, N., Zhou, H., Mitchell, A. D., Ashwell, C. M. and Lamont, S. J. (2003), 'Chicken quantitative trait loci for growth and body composition associated with transforming growth factor-β genes', *Poult. Sci.*, 82(3), 347–56.

Li, H., Deeb, N., Zhou, H., Ashwell, C. M. and Lamont, S. J. (2005), 'Chicken quantitative trait loci for growth and body composition associated with the very low density apolipoprotein-II gene', *Poult. Sci.*, 84(5), 697–703.

Li, G., Sun, D. X., Yu, Y., Liu, W. J., Tang, S. Q., Zhang, Y., Wang, Y. C., Zhang, S. L. and Zhang, Y. (2011), 'Genetic effect of the follicle-stimulating hormone receptor gene on reproductive traits in Beijing You chickens', *Poult. Sci.*, 90(11), 2487–92.

Lian, L., Qu, L. J., Sun, H. Y., Chen, Y. M., Lamont, S. J., Liu, C. J. and Yang, N. (2012), 'Gene expression analysis of host spleen responses to Marek's disease virus infection at late tumor transformation phase', *Poult. Sci.*, 91(9), 2130–8.

Liu, H. C., Kung, H. J., Fulton, J. E., Morgan, R. W. and Cheng, H. H. (2001a), 'Growth hormone interacts with the Marek's disease virus SORF2 protein and is associated with disease resistance in chicken', *PNAS*, 98(16), 9203–8.

Liu, H. C., Cheng, H. H., Softer, L. and Burnside, J. (2001b), 'A strategy to identify positional candidate genes conferring Marek's disease resistance by integrating DNA microarrays with genetic mapping', *Anim. Genet.*, 32(6), 351–9.

Liu, H. C., Niikura, M., Fulton, J. and Cheng, H. H. (2003), 'Identification of chicken stem lymphocyte antigen 6 complex, locus E (LY6E, alias SCA2) as a putative Marek's disease resistance gene via a virus-host protein interaction screen', *Cytogenet. Genome Res.*, 102(1–4), 304–8.

Liu, P., Yang, F., Zhuang, Y., Xiao, Q., Cao, H., Zhang, C., Wang, T., Lin, H., Guo, X. and Hu, G. (2016), 'Dysregulation expression of microRNAs and mRNAs in pulmonary artery remodeling in ascites syndrome in broiler chickens', Oncotarget, doi:10.18632/oncotarget.12888 [Epub ahead of print].

Liu, W., Kaiser, G. and Lamont, S. J. (2003), 'Natural resistance-associated macrophage protein 1 gene polymorphisms and response to vaccine against or challenge with Salmonella enteritidis in young chicks', *Poult. Sci.* 82(2), 259–66.

Liu, W., Li, D., Liu, J., Chen, S., Qu, L., Zheng, J., Xu, G. and Yang, N. (2011), 'A genome-wide SNP scan reveals novel loci for egg production and quality traits in white leghorn and brown-egg dwarf layers', *PLoS One*, 6(12), e28600, doi:10.1371/journal.pone.0028600.

Liu, W., Yu, Y., Li, G., Tang, S., Zhang, Y., Wang, Y., Zhang, S. and Zhang, Y. (2012), 'Single-nucleotide polymorphisms in the promoter of the growth hormone-releasing hormone receptor gene are associated with growth and reproduction traits in chickens', *Anim. Genet.*, 43(5), 564–9.

Loveridge, N., Farquharson, C., Hesketh. , J. E., Jakowlew, S. B., Whitehead, C. C. and Thorp, B. H. (1993), 'The control of chondrocyte differentiation during endochondral bone growth in vivo: changes in TGF- and the proto-oncogene c-myc', *J. Cell Sci.*, 105(Pt. 4), 949–56.

Lui, X. Q., Wang, F., Jin, J., Zhou, Y. G., Ran, J. S., Feng, Z. Q., Wang, Y. and Liu, Y. P. (2015), 'MyD88 polymorphisms and association with susceptibility to Salmonella Pullorum', *Biomed. Res. Int.*, 2015, 692973, doi:10.1155/2015/692973.

MacEachern, S., Muir, W. M., Crosby, S. and Cheng, H. H. (2012), 'Genome-wide identification and quantification of cis- and trans-regulated genes responding to Marek's disease virus infection via analysis of allele-specific expression', *Front. Genet.*, 2, 113, doi:10.3389/fgene.2011.00113.

Malek, M. and Lamont, S. J. (2003), 'Association of INOS, TRAIL, TGF-beta2, TGF-beta3, and IgL genes with response to Salmonella enteritidis in poultry', *Genet. Sel. Evol.*, 35(Suppl. 1), S99–111.

Malo, D., Vogan, K., Vidal, S., Hu, J., Cellier, M., Schurr, E., Fuks, A., Bumstead, N., Morgan, K. and Gros, P. (1994), 'Haplotype mapping and sequence analysis of the mouse Nramp gene predicts susceptibility to infection with intracellular parasites', *Genomics*, 23(1), 51–61.

Malo, D. and Skamene, E. (1994), 'Genetic control of host resistance to infection', *Trends Genet.*, 10(10), 365–71.

McMutry, J. P., Francis, G. L. and Upton, Z. (1997), 'Insulin-like growth factors in poultry', *Domest. Anim. Endocrinol.*, 14(4), 199–229.

Muir W. H., Chen, H.-W. and Croney, C. (2014) 'Methods to address poultry robustness and welfare issues through breeding and associated ethical considerations', *Front. Genet.* 5, 407. doi:10.3389/fgene.2014.00407.

Nagaraja, S. C., Aggrey, S. E., Yao, J., Zadworny, D., Fairfull, R. W. and Kuhnlein, U. (2000), 'Trait association of a genetic marker near the IGF-I gene in egg-laying chickens', *J. Hered.*, 91(2), 150–6.

National Chicken Council (2012), Per Capita Consumption of Poultry and Livestock, 1965 to Estimated 2016, in Pounds. Available at: http://www.nationalchickencouncil.org/about-the-industry/statistics/per-capita-consumption-of-poultry-and-livestock-1965-to-estimated-2012-in-pounds/ (Accessed 21 Sept. 2016).

Nestor, K. E., Nobel, D. O., Zhu, J. and Moritsu, Y. (1996), 'Direct and correlated responses to long-term selection for increased body weight and egg production in turkeys', *Poult. Sci.*, 75(10), 1180–91.

Olkowski, A. A. and Classen, H. L. (1997), 'Malignant ventricular disrhythmia in broiler chicken which die of sudden death syndrome', *Vet. Rec.*, 140(7), 177–9.

Olkowski, A. A., Wojnarowicz, C., Nain, S., Ling, B., Alcon, J. M. and Laarveld, B. (2008), 'A study on pathogenesis of sudden death syndrome in broiler chickens', *Res. Vet. Sci.*, 85(1), 131–40.

Ou, J. T., Tang, S. Q., Sun, D. X. and Zhang, Y. (2009), 'Polymorphisms of three neuroendocrine-correlated genes associated with growth and reproductive traits in the chicken', *Poult. Sci.* 88(4), 722–7.

Pelleteir, N., Ibarburu, M. and Xin H. (2014), 'Comparison of the environmental footprint of the egg industry in the United States in 1960 and 2010', *Poult. Sci.* 93(2), 241–55.

Perumbakkam, S., Muir, W. M., Black-Pyrkosz, A., Okimoto, R. and Cheng, H. H. (2013), 'Comparison and contrast of genes and biological pathways responding to Marek's disease virus infection using allele-specific expression and differential expression in broilers and layer chickens', *BMC Genomics*, 14, 64, doi:10.1186/1471–2164-14–64.

Podisi, B. K., Knott, S. A., Dunn, I. C., Law, A. S., Burt, D. W. and Hocking, P. M. (2011), 'Overlap of quantitative trait loci for early growth rate, and for body weight and age at onset of sexual maturity in chickens', *Reproduction*, 141(3), 381–9.

Qin, N., Liu, Q., Zhang, Y. Y., Fan, X. C., Xu X. X., Lv, Z. C., Wei, M. L., Jing, Y., Mu, F. and Xu, R. F. (2015a), 'Association of novel polymorphisms of forkhead box L2 and growth differentiation factor-9 genes with egg production traits in local Chinese Dagu hens', *Poult. Sci.* 94(1), 88–95.

Qin, N., Fan, X. C., Zhang, Y. Y, Xu, X. X., Tyasi, T. L., Jing, Y., Mu, F., Wei, M. L. and Xu, R. F. (2015b), 'New insights into implication of the SLIT/ROBO pathway in the prehierarchical follicle development in the hen ovary', *Poult. Sci.*, 94(9), 2235–46.

Rath, N. C., Huff, G. R., Huff, W. E., and Balog, J. M. (2000), 'Factors regulating bone maturity and strength in poultry', *Poult. Sci.*, 79(7), 1024–32.

Rath, N. C., Huff, W. E., Balog, J. M. and Huff, G. R. (2004), 'Comparative efficacy of different dithiocarbamates to induce tibial dyschondroplasia in poultry', *Poult. Sci.*, 83(2), 266–74.

Rath, N. C., Richards, M. P., Huff, W. E., Huff, G. R. and Balog, J. M. (2005), 'Changes in the tibial growth plates of chickens with thiram-induced dyschondroplasia', *J. Comp. Pathol.*, 133(1), 41–52.

Rath, N. C., Huff, W. E. and Huff, G. R. (2007), 'Thiram-induced changes in the expression of genes relating to vascularization and tibial dyschondroplasia', *Poult. Sci.*, 86(11), 2390–5.

Reich, A., Jaffe, N., Tong, A., Lavelin, I., Genina, O., Pines, M., Sklan, D., Nussinovitch, A. and Monsonego-Ornan, E. (2005), 'Weight loading young chicks inhibits bone elongation and promotes growth plate ossification and vascularization', *J. Appl. Physiol.*, 98(6), 2381–9.

Reiner, G., Hartmann, J. and Dzapo, V. (1995), 'Skeletal muscle sarcoplasmic calcium regulation and sudden death syndrome in chickens', *Br. Poult. Sci.*, 36(4), 667–75.

Rekaya, R., Sapp, R. L., Wing, T. and Aggrey, S. E. (2013), 'Genetic evaluation for growth, body composition, feed efficiency, and leg soundness', *Poult. Sci.* 92(4), 923–9.

Robertson, A. (1966), 'A mathematical model of the culling process in dairy cattle', *Anim. Prod.*, 8(1), 95–108.

Romanov, M. N. (2001), Genetics of broodiness in poultry-A Review, Asian-Aust. *J. Anim. Sci.*, 14(11), 1647–54.

Rubin, C., Lindberg, J., Fitzsimmons, C., Savolainen, P., Jensen, P., Lundeberg, J., Andersson, L. and Kindmark, A. (2007), 'Differential gene expression in femoral bone from red junglefowl and domestic chicken, differing for bone phenotypic traits', *BMC Genomics*, 8, 208, doi:10.1186/1471–2164-8-208.

Sasaki, O., Odawara, S., Takahashi, H., Nirasawa, K., Oyamada, Y., Yamamoto, R., Ishii, K., Nagamine, Y., Takeda, H., Kobayashi, E. and Furukawa, T. (2004), 'Genetic mapping of quantitative trait loci affecting body weight, egg characterand egg production in F2 intercross chickens', *Anim. Genet.*, 35(3), 188–94.

Schat, K. A., Taylor, Jr.R. L. and Briles, W. E. (1994), 'Resistance to Marek's disease chickens with recombinant haplotypes to the major histocompatibility (B) complex', *Poult. Sci.* 73(4), 502–8.

Shen, S., Berry, W., Jaques, S., Pillai, S. and Zhu, J. (2004), 'Differential expression of iodothyronine deiodinase type 2 in growth plates of chickens divergently selected for incidence of tibial dyschondroplasia', *Anim. Genet.*, 35(2), 114–18.

Shen, X., Zeng, H., Xie, L., He, J., Li, J., Xie, X., Luo, C., Xu, H., Zhou, M., Nie, Q. and Zhang, X. (2012), 'The GTPase activating Rap/RanGAP domain-like 1 gene is associated with chicken reproductive traits', *PLoS One*, 7(4), e33851, doi:10.1371/journal.pone.0033851.

Shen, X., Bai, X., Xu, J., Zhou, M., Xu, H., Nie, Q., Lu, X. and Zhang, X. (2016), 'Transcriptome sequencing reveals genetic mechanisms underlying the transition between laying and brooding phases and gene expression changes associated with divergent reproductive phenotypes in chickens', *Mol. Biol. Rep.*, 43(9), 977–89.

Shi, S., Shen, Y., Zhao, Z., Hou, Z., Yang, Y., Zhou, H., Zou, J. and Guo, Y. (2014), 'Integrative analysis of transcriptomic and metabolomic profiling of ascites syndrome in broiler chickens induced by low temperatures', *Mol. Biosyst.*, 10(11), 2984–93.

Shim, M. Y., Karnuah, A. B., Anthony, N. B., Pesti, G. M. and Aggrey, S. E. (2012), 'The effects of broiler chicken growth rate on valgus, varus, and tibial dyschondroplasia', *Poult. Sci.*, 91(1), 62–5.

Steadham, E. M., Lamont, S. J., Kujdych, I. and Norskog, A. W. (1987), 'Association of Marek's disease with Ea-B and immune response genes in sublines and F2 populations of the Iowa State S1 Leghorn line', *Poult. Sci.*, 66(4), 571–5.

Stewart, A. J., Roberts, S. J., Seawright, E., Davey, M. G., Fleming, R. H. and Farquharson, C. (2006), 'The presence of PHOSPHO1 in matrix vesicles and its developmental expression prior to skeletal mineralization', *Bone*, 39(5), 1000–7.

Subramaniam, S., Preeyanon, L. and Cheng, H. H. (2013), 'Transcriptional Profiling of MEq-Dependent Genes in Marek's Disease Resistant and Susceptible Inbred Chicken Lines', *PLoS ONE*, 8(10), e78171. doi:10.1371/journal.pone.0078171.

Swaggerty, C. L., Pevzner, I. Y. and Kogut, M. H. (2014), 'Selection for pro-inflammatory mediators yields chickens with increased resistance against *Salmonella enterica serovar Enteritidis*', *Poult. Sci.*, 93(3), 535–44.

Takahashi, H., Sasaki, O., Nirasawa, K. and Furukawa, T. (2010), 'Association between ovocalyxin-32 gene haplotypes and egg shell quality in an F2 intercross between chicken divergently selected for egg shell strength', *Anim. Genet.*, 41(5), 541–4.

Teixeira, C. C., Xiang, J., Roy, R., Kudrashov, V., Binderman, I., Mayer-Kuckuk, P. and Boskey, A. L. (2011), 'Changes in matrix protein gene expression associated with mineralization in the differentiating chick limb-bud micromass culture system', *J. Cell Biochem.*, 112(2), 607–13.

Tian, W. X., Zhang, W. P., Li, J. K., Bi, D. R., Guo, D. Z., Pan, S. Y., Zhang, Y. H. and Qin, P. (2009), 'Identification of differentially expressed genes in the growth plate of broiler chickens with thiram-induced tibial dyschondroplasia', *Avian Pathol.*, 38(2), 161–6.

Tilquin, P., Barrow, P. A., Marly, J., Pitel, F., Plisson-Petit, F., Velge, P., Vignal, A., Baret, P. V., Bumstead, N. and Beaumont, C. (2005), 'A genome scan for quantitative trait loci affecting the Salmonella carrier-state in chicken', *Genet. Sel. Evol.*, 37(5), 539–61.

Tohidi, R., Idris, I. B., Panandam, J. M. and Bejo, M. H. (2012), 'The effects of polymorphisms in IL-2, IFN-γ, TGF-β2, IgL, TLR-4, MD-2, and iNOS genes on resistance to Salmonella enteritidis in indigenous chickens', *Avian Pathol.*, 41(6), 605–12.

Tohidi, R., Idris, I. B., Malar, P. J. and Hair, M. B. (2013), 'The effects of polymorphism in 7 candidate genes on resistance to Salmonella Enteritidis in native chickens', *Poult. Sci.*, 92(4), 900–9.

Tong, A., Reich, A., Genin, O., Pines, M. and Monsonego-Ornan, E. (2003), 'Expression of chicken 75-kDa gelatinase B-like enzyme in perivascular chondrocytes suggests its role in vascularization of the growth plate', *J. Bone Miner. Res.*, 18(8), 1443–52.

Tuiskula-Haavisto, M., Honkatukia, M., Vilkki, J., de Koning, D. J., Schulman, N. F. and Maki-Tanila, A. (2002), 'Mapping of quantitative trait loci affecting quality and production traits inegg layers', *Poult. Sci.*, 81(7), 919–27.

Tuiskula-Haavisto, M., de Koning, D. J., Honkatukia, M., Schulman, N. F., Maki-Tanila, A. and Vilkki, J. (2004). 'Quantitative trait loci with parent-of-origin effects in chicken', *Genet. Res.*, 84(1), 57–66.

Uemoto, Y., Suzuki, C., Sato, S., Sato, S., Ohtake, T., Sasaki, O., Takahashi, H. and Kobayashi, E. (2009), 'Polymorphism of the ovocalyxin-32 gene and its association in with egg production traits in the chicken', *Poult. Sci.*, 88(12), 2512–17.

Vallejo, R. L., Bacon, L. D., Liu, H. C., Witter, R. L., Groenen, M. A. M., Hillel, J. and Cheng, H. H. (1998), 'Genetic mapping of quantitative trait loci affecting susceptibility to Marek's disease virus induced tumors in F2 intercross chickens', *Genetics*, 148(1), 349–60.

Van der Most, P. J., de Jong, B., Parmentier, H. K. and Verhulst, S. (2011), 'Trade-off between growth and immune function: a meta-analysis of selection experiments', *Funct. Ecol.* 25(1), 74–80.

Velada, I., Capela-Silva, F., Reis, F., Pires, E., Egas, C., Rodrigues-Santos, P. and Barros, M. T. (2011), 'Expression of genes encoding extracellular matrix macromolecules and metalloproteinases in avian tibial dyschondroplasia', *J. Comp. Pathol.*, 145(2–3), 174–86.

Velleman, S. G. (2000), 'The role of the extracellular matrix in skeletal development', *Poult. Sci.*, 79(7), 985–9.

Warman, M. L., Abbot, M., Apte, S. S., Hefferon, T., McIntosh, I., Cohn, D. H., Hecht, J. T., Olsen, B. R. and Francomano, C. A. (1993), 'A type X collagen mutation causes Schmid metaphyseal chondrodysplasia', *Nat. Gen.*, 5(1), 79–82.

Webster, S. V., Farquharson, C., Jefferies, D. and Kwan, A. P. L. (2003), 'Expression of type X collagen, Indian hedgehog and parathyroid hormone related-protein in normal and tibial dyschondroplastic chick growth plates', *Avian Pathol.*, 32(1), 69–80.

Winter, W. P., Buss E. G., Clagett C. O. and Boucher R. V. (1967), 'The nature of the biochemical lesion in avian renal riboflavinuria -II. The inherited change of a riboflavin-binding protein from blood and eggs', *Comp. Biochem. Physiol.*, 22(3), 897–906.

Wolc, A., Arango, J., Jankowski, T., Dunn, I., Settar, P., Fulton, J. E., O'Sullivan, N. P., Preisinger, R., Fernando, R. L., Garrick, D. J. and Dekkers, J. C. M. (2014), 'Genome-wide association study for egg production and quality in layer chickens', *J. Anim. Breed. Genet.*, 131(3), 173–82.

Wu, X., McKenna, M. A., Feng, X., Nagy, T. R. and McDonald, J. M. (2003), 'Osteoclast apoptosis: the role of Fas *in vivo* and *in vitro*', *Endocrinology*, 144(12), 5545–55.

Xu, H. P., Shen, X., Zhou, M., Fang, M., Zeng, H., Nie, Q. and Zhang, X. (2010a), 'The genetic effects of the dopamine D1 receptor gene on chicken egg production and broodiness traits', *BMC Genetics*, 2010, 11, 17, doi:10.1186/1471–2156-11–17.

Xu, H. P., Shen, X., Zhou, M., Luo, C. L., Kang, L., Liang, Y., Zeng, H., Nie, Q. H., Zhang, D. X. and Zhang, X. Q. (2010b), 'The dopamine D2 receptor gene polymorphisms associated with chicken broodiness', *Poult. Sci.*, 89(3), 428–38.

Xu, H. P., Zeng, H., Zhang, D. X., Jia, X. L., Luo, C. L., Fang, M. X., Nie, Q. H. and Zhang, X. Q. (2011), 'Polymorphisms associated with egg number at 300 days of age in chickens', *Genet. Mol Res.*, 10(4), 2279–89.

Yang, F., Cao, H., Xiao, Q., Guo, X., Zhuang, Y., Zhang, C., Wang, T., Lin, H., Song, Y., Hu, G. and Liu, P. (2016), 'Transcriptome analysis and gene identification in the pulmonary artery of broilers with ascites syndrome', *PLoS One*, 11, 6, e0156045, doi:10.1371/journal.pone.0156045.

Yonash, N., Bacon, L. D., Witter, R. L. and Cheng, H. H. (1999), 'High resolution mapping and identification of new quantitative trait loci (QTL) affecting susceptibility to Marek's disease', *Anim. Genet.*, 30(2), 126–35.

Yuan, J., Sun, C., Dou, T., Yi, G., Qu, L., Qu, L., Wang, K. and Yang, N. (2015), 'Identification of promising mutants association with egg production traits revealed by genome-wide association study', *PLoS One*, 10, 10, e0140615, doi:10.1371/journal.pone.0140615.

Zhang, G., Zhang, L., Wei, Y., Wang, J., Ding, F., Dai, G. and Xie, K. (2012), 'Polymorphisms of the myostatin gene and its relationship with reproduction traits in the Bian chicken', *Anim. Biotechnol.*, 23(3), 184–93.

Zhang, H., Liu, S. H., Zhang, Q., Zhang, Y. D., Wang, S. Z., Wang, Q. G., Wang, Y. X., Tang, Z. Q. and Li, H. (2011), 'Fine-mapping of quantitative trait loci for body weight and bone traits and positional cloning of the RB1 gene in chicken', *J. Anim. Breed. Genet.*, 128(5), 366–75.

Zhou, H., Deeb, N., Evock-Clover, C. M., Mitchell, A. D., Ashwell, C. M. and Lamont, S. J. (2007), 'Genome-wide linkage analysis to identify chromosomal regions affecting phenotypic traits in the chicken. III. Skeletal integrity', *Poult. Sci.*, 86(2), 255–66.

Zhou, M., Du, Y., Nie, Q., Liang, Y., Luo, C., Zeng, H. and Zhang, X. (2010), 'Associations between polymorphisms in the chicken VIP gene, egg production and broody traits', *Br. Poult. Sci.*, 51(2), 195–203.

Chapter 2

A balanced approach to commercial poultry breeding

Nicholas B. Anthony, University of Arkansas, USA

1 Introduction

Per capita poultry production, on the global stage, has far exceeded expectations. In fact, poultry's share of global meat production has more than doubled over the past half century. This growth is expected to continue as reflected in the world food consumption figures and projections to the year 2030 (FAO, 2003) and 2050 (FAO, 2012). Poultry production in the United States has continued to experience ongoing growth in both production and per capita consumption (National Chicken Council, 2016). Chicken meat consumption surpassed pork consumption in the mid-1980s and beef in the early 1990s. Total poultry consumption (chicken + turkey) is currently poised to surpass total combined red meat consumption (beef + pork). Much of this success has been due to the ability of the poultry industry to produce an affordable, quality protein product in a cost-effective fashion. Per capita egg consumption trends have been stable since 2000, averaging 253 eggs (American Egg Board, 2016). Egg prices have increased but have remained very competitive as a cost-effective protein/energy source.

Success in the development of the egg and meat sectors of the poultry industry required the development of separate, unique selection programmes. The concept of the dual-purpose bird that could produce both eggs and meat was challenged primarily because of the negative correlation between the two traits. Therefore, in the late 1940s selection programmes emerged that focused specifically on the development of either egg- or meat-type chickens. Beyond the basic understanding of this negative relationship has been the exploitation of genotype, environment and interaction of the two as a means to advance the phenotype of the bird. Progress to this point has been accomplished through the implementation of traditional quantitative and population genetic practices. The ability

http://dx.doi.org/10.19103/AS.2016.0011.03

to forecast and adapt to consumer preference and adapt technological changes has led to continued efficient poultry production.

2 Balance of supply and demand

The concept of 'supply and demand' or 'resource allocation' seems to be relatively simple but it is often forgotten. The resources available to an animal are allocated towards the immediate needs. It is clear, however, that these needs change as the animal ages. The growth of the avian embryo has a clear path based on developmental priorities (Lillie, 1908; Hamburger and Hamilton, 1951; Lilja, 1983). For the embryo, neurological development precedes skeletal development which is followed by muscle formation. Post-hatch growth is having a similar pattern of priority with deposition of carcass fat having the lowest priority (Katanbaf et al., 1988; Anthony et al., 1989, 1991). It is easy to visualize growth as the stacking of respective growth curves for neurological, skeletal, muscle and fat deposits each having different timing of accumulation, inflection points and asymptotic contribution (Fig. 1). In an environment that has unlimited resources and no environmental challenges, the animal is able to achieve its genetic potential for all aspects of growth. Unfortunately, achievement of commercial genetic potential for growth and yield characteristics does not always result in the most fit individual as indicated by the negative correlation between growth and reproductive performance.

Growth will not occur until the maintenance requirements for bird are met. Maintenance was defined by Arnsby and Moulton (1925) as the conservation status of an animal that is not performing any work or producing any product. When nutrient intake and excretion are perfectly balanced the animal does not experience any weight gain or loss. See Leveille and Fisher (1960), Leveille and co-workers (1960), Sakomura and Coon (2003), and Bonato et al. (2011) for other definitions of maintenance. It is very difficult to calculate maintenance on a commercial level because once it is exceeded growth occurs which adds to the maintenance cost for the following day. Both nutritionists and geneticists have attempted to separate the efficiencies of maintenance and growth through partitioning the components of feed utilization. Residual feed intake is one such measure (Koch et al., 1963; van Bebber and Mercer, 1994; Aggrey et al., 2010; Willems et al., 2013).

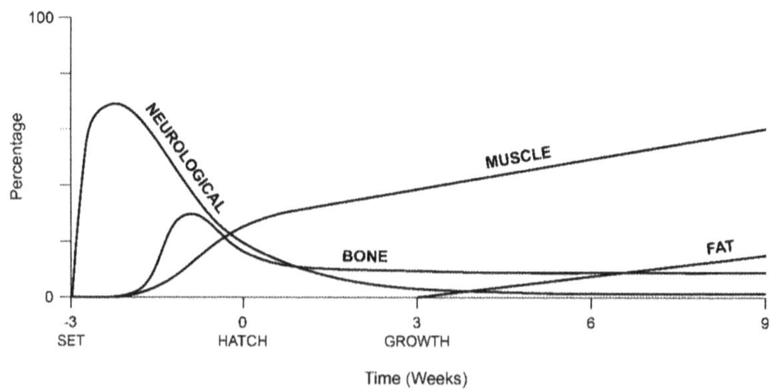

Figure 1 Simulated growth curves for neurological, bone, muscle and fat development.

The growth of an individual in a resource-restricted environment will result in the failure to meet genetic potential of growth-related traits. The timing and duration of the restriction period will, however, impact the ability of the bird to recover from the restriction. This becomes particularly important when considering the nutrition of the breeder hen and what she is able to deposit in the egg in support of embryonic and post-embryonic development (Kidd, 2003; Calini and Sirri, 2007). Similarly, delay in chick placement immediately post hatch has been shown to have a detrimental effect on processing yields. In some instances, commercial broilers have been shown to recover from short-term restrictions if they occur in the periods of hypotrophic growth after the first week post hatch (Velleman et al., 2014). In other instances, the growth recovery from a short-term early restriction environment is variable (Cherry et al., 1978; Malone et al., 1980; Plavnik and Hurwitz, 1988). This has contributed to the ongoing discussion regarding restriction, recovery and compensatory gain (Zubair and Lesson, 1996). The broiler breeder industry manages breeder weights as a means of reducing lean and fat tissue resulting in a recovery of fitness traits important for onset, synchrony and persistence of lay. The criteria for restriction is breed specific and defined by the company protocol for which the breeder stock was obtained.

Resource allocation is more than simply meeting the needs associated with the timed deposition of tissues in a full or restricted fed environment. Activation of the immune system in response to a disease challenge has an associated resource cost (Rauw, 2012). Cold or heat stress will also have an impact on resource allocation because there is an energy cost to maintenance of body temperature (Siegel, 1995). Energy expenditures related to bird-to-bird interactions and injury will impact resource allocation (Craig, 1982; Shini et al., 2010; Lay et al., 2011). Regardless of how it happens, the shifting of resources will be at the expense of growth and in some cases will result in a loss of body weight. At the end of the challenge, growth will resume when resources once again exceed maintenance. The timing and the duration of the challenge could have a tremendous impact on the ability of the bird to return to pre-challenge condition. Confounding this is the pre-programming of the bird and the ability to shift resources away from one demand to a secondary demand associated with a challenge condition. If an animal is unable to maintain developmental homeostasis they will likely experience slowed growth or even succumb to the challenge.

One must consider a population to be under the same constraints that individuals may experience. An individual manages its available resources as a means of survival and growth to maturity. Siegel and Dunnington (1997) described the importance of genetic homeostasis of a population as compared to the developmental homeostasis of an individual. For genetic homeostasis, the population must maintain enough genetic variation to moderately perform with regard to growth, reproduction and immune response. In addition, the population must maintain enough heterozygosity to buffer against challenge and change in environments. These concepts are important considerations for a balanced selection approach and will be discussed later in this chapter.

3 Evolution of selection programmes

Poultry breeding and genetics found its roots many years ago when selection was for qualitative traits focused on plumage colour, pattern, structure and distribution (Smyth, 1990). Other major gene variants included musculature, skeletal, nervous, skin and immunity. Small population size and close mating structure exposed mutations through increased homozygosity. Often discovered as a deviation from 'normal', mutations were concentrated and exaggerated in populations to create the early breeds. In 1874 the breeds and standards were formalized when the American Poultry Association published the first official breed standard for the fancy poultry of North America. In addition to the novelty component of early poultry breeding was the contribution to the understanding of Mendelian heredity (Punnett, 1923). The contributions of poultry to genetics from Neolithic times to the present is reviewed (Siegel et al., 2006; Siegel, 2014). Today the modern recognized poultry breeds serve as a reservoir of conserved genetic variation.

Selection practices during the twentieth century evolved from pure breed populations and single trait selection to a more sophisticated process that manages phenotypic and pedigree information in a sustainable way. Early quantitative selection practices utilized mass selection of individuals based on individual phenotype. Truncation selection was applied. For the most part, selection targeted high final body weights with little regard for other production traits. Qualitative traits continued to be identified in the hatchery, recorded as a defect and eliminated. Defects included colour slips, curled toes and cross beak phenotypes. The primary selection, however, focused on body weight to a specific age. It was not long before the traits correlated with body weight became apparent. These included the reduction of general fitness traits and increases in fat deposition. Processing ages reduced as birds were able to achieve targeted processing body weights at a younger age. This, in turn, resulted in the shift of weight selection to younger ages. Feed conversion became part of the equation as feed represent approximately 70% of the input cost of rearing a broiler or leghorn chicken. The complexity of selection programmes increased.

A modern commercial pedigree line is a population defined by the genetic constitution and transmission of the genes from one generation to the next (Falconer and MacKay, 1996). In the absence of mutation, migration, artificial and natural selection, and genetic drift a population will achieve equilibrium. The commercial geneticist, working in concert with the outside forces will manage the evolution of the population with decisions including but not limited to selection intensity, mating structure, population size and introduction of new genetic material. The poultry geneticist even attempts to manage the rearing environment as to minimize the impact of environmental and genotype by environment interactions on phenotypic variation. Decisions made will literally determine the fate of the population. Therefore, a balanced approach to selection is more than properly weighting a multitude of traits in a selection index.

Artificial selection will impose on the genetic homeostasis of a population. This can be realized as a shift in the fitness of the selected population as it attempts to re-establish homeostasis. Dunnington and Siegel (1996) discussed this with regard to correlated responses as influenced by reallocation of resources redirected to 8-week body weight. In fact, change due to divergent selection for 8-week body weight was reported for growth, metabolic, reproduction and immunological characteristics. It is clear that even early in a selection programme there are individual bird and population changes that reflect the change in resource allocation due to, in this case, achieving a body weight that is different from that found in the founder population. Relaxed selection pressure often results in the population holding or experiencing a slight drift back towards the original base population. This is likely associated with fertility differences across the distribution of the selection trait. Natural selection would not typically favour extreme phenotypes. Reverse selection applied early in a selection programme will allow for the return to base population values.

Long-term selection studies have shown that it is possible to approach genetic and physical selection limits for chickens (Dunnington and Siegel, 1996), Japanese quail (Marks, 1996; Nestor et al., 1996), turkeys selected for body weight (Nestor et al., 2008) and turkeys selected for egg production (Emmerson et al., 2002; Nestor et al., 2004). Long-term selection is often accompanied with selection plateaus that are several generations in duration but often are disrupted by a sudden resumption of response. Favourable mutations and their interaction with other genes may be responsible for breaking a selection plateau. Subtle environmental shifts and adaptation to these changes may also help break a selection plateau. Marks (1996) reported that sufficient genetic variation was present during the pause periods for his quail lines but the line failed to respond to forward selection. Reverse selection is used to determine if there has been depletion of additive genetic variation. The failure of a population to respond to reverse selection is discussed (Falconer and MacKay, 1996).

It is difficult to evaluate the selection response observed in the field since it is confounded with adaptation to new resource priorities and approach to selection limits. This becomes even more complex as expanded lists of traits are incorporated in index selection. Many factors influence the ability of the population to buffer the impact of genetic change. These factors include the maintenance of heterozygosity through dominance, over-dominance and epistatic effects. Polygenic networks may be in place that maintain genetic homeostasis. If these networks are modified through selection the population may collapse. This would be similar to a threshold trait (ascites, tibial dyschondroplasia, femoral head necrosis) that requires an accumulation of environmental challenges to occur before the developmental homeostasis is lost allowing the expression of the negative phenotype. It is clear that many of the decisions regarding selection practices and mating structure and environment change can result in unintended consequences and population extinction. Maintenance of physiological homeostasis is further discussed (Rauw et al., 1998; Cheng and Muir, 2005; Cheng, 2010).

Quoting from Siegel and Dunnington (1997) 'Many of the foundation theories of population genetics in poultry and livestock breeding are based on the early writings of Fisher, Wright, and Haldane as well as those of Lush (1945), Lerner (1950), Li (1955), Pirchner (1969) and Falconer and MacKay (1996)'. The fundamentals of selection today are in common with the ways of the past. Although our tools are more sophisticated and levels of accuracy have greatly improved, we are still trying to move genes in a way that tease out the genotypes that are desirable for humane, efficient and economical poultry production. Selection indexes were developed to incorporate simultaneous selection of several traits weighted based on perceived relevance.

It is important to recognize that many of the mating strategies that we still use today were influenced by plant breeding strategies being implemented at the time. The importance of accelerating genetic gain was key. Therefore, strategies for reducing generation interval, increasing selection intensity and exploiting additive variation were developed (Hill, 2014). In addition, line-cross strategies were developed to exploit the non-additive genetic variation. This resulted in the development of the three and four-way cross system currently used by today's poultry industry and outlined in the next section.

Genetic management of growth rate and body weight continue to be a major focus for essentially all commercial breeding programmes. Growth-related traits tend to be moderately to highly heritable and therefore respond readily to selection. See Chambers (1990) for a comprehensive summary of genetic parameters associated with production traits. Truncation selection is applied at industry-defined ages primarily to reduce bird numbers throughout the growth phase. If desired, selection at certain points along the growth curve could allow for the adjustment of growth curve parameters. Anthony (1995) reported that trait selection should be applied at the age that processing will occur in order to maximize the selection response. Feed conversion ratio, as previously mentioned, carries an economic as well as physiological importance. Just as body weight is the composite of the bone, fat and organ growth curves, feed conversion ratio is as complex (Emmerson, 1997). Both traits are clearly intertwined, especially since body weight gain during the feed conversion period is the denominator of the feed conversion ratio. There is an ongoing struggle to resolve the conflict between the selection criteria and breeding goal. Emmerson (1997) clearly described this conflict: 'The commercial goal of feed conversion selection is to reduce the amount of feed required to grow birds to a constant market weight. This goal favours fast-growing individuals, as they will reach market weight at an earlier age and consequently, energy requirements for maintenance will make up a smaller percentage of total energy intake. However, feed conversion can only be practically evaluated for a fixed age period, which penalizes heavier individuals because their maintenance requirements are higher and energy consumption of maintenance makes up a larger percentage of feed consumed during the test period.'

The poultry selection programme of the twenty-first century continues to advance in sophistication as technologies and computation power expands at exponential rates. Traits of economic importance such as growth, yield and feed conversion ratio will continue to be the primary focus of every primary breeder. However, new challenges, many driven by the consumer, will test the ability of the commercial broiler and layer to maintain genetic homeostasis. For example, the elimination of antibiotics from the feed, climate change, the development of ethanol industry (corn price and distillers' grains) and animal welfare concerns are just a few of the changes that will directly impact the industry and alter the scope and complexity of selection goals and pressure.

Fortunately, the poultry industry has matured from single trait or simple index selection to the present-day selection programme that monitors more than fifty traits (Katanbaf and Hardiman, 2010). It is a hallmark of the poultry industry to produce a product focused on improved health and well-being. Therefore, half of the fifty traits are directed towards fitness. The remaining traits are focused on economically producing a rapidly growing, high-yielding feed efficient product. It should be understood that not all of the traits are used directly in the selection of breeders. However, monitoring these traits allow for fine-tuning of breeding decisions made at final selection. In addition, monitoring these traits has allowed for the development of massive historic records documenting selection and correlated responses.

4 Pipeline genetics

The primary poultry breeder must have a clear understanding of the consumer needs because the product must go through several generations of managed multiplication and breeding (Anthony, 1998; Pollock, 1999). It is therefore critical that breeder goals be set and managed accordingly. The flow of product through the programme not only takes advantage of additive genetic variation but also non-additive traits that are exposed when lines are eventually crossed to produce parent stock that yields our commercial bird. Most of the commercial product is the composite of a four-way cross between specialized male

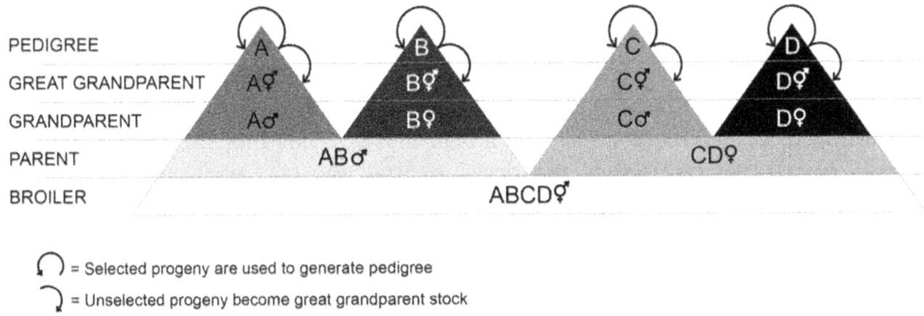

Figure 2 Pedigree structure for commercial broilers; 4-way cross.

and female lines selected for line-specific traits (Fig. 2). This process is often referred to as the pipeline by commercial geneticist.

The pedigree level represents the position in the pipeline where permanent genetic progress is made. Therefore, if the end product is a four-way cross, as diagrammed, then four separate pedigree lines must contribute. A primary breeder will rely on the development of pedigree lines that will focus independently on traits associated with the breeding goals set for the respective line. The position that these lines are used in the mating structure helps guide the selection focus. For example, the 'A position' pedigree population will provide the males that generate the male side of the parent stock package. The 'B position' pedigree population will contribute the females that generate the male side of the parent package. The 'C position' will contribute the males that generate the female side of the parent package. The 'D position' contributes the females that generate the female side of the parent package. Male and female parent packages may be purchased directly from the breeding company or novel combinations can be created by combining male parent stock from one company with a female parent stock from a competitor. The assumption, however, is that the primary breeder has designed a breeder package that will meet the respective consumer and or market needs.

The A and B position male lines focus their selection opportunity on growth- and yield-related traits. These would include body weight, feed conversion ratio, liveability, leg health and breast yield. A trait such as breast yield may become a composite trait that is built off of a variety of measures. For example, direct measures such as the subjective conformation scoring; quantitative measures including breast length, width and depth measures; ultrasound; DEXA; and NMR are considered. Also included would be indirect measures from sibling carcass processing data. All of these traits are descriptive of the same breast yield phenotype but have different heritability's and correlations associated with them. The geneticist must balance the pros and cons of these measures in describing the breast yield trait. The C position population will also have considerable selection pressure on body weight, feed conversion and breast meat yield. Since this population will contribute to the female parent stock, fitness traits including egg production, hatchability and liveability are also weighted into the index. The D position population will focus primarily on egg production, fertility and hatchability. Although secondary, bodyweight, feed conversion and breast yield are included in the index.

The overarching theme of balanced selection is to move all of the respective traits in the direction of the preferred response (i.e. increased percentage yield and decreased feed conversion ratio). This requires a sound understanding of the heritabilities and correlations between the traits for the respective populations. These relationships are not 'one size fits all' but rather population specific. It is also important to gather sufficient phenotypic sampling in the pedigree populations to identify desirable birds. Often there is the temptation to overreact to weighting a trait that you have fallen behind on. This must be avoided because it diminishes the pressure that you can apply for other traits. This can create a see-saw effect in response and product variation throughout the pipeline. Greater selection pressure is placed on males since mating ratio is typically one male to ten females.

Birds not selected at the pedigree level move down the pipeline to the great grandparent (GGP) level. The mating ratio of one male to ten females is the same for the GGP level. Selection pressure is on the male side and based on the performance traits measured in the original pedigree. GGP female selection is minor and generally based on defects that may interfere with reproductive success. Progeny from GGP breeders generate Grandparent (GP) offspring. Male and female GP are selected in the hatchery for general hatchery defects. Selection for GP males is through truncation of males that do not achieve minimal body weight. Parent stocks are produced by crossing A position GP males with B position GP females. Similarly, the C position GP male is mated to the D position GP female. Reciprocal crosses are not generated at the GP level. The terminal cross is the commercial broiler. Strategies can be employed in the pipeline to take advantage of sex-linked traits that can be used to sex day-old chicks at hatch. Specific mating's produce sexable chicks at the parent or broiler levels. Occasionally, the sex-linked dwarfing gene is implemented in the pipeline; however, the performance loss associated with the dwarfing allele offsets the benefits of its inclusion.

It takes 4–5 generations (usually 4–5 years) for change at the pedigree level to work its way down to the commercial level. Often the pipeline generations are referred to as multiplication steps. Although this is true, the pipeline also presents the opportunity to recover non-additive variation lost in pedigree selection. The pipeline contains a specific line cross to produce the male parent stock and an independent line cross to produce the female parent stock. This is followed by a terminal cross to generate the commercial product. Through these crosses the geneticist attempts to take advantage of non-additive genetic variation while managing to retain sufficient additive variation for the traits of interest. The identification of a desirable breed combinations or 'Nicks' between parent packages can elevate a breeding company above their competitors. Just as significant, however, is if negative attributes appear at the parent stock or broiler level and your product is no longer desired. One additional benefit of the four-way cross is the blending of pure line genetics, making it impossible to acquire pure lines from cross bred commercial parent or broiler stock.

The breeder industry has morphed in concert with the changing market sectors. Once a market requiring efficient production of birds for the whole bird market has changed to one that is predominately further processed for parts production or the debone market (Fig. 3). With that came the need for bird uniformity to minimize cutting loss by mechanized processes. Emmerson (2003) divided the industry into three sectors. The first is designed around the developing markets that prefer a slower growing bird that is balanced with reproductive performance. In the United States this would serve the emerging organic and free-range markets. In developing countries this would serve whole bird markets. A

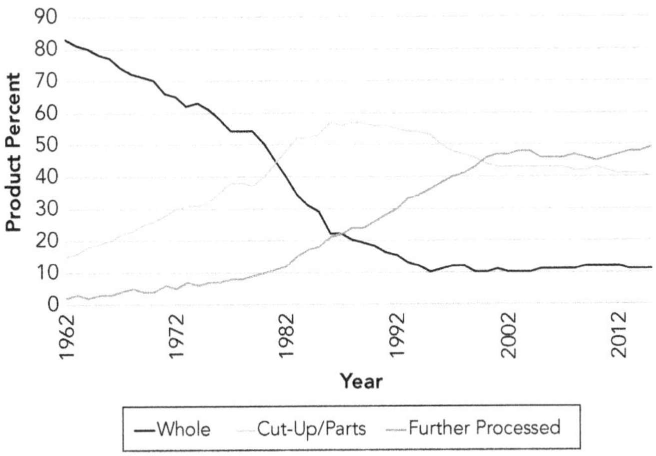

Figure 3 Allocation of processed poultry products over the past half century. Source: Data from National Broiler Council, 2014; 2010–2015 are estimated values.

second sector describes the parts market which would target broiler characteristics with minor emphasis on reproductive efficiency. The parts market serves both fast food and service markets. The third sector represents the debone market which focuses on yield and conformation. These birds tend to be reared to heavier weights and have taken advantage of increasing processing plant yields by increasing the per bird weight per shackle. This product is directed towards the fast food, food service and reformed poultry products.

With these sectors of the industry in mind the commercial geneticist has to develop a breeding programme that is both current and visionary to the future consumer preferences. Most recently the poultry industry has experienced some seismic shifts in regulations that will certainly test the resilience of commercial populations. The cage-free system for commercial egg production will likely expose behavioural issues since selection has been for close rearing conditions and adaptability in close confinement. In a cage-free system, there is increased opportunity for more bird-to-bird interactions in establishing the pecking order (Muir and Craig, 1998). This will lead to a shift in resource allocation away from production traits as birds expend energy protecting territory. The elimination of antibiotics in poultry production will present new challenges with disease organisms that have been kept in check. There will be a shift in resource allocation to address immune challenges that are certain to occur. The development of organic and free-range markets creates yet another set of environmental shifts that will redirect resources from growth- and yield-related traits. As previously mentioned in this chapter, some commercial lines may not be competitive under these new rearing environments. In effect the line becomes extinct because it is no longer relevant.

The movement towards slow-growing chicken reared in a free-range or organic environment has created a renewed interest in the use of heritage breeds. Successful use of modern commercial broiler strains in this 'more natural' environment is unreasonable considering they have been selected for performance in a tightly controlled commercial environment. Rearing the commercial broiler in a free-range environment would be akin to moving the human population of New Jersey to a third-world country and the vice versa. The additional challenge associated with the elimination of antibiotics would add further complications. Environmental challenges associated with diet and disease would have a huge impact on survival and growth of these human transplants. It does not mean that the human population is weak but rather exposed to different environmental conditions.

The modern commercial chicken can, however, play an important role in the slow growth market. One could utilize commercial parent stock crossed on an improved heritage breed in a mating structure similar to that previously described. That could result in an intermediate product that would be more tolerant of the dynamic free-range or organic environment. The intermediate product would express slowed growth rate, poor feed conversion and reduced yield as compared to the commercial broiler grown in confinement but improved relative to a pure heritage breed. Additional increases should be expected with regard to the carbon footprint as slow growth strains and layers reared in free-range and/or organic conditions will place a higher demand on the environment because of the efficiency loss relative to the modern commercial chicken (Williams et al., 2006; Pelletier et al., 2014).

5 Balanced selection, molecular methods and animal well-being

5.1 Balanced selection and molecular methods

As technology advances, there will be a plenty of new tools available for the commercial geneticist to 'simplify' the selection process or improve on the accuracy of breeding values. Siegel and co-workers (2006) and Fulton (2014) reviewed the strengths and limitations of many of molecular tools available over the past two decades. Needless to say we are in a discovery-rich environment with molecular genetics. In fact, Fulton describes the evolution of ten different molecular applications to the poultry model. It appears that the most efficient applications for molecular methods today would be for direct selection for sex-limited and sex-influenced traits, traits that are difficult, expensive or have to be measured late in production (Fulton, 2014). Molecular methods could also be used to genotype day-old chicks as a means to eliminate birds that will not be desirable later. Molecular focus on traits that are highly heritable and relatively easy to measure should be avoided at this point in time because the existing infrastructure and cost advantage of a traditional breeding programme makes it difficult to justify. Highly heritable traits are easy to move using traditional methods. Gene editing, gut microbiome and the development of primordial germ cells are the latest 'opportunities' for the modern breeding programme.

Opportunities to move towards molecular methods and away from traditional selection methods will occur. The thought of collecting pedigreed blood samples in the hatchery and rearing only the candidates for replacement is an attractive concept. Unfortunately, the molecular approach at this point in history is like peeling an onion. Pulling back one layer exposes another layer of challenges. Basic research has not been sufficiently married to the applied, thus leaving a disconnect between genotype and phenotype and genotype and environment. One has to understand the impact of resource reallocation that will occur when preferentially selecting for genotype alone.

Although molecular methods are developing rapidly, the importance of collecting phenotypic data cannot be undervalued. If not used directly in the breeding programme, phenotype is a means of monitoring selection progress and impact on correlated traits. It assures a physical connection with the population that allows the geneticist to gauge selection progress and evaluate the genetic stability of the population. This leads to product awareness that supports product promotion. Finally, there is the opportunity for discovery of new traits and challenges that may impact the future product.

5.2 Balanced selection and animal well-being

Because commercial breeding of poultry has led to rapid production advances, we have seen tremendous industry growth and increased per capita consumption. Unfortunately, this has led to a number of consumer misconceptions which include, but are not limited to, environment, health and animal welfare. Fortunately, all of these areas are priorities for the poultry industry because they have a responsibility to be stewards for these areas, and the optimization of the former leads to sustainability. The breeding objectives with regard to feed conversion and growth have reduced the carbon footprint of poultry production by 23% over the past 20 years (Williams et al., 2006). Both health and animal welfare continue to be addressed in current selection programmes as the balanced selection approach not only focuses on the performance traits of growth, yield and feed conversion but also fitness traits including fertility, hatchability and early chick survivability. Mortality records of processed flocks dating back to the mid-1920s clearly show, when confinement rearing was initiated, rapid declines in mortality were observed. Beyond the 1950s progress in bird survivability continued to be realized along with concomitant progress in market weight and feed conversion (National Chicken Council, 2016). Hence, commercial primary breeding of broilers has improved health on a yearly basis.

The balanced approach to genetic selection has resulted in a broiler that is adaptable and thrives in the modern commercial production environment. The commercial geneticist is responsible for managing the evolution through artificial selection and breeding strategies applied in conjunction with natural selection. This process accelerates permanent genetic change within the environmental constraints of the poultry industry. Certainly new challenges appear as a population matures from a heterogeneous population to one that is uniform. Inbreeding is managed at the pedigree level in a primary breeding programme while the final commercial product is not inbred because it is a product of a three- or four-way cross. Inbreeding at the pedigree level will expose the good and bad non-additive traits but natural selection did the same as was recognized by Darwin in the *Origin of Species*. The finches of the Galápagos islands are specific to the island environment they were selected. The commercial broiler is specific to the environment they were selected. The commercial geneticist has developed a bird that is 'fit' in modern commercial environment, more fit than any other chicken breeds reared in the commercial environment.

6 Conclusion

It is projected that the current world population of 7.5 billion people will grow to 9.7 billion by 2050. In order to support this population growth, sustainable world food supply will have to increase by 70%. The food production in developing countries will need to double (FAO, 2012). Current breeding practices have greatly improved the yields and efficiencies of grain, livestock and poultry production. Future food security will require continued progress through a balanced approach to selection while implementing new technology and computing capacities. Globalization has created opportunities and challenges to expand our production capabilities. It is important that the geneticist utilize a balanced approach to selection to continue to improve on phenotype while generating a safe wholesome product in a programme having a continued focus on fitness and well-being.

7 References

Aggrey, S. E., A. B. Karnuah, B. Sebastian and N. B. Anthony. 2010. Genetic properties of feed efficiency parameters in meat-type chickens. *Genet. Sel. Evol.* 42:25. doi:10.1186/1297-9686-42-25.

Anthony, N. B. 1995. Selection for increased body weight in poultry. What role does selection age have on response? pp. 61–77. In *Proceedings of the National Breeders Roundtable*, St. Louis, MO.

Anthony, N. B. 1998. A review of genetic practice in poultry: efforts to improve meat quality. *J. Muscle Foods.* 9:25–33.

Anthony, N. B., E. A. Dunnington and P. B. Siegel. 1989. Embryo growth of normal and dwarf chickens from lines selected for high and low 56-day body weight. *Arch. Geflugelk.* 53:116–22.

Anthony, N. B., D. E. Jones, E. A. Dunnington and P. B. Siegel. 1991. Allometric growth and DNA, RNA and total protein of brains from white rock chickens selected for 56-day body weight. *J. Animal. Breed. Genet.* 108:54–60.

Armsby, H. P., and C. R. Moulton. 1925. *The Animal as a Converter of Matter and Energy*. The Chemical Catalog Co. Inc., New York.

Bonato, M. A., N. K. Sakomura, J. C. Siqueira, J. B. K. Fernandes and R. M. Gous. 2011. Maintenance requirements for methionine and cysteine, and threonine for poultry. *S. Afr. J. Anim. Sci.* 41:209–22. https://doi.org/10.4314/sajas.v41i3.3

Calini, F., and F. Sirri. 2007. Breeder nutrition and offspring performance. *Braz. J. Poult. Sci.* 9:77–83. https://doi.org/10.1590/s1516-635x2007000200001

Chambers, J. R. 1990. Genetics of growth and meat production in chickens. In Crawford, R. D. (ed.), *Poultry Breeding and Genetics*, Elsevier, Amsterdam: The Netherlands, pp. 599–643.

Cheng. H., and W. M. Muir. 2005. The effects of genetic selection for survivability and productivity on chicken physiological homeostasis. *Worlds Poult. Sci. J.* 61:383–97.

Cheng, H. W. 2010. Breeding of tomorrow's chickens to improve well-being. *Poult. Sci.* 89:805–13.

Cherry, J. A., P. B. Siegel and W. L. Beane. 1978. Genetic-nutritional relationships in growth and carcass characteristics of broiler chickens. *Poult. Sci.* 57:1482–7.

Craig, J. V. 1982. Behavioral and genetic adaptation of laying hens to high density environments. *Bioscience* 32:33–7.

Dunnington, E. A., and P. B. Siegel. 1996. Long-term divergent selection for eight-week body weight in White Plymouth Rock chickens. *Poult. Sci.* 75:1168–79.

Emmerson, D. A. 1997. Commercial approaches to genetic selection for growth and feed conversion in domestic poultry. *Poult. Sci.* 76:1121–5.

Emmerson, D. A. 2003. Breeding objectives and selection strategies for broiler production. In Muir, W. E. and Aggrey, S. E. (eds), *Poultry Genetics, Breeding and Biotechnology*, CAB Publishing, Wallingford: Oxford, UK, pp. 113–26. https://doi.org/10.1079/9780851996608.0113

Emmerson, D. A., S. G. Velleman and K. E. Nestor. 2002. Genetics of growth and reproduction in the turkey. 15. Effect of long-term selection for increased egg production on the genetics of growth and egg production traits. *Poult. Sci.* 81:316–20.

Falconer, D. S., and T. Mackay, 1996. *Introduction to Quantitative Genetics*. 4th Edition. Pearson Education: Harlow, UK.

FAO. 2003. World agriculture: Towards 2015/2030 – An FAO Perspective. In FAO, editor. Rome: Earthscan. http://www.fao.org/3/a-y4252e.pdf.

FAO. 2012. World agriculture: Towards 2030/2050 – The 2012 revision. In FAO, editor. Rome: Earthscan. http://www.fao.org/docrep/016/ap106e/ap106e.pdf.

Flock, D. K., K. F. Laughlin and J. Bentley. 2006. Minimizing losses in poultry breeding and production: How breeding companies contribute to poultry welfare. *Worlds Poult. Sci. J.* 61:227–37.

Fulton, J. E. 2014. Genomic selection for poultry breeding. *Anim. Front.* 2:11–28.

Hamburger, V., and H. L. Hamilton. 1951. A series of normal stages in the development of the chick embryo. *J. Morphol.* 88: 49–92.

Hill, W. G. 2014. Applications of population genetics to animal breeding from Wright, Fisher and Lush to genomic prediction. *Genetics* 196:1–16.

Hunton, P. 1984. Selection limits: Have they been reached in the poultry industry? *Can. J. Anim. Sci.* 64:217–21.

Katanbaf, M. N., P. B. Siegel and E. A. Dunnington. 1988. Organ growth of selected lines of chickens and their F1 crosses to a common body weight or age. *Theor. Appl. Genet.* 76:540–4. https://doi.org/10.1007/bf00260904

Katanbaf, M. N., and J. W. Hardiman. 2010. Primary broiler breeding – striking a balance between economic and well-being traits. *Poult. Sci.* 89:822–4.

Kidd, M. T. 2003. A treatise on chicken dam nutrition that impacts on progeny. *Worlds Poult. Sci. J.* 59:475–94.

Koch, R., Swiger, L., Chambers, D. and Gregory, K. 1963. Efficiency of feed use in beef cattle. *J. Anim. Sci.* 22:486–94.

Lay Jr., D. C., R. M. Fulton, P. Y. Hester, D. M. Karcher, J. B. Kjaer, J. A. Mench, B. A. Mullens, R. C. Newberry, C. J. Nicol, N. P. O'Sullivan and R. E. Porter. 2011. Emerging issues: Social sustainability of egg production symposium: hen welfare in different housing systems. *Poult. Sci.* 90:1–17.

Lerner, I. M. 1950. *Population Genetics and Animal Improvement*. Cambridge University Press, New York, NY.

Lerner, I. M. 1954. *Genetic Homeostasis*. John Wiley and Sons, Inc., New York, NY.

Leveille, G. A., and H. Fisher. 1960. Amino acid requirements for maintenance in the adult rooster. III. The requirements for leucine, isoleucine, valine and threonine, with reference also to the utilization of the d-isomers of valine, threonine and isoleucine. *J. Nutr.* 70:135–40.

Leveille, G. A., R. Shapiro and H. Fisher 1960 Amino acid requirements for maintenance in the adult rooster. IV. The requirements for methionine, cystine, phenylalanine, tyrosine and tryptophan: The adequacy of the determined requirements. *J. Nutr.* 72:8–15.

Li, C. C. 1955. *Population Genetics*. University of Chicago Press, Chicago, IL.

Lush, J. L. 1945. *Animal Breeding Plans*. Iowa State University Press, Ames, IA.

Lilja, C. 1983. A comparative study of postnatal growth and organ development in some species of birds. *Growth* 47:317–39.

Lillie, F. R. 1908. *The Development of the Chick*. Holt, New York.

Lush, J. L. 1945. *Animal Breeding Plans*. Iowa State University Press, Ames, IA.

Katanbaf, M. N., and J. W. Hardiman. 2010. Primary broiler breeding – striking a balance between economic well-being traits. *Poult. Sci.* 89:822–4.

Malone, G. W., G. W. Chaloupka, E. W. Walpole and L. L. Littlefield. 1980. The effect of dietary energy and light treatment on broiler performance. *Poult. Sci.* 59:567–81. https://doi.org/10.3382/ps.0590576

Marks. H. L. 1996. Long-term selection for body weight in Japanese quail under different environments. *Poult. Sci.* 75:1198–203.

Muir, W. M. and J. V. Craig. 1998. Improving animal well-being through genetic selection. *Poult. Sci.* 77:1781–8.

National Chicken Council. 2016. U.S. Broiler performance, 1925 to Present. http://www.nationalchickencouncil.org/about-the-industry/statistics/u-s-broiler-performance/.

Nestor, K. E., J. W. Anderson, R., A. Patterson and S. G. Velleman. 2004. Genetic variation in body weight and egg production in an experimental line selected long term for increased egg production, a commercial dam line, and reciprocal crosses between lines. *Poult. Sci.* 83:1055–9.

Nestor, K. E., W. L. Bacon, N. B. Anthony and D. O. Noble. 1996. Divergent selection for body weight and yolk precursor in Coturnix coturnix japonica. 10. Response to selection over thirty generations. *Poult. Sci.* 75:303–10.

Nestor, K. E., J. W. Anderson, R. A. Patterson and S. G. Velleman. 2008. Genetics of growth and reproduction in the Turkey. 17. Changes in genetic parameters over forty generations of selection for increased sixteen-week body weight. *Poult. Sci.* 87:1971–9. doi:10.3382/ps.2008-00137

Pelletier, N., M. Ibarburu and H. Xin. 2014. Comparison of the environmental footprint of the egg industry in the United States in 1960 and 2010. *Poult. Sci.* 93:241–55.

Pirchner, F. 1969. *Population Genetics in Animal Breeding*. W. H. Freeman Co., San Francisco, CA.

Plavnik, I., and S. Hurwitz. 1988. Early feed restriction in chicks: Effect of age, duration and sex. *Poult. Sci.* 67:384–90.

Pollock D. L. 1999. A geneticist's perspective from within a broiler primary breeder company. *Poult. Sci.* 78:414–18.

Punnett, R. C. 1923. *Heredity in Poultry*. Macmilllan & Co., London.

Rauw, W. M., E. Kanis, E. N., Noordhuizen-Stassen and F. J. Grommers. 1998. Undesirable side effects of selection for high production efficiency of farm animals: a review. *Livest. Prod. Sci.* 56:15–33.

Rauw, W. M. 2012. Immune response from a resource allocation perspective. *Front. Genet.* 3:267. https://doi.org/10.3389/fgene.2012.00267

Sakomura, N. K., and Coon, C. 2003. Amino acid requirements for maintenance of broiler breeder pullets. *Proceedings of European Symposium on Poultry Nutrition*, Lillehammer, Norway, pp. 280–1.

Shini, S., G. R. Huff, A. Shini and P. Kaiser. 2010. Understanding stress-induced immunosuppression: Exploration of cytokine and chemokine gene profiles in chicken peripheral leukocytes. *Poult. Sci.* 89:841–51.

Siegel, H. S. 1995. Gordon Memorial Lecture. Stress, strains and resistance. *Br. Poult. Sci.* 36:3–22.

Siegel, P. B. 2014. Evolution of the modern broiler and feed efficiency. *Annu. Rev. Anim. Biosci.* 2:375–85.

Siegel, P. B., J. B. Dodgson and L. Anderson, 2006. Progress from chicken genetics to the chicken genome. *Poult. Sci.* 85(12):2050–60.

Siegel, P. B., and E. A. Dunnington. 1997. Genetic selection strategies-population genetics. *Poult. Sci.* 76:1062–5.

Smyth, R. J. 1990. Genetics of plumage, skin and eye pigmentation in chickens. Chapter 5. In Crawford, R. D. (ed.), *Developments in Animal and Veterinary Sciences, 22; Poultry Breeding and Genetics.* Elsevier Science Publisher: Amsterdam, The Netherlands.

van Bebber, J., and Mercer, J. T. 1994. Selection for efficiency in broilers: A comparison of residual feed intake with feed conversion ratio. *Proceedings of the 5th World Congress of Genetics Applied to Livestock Production*. 7–12 August 1994, Guelph, pp. 53–6.

Velleman, S. G., C. S. Coy and D. A. Emmerson. 2014. Effect of the timing of posthatch feed restrictions on broiler breast muscle development and muscle transcriptional regulatory factor gene expression. *Poult. Sci.* 93:1484–94. http://dx.doi.org/10.3382/ps.2013-03813.

Williams, A. G., E. Audsley and D. L. Sandars. 2006. Determining the environmental burdens and resource use in the production of agricultural and horticultural commodities. Main Report. Defra Research Project IS0205.

Willems O. W., S. P. Miller and B. J. Wood, 2013. Aspects of selection for feed efficiency in meat producing poultry. *Worlds Poult. Sci. J.* 69:77–87

Zubair, A. K., and S. Lesson. 1996. Compensatory growth in the broiler chicken: a review. *Worlds Poult. Sci.* 52:189–201.

Chapter 3

Marker-assisted selection in poultry

P. M. Hocking and J. Hickey, University of Edinburgh, UK

1 Introduction

Marker-assisted selection (MAS) is a form of indirect selection for a desired change in a trait based on a marker as distinct from direct phenotypic selection on the trait itself. Enzymes, proteins and blood group polymorphisms were investigated as suitable markers for quantitative traits in early research. In general, these studies were not very successful in detecting quantitative trait loci (QTL) because there were a limited number of markers, and tight linkages between marker and trait were rare (Bovenhuis et al., 1997). Nevertheless, blood group markers for the B^{21} and B^{12} haplotypes have been used as selection criteria to decrease susceptibility to Marek's disease (McKay, 1998).

Whereas there are several important single-gene mutations that are commercially important in poultry breeding, such as dominant white, the dwarfing gene (*dw*) and colour sexing genes, genetic markers are not required to change their frequency in a population. A bird may have one or two copies of the gene, or none at all, and the overall merit of the individual will be a balance between the economic value of removing one or two copies of the gene on the one hand and the merit of the bird for production traits such as feed conversion on the other. The application of such gene-based selection to supplement

http://dx.doi.org/10.19103/AS.2016.0011.04

the conventional determination of estimated breeding values (EBVs) is available in the literature (Fernando and Grossman, 1989). If a DNA marker was closely linked to an important gene (more accurately the mutation underlying a phenotypic difference in the population), this information could also be used in this method. The development of single nucleotide polymorphism (SNP) chips (described in Section 4), however, has made single-gene-based selection of largely theoretical interest only. However, there are several practical situations where indirect selection for a quantitative trait may be preferable to direct selection: specifically, if the desired trait is difficult to measure with precision (e.g. feather pecking and cannibalism); is expensive to measure (e.g. feed conversion efficiency); is observable in only one sex (e.g. semen quality); is measured late in life (total egg production and longevity); or requires that the animal is killed (e.g. carcass traits). This has led to the continued search for DNA markers with the required attributes for successful implementation of MAS. This search culminated in recent years with the development of microsatellites and latterly of SNP markers that have made the implementation of MAS possible in commercial pedigree breeding.

The general principles governing the success of MAS have been known for many years and are readily available in standard textbooks of quantitative genetics. The theoretical background surrounding the successful application of MAS will be briefly described in the next section where it will be shown that a single basic criterion largely determines the potential of MAS whatever marker is chosen.

2 Traditional marker-assisted selection theory

The conditions under which indirect selection are beneficial are well known (Falconer and MacKay, 1996: 219–20): indirect selection is better than direct selection for the trait (X) when the product of the genetic correlation between the marker (y) and the breeding value of the trait ($r_{X,y}$) and the accuracy of the marker (h_y, the square root of the heritability of the marker Y) is greater that the accuracy of direct selection (h_X), that is, when $h_y r_{X,y} > h_X$. This is a pretty tough criterion, requiring both the heritability of the marker (h_y^2) and the genetic correlation to be relatively high. Heritability (h_y^2) is a function of genetic variability for the trait that can be modified to some extent by taking steps to reduce environmental variation and by incorporating information from close relatives. These steps can be taken relatively easily and are readily achievable, whereas increasing the genetic correlation ($r_{X,y}$) is essentially a function of the number of markers: if a very large number of markers are nearly randomly distributed across the genome, a few are inevitably found to be close to a causative mutation – a difference in the DNA responsible for variation in the phenotype of an animal. Markers with the required genetic characteristics could then be used to supplement the phenotypic information used to calculate breeding values including traits that were not measured (Fernando and Grossman, 1989; van Arendonk et al. 1994).

3 Microsatellite markers

Microsatellites were among the earliest of the DNA markers that were potentially useful for MAS. Microsatellites were the first widely used markers that had the required attributes: they were numerous, randomly distributed in the genome and were detectable in common

a. A microsatellite in the nucleotide sequence of a double DNA strand

First copy:	GTACCATATATATCGTAAC
Second copy:	GTACCATATATCGTAAC

b. An SNP in the nucleotide sequence of a DNA double strand

First copy:	CCTTAGCTAATCGA
Second copy:	CCTTAGGTAATCGA

c. Genome selection using SNP effects estimated for a specific bird

SNP:	S1	S2	S3	S4	S5	S6	S7	S8
DNA:								
Phenotype:	+2	+1	0	+5	+3	-3	-1	+1

Overall merit: sum of SNP effects = +8

Figure 1 Genetic markers and SNP-based genome selection. (a) A microsatellite marker is a short tandem repeat motif (here ATATAT, shown by the red font) that is anchored to a specific location of the genome by the unique flanking sequence. The number of repeats varies and is detectible in sequencing gels or machines that determine small differences in the lengths of DNA fragments. (b) A DNA code change from cytosine (C) to guanine (G), indicated here by the red font, creates an SNP in a DNA (double) strand. Each SNP has a unique location identified by the flanking DNA sequence consisting of the four DNA bases, C, G, T (thymine) and A (adenine). The flanking sequences have to be sufficiently long to provide a unique identification for each SNP. (c) Whole-genome selection (WGS) based on SNPs is illustrated for 8 SNP markers with phenotypic effects shown. The sum of the SNP effects gives an estimate of the relative merit of the individual.

DNA-sequencing machines. Microsatellites are short tandem repeat sections of DNA that are anchored at specific locations (DNA sequence) in the genome (Fig. 1). The main disadvantage of microsatellites for the purpose of MAS was that they required technical expertise to determine and they were not sufficiently numerous to mark a sufficient number of useful genes. Nevertheless, these techniques were widely used and they led to the initial identification of many QTL in both experimental crosses (Sewalem et al., 2002) and commercial flocks (de Koning et al., 2003): this was an important development as it demonstrated that markers could be used to identify areas of the genome that contained genes with a detectable effect on economic traits (Hocking, 2005). The limitations of microsatellite markers were overcome with the development of techniques to rapidly genotype single SNPs, which are abundant in the genome, and are easy and relatively cheap to measure using robotic machines.

4 Single nucleotide polymorphism (SNP) markers

An SNP is a single base pair mutation at a specific locus, usually consisting of two alleles (Fig. 1). SNPs are widely distributed in very large numbers across the genome (Wong et al., 2004), and the ease and cost of genotyping large numbers of SNPs has made them the marker of choice for commercial application. A chicken SNP chip consisting of 600K SNPs covering most of the chicken genome (Kranis et al., 2013) is now available commercially and facilitates the genotyping of individual birds relatively cheaply and quickly. This and the associated statistical and computer storage technology have made previous methods largely redundant. Combining information from different sources (traits, markers, relatives and pedigree) is theoretically the most effective method for maximising genetic progress, and in the sections below we summarise published work on optimising MAS in poultry breeding using SNP markers.

5 Whole-genome selection (WGS)

WGS was proposed by Meuwissen et al. (2001) and has revolutionised many breeding programmes, including poultry breeding programmes. In simple terms, WGS involves establishing a training set that is genotyped and phenotyped, using this training set to calculate a prediction equation and, then, combining this prediction equation with marker genotypes of selection candidates to enable the breeding values of these birds to be calculated. Early assessments of the method using relatively small SNP chips for WGS in broiler populations (see Table 1) were encouraging (Gonzalez-Recio et al., 2009; Chen et al., 2011) and have led to the implementation of MAS for genetic improvement of pedigree broiler flocks and increasingly in layers. The breeder's equation models the rate of genetic gain that selection achieves and has four parameters – generation interval, selection accuracy, selection intensity and genetic variation. The process of being able to calculate breeding values of candidates based on their marker genotypes and a previously calculated prediction equation enables genomic selection to directly impact three of the four components of the breeders' equation. First, phenotypic information on the selection candidates themselves, their sibs or progeny is not required for accurate calculation of breeding values. By removing the requirement for such phenotypes the generation interval can be shortened. Second, in some circumstances, the accuracy of genomic prediction can be greater than the accuracy of breeding values calculated on an individual's own phenotypes, on the phenotypes of its sibs and other relatives, and on the phenotypes collected on its descendants. Thus, genomic selection can be more accurate than conventional selection. Third, collecting marker genotypes on a selection candidate is often cheaper than collecting sufficient phenotypic information pertaining to a selection candidate. Thus, when resources are limited, as is the case in practice, genomic selection can enable greater selection intensity than conventional selection. Finally, although genomic selection does not directly address available genetic variance, which is the fourth parameter in the breeders' equation, the genomic relationship matrix, which is a key component of genomic prediction methods, can be used to enable genome-driven optimal contributions for the management of utilisation of genetic variation.

Table 1 Genetic gain (%) from using genome SNP information in a pedigree broiler breeding programme (G-BLUP) over conventional BLUP estimation of breeding values. A total of 184 000 broiler chickens were assessed and 3000 birds were genotyped with a 58K SNP chip (from Chen et al., 2011)

		Accuracy of BV		
Trait	Heritability	BLUP	G-BLUP	Gain, %
Weight at 6 weeks, g	0.20	0.51	0.61	20
Breast muscle area, cm^2	0.30	0.34	0.40	18
Leg score/1	0.11	0.28	0.37	32

The key difference between traditional MAS and WGS is that WGS skips the significance testing step and uses all markers when deriving a prediction equation. In contrast, traditional MAS involves, first, identifying the markers that have significant effects and, second, building a prediction equation based only on those significant markers. The problem with the traditional MAS approach is that the significant markers typically only explained a very small percentage of the total genetic variation that affects a trait, and thus the accuracy of the resulting prediction equation was typically small. By using all of the markers, WGS-derived prediction equations can capture much more of the total genetic variation that affects a trait and consequently can be much more accurate. However, the accuracy of genomic prediction is typically only high if the training set used to train the prediction equation is well constructed. In practical breeding programmes, well-constructed genomic selection training sets are large (several thousand individuals), contain sufficient numbers of individuals that are close relatives to the selection candidates who are having breeding values calculated, and are sufficiently genetically diverse to enable the statistical model implicitly or explicitly model the underlying QTL.

In different species and in different breeding programmes, WGS has delivered breeding benefit in different ways. In dairy cattle, genomic selection has typically been used to reduce the generation interval considerably and to increase the selection intensity in some cases. In pig breeding, genomic selection has typically been used to increase the accuracy of selection for traits that are measured on an individual after the point of selection, such as traits related to female fertility. This has led to a better alignment between the breeding goal and the accuracy of selection.

6 Application of WGS to poultry breeding

Poultry breeding programmes have a number of features that impact how genomic selection can be beneficial (Wolc et al., 2016; Meuwissen et al., 2001). Traditional breeding programmes in poultry already have short generation intervals. Broilers had multiple overlapping generations per year with selection every six weeks and non-overlapping annual generations in layers. There is some scope for shortening the generation interval in layers but not as much as was the case for dairy cattle (Wolc et al., 2016). Poultry breeding programmes have very large numbers of selection candidates and high selection intensities. Whereas each selection candidate delivers low marginal revenue, the very high reproductive rate, combined with an efficient multiplication phase during the formation of line crosses, results in the costly marginal improvements at the pedigree level being spread over a very large number of commercial progeny. Nevertheless, to be economically viable, genotyping has to be relatively inexpensive and also requires robust procedures for collection of blood samples, processing for DNA extraction, accurate sample tracking and large-scale data storage and processing that are also costly to develop and implement. Taken together, the primary current role for genomic selection in poultry breeding programmes is to increase the accuracy of selection, especially for traits that are expensive to measure (e.g. feed conversion ratio), measured late in the life of an individual (e.g. female fertility) or in one sex (e.g. egg production in male egg-laying strains). For traits which are not available on selection candidates, genomic predictions capture information on Mendelian sampling terms, that is, they estimate more accurately which alleles an individual selection candidate inherits from its sire and dam. This opens up the 'black box' and allows selection among full sibs, for example, in contrast to conventional pedigree-based predictions. It has also been shown that this can have a considerable effect on reducing the rate of inbreeding or conversely to substantially reduce the size of the breeding programme for the same rate of inbreeding, especially in layers (Wolc et al., 2015).

One of the essential criteria of indirect selection is that the correlation between marker and trait is high and stable over successive generations, as outlined in Section 2. Experience of applying WGS in poultry breeding programmes has shown that the correlation between the effects estimated in the training set erodes quickly over two or three generations, even for high-density SNP chips, and has to be re-estimated. This occurs because recombination breaks down the linkages between the markers and their phenotypic effects. Whereas recombination in a single bird may not have too serious an impact on its genomic EBV (GEBV), when integrated over a large flock, the effect is to reduce the overall correlation and prediction accuracy. Considerations of the optimum size of the training data set and the closeness to the validation set to the training set are therefore important criteria in the efficient implementation of MAS by WGS in poultry breeding programmes. In addition, it is essential that changes in traits that are not primary selection criteria, such as skeletal disorders, muscle quality and immunity (so-called 'unexpected consequences of selection' (Hocking, 2014)), are not compromised.

7 Future trends

The paper by Wolc et al. (2016) should be referred to for a recent and comprehensive account of the application of WGS in poultry breeding programmes and a discussion of the immediate future from leaders in the poultry breeding industry and associated scientific research. According to these authors, recent developments have included the use of low- and medium-density SNP chips with imputation of SNPs that were not genotyped to reduce the relatively high costs of using high-density alternatives. This process can be combined with SNP chips that target specific areas of the genome containing genes that affect economically important traits of interest to increase precision of EBVs. Genotyping by random sequencing has been proposed because it can remove the ascertainment bias that is inherent to most SNP arrays and the implication that most causative variants will be marked. Sequencing costs have been considerably reduced and this may become a viable alternative to SNP chips as it would capture all genetic variations. Combined with a large training data set, genome sequencing could lead to identification of the causal mutations and a considerable increase in accuracy of estimating GEBVs. Looking forward, as genomic data becomes cheaper it may be possible to use genomic prediction methods to enable phenotypes and genotype data collected on commercial crossbred animals to be used when calculating breeding values for purebred selection candidates. Extension of these methods to new phenotypes (traits) such as disease resistance or meat quality that preclude the bird from being used in selection flocks could further increase the alignment between the accuracy of selection and the breeding goal, which ultimately is performance on commercial farms.

8 Conclusions

The development of technology – novel chemistry, robots and computer software – for rapid genotyping of large numbers of DNA markers as SNPs cheaply and on large numbers of individual birds has made the application of MAS to commercial poultry breeding programmes both feasible and practical. The benefits are considerable in terms of improved accuracy of estimation of breeding values in combination with phenotypic measurements. Considerable benefits have been realised for traits with low heritability or measured in one sex such as egg production. The latter could also lead to a reduction in the generation interval in males of egg-laying strains, for example, where early selection decisions based on predicted merit before sexual maturity could be made. It is very likely that further technical developments such as genome sequencing could lead to a further step change in the rate of genetic improvement. Additional benefits could include the incorporation of welfare traits with no loss of gains in production traits because of the greater accuracy of estimating breeding values. The rapid multiplication rate of poultry is likely to mean that these costly developments will become feasible in poultry breeding programmes, whereas in the short term they might not do so for other species, because costs can be apportioned to the huge numbers of commercial poultry emanating from relatively few genetically improved pedigree birds in a relatively short period of time.

9 Where to look for further information

A fuller account of the technical background to MAS theory and application to poultry before the development of SNP chips can be found in book chapters by Dekkers and van der Werf (2007) and de Koning and Hocking (2007), respectively. The seminal papers by Fernando and Grossman (1989) for single markers and van Arendonk et al. (1994) for multiple markers show how markers can be incorporated into conventional best linear unbiased prediction (BLUP) estimation of breeding values. Lastly, a paper by Meuwissen et al. (2016) is recommended as a very readable and comprehensive account of whole genome selection in farm animals.

10 Acknowledgement

The Roslin Institute is supported by a BBSRC Institute Strategic Programme Grant (BB/J004235/1).

11 References

Bovenhuis, H., van Arendonk, J. A. M., davis, G., Elsen, J. M., Haley, C. S., Hill, W. G., Baret, P. V., Hetzel, D. J. S. and Nicholas, F. W. 1997. Detection and mapping of quantitative trait loci in farm animals. *Livestock Production Science*, 52, 135–44.

Chen, C. Y., Misztal, I., Aguilar, I., Tsuruta, S., Meuwissen, T. H. E., Aggrey, S. E., Wing, T. and Muir, W. M. 2011. Genome-wide marker-assisted selection combining all pedigree phenotypic information with genotypic data in one step: An example using broiler chickens. *Journal of Animal Science*, 89, 23–8.

de Koning, D. J., Windsor, D., Hocking, P. M., Burt, D. W., Law, A., Haley, C. S., Morris, A., Vincent, J. and Griffin, H. 2003. QTL detection in commercial broiler lines using candidate regions. *Journal of Animal Science*, 203, 1158–65.

Dekkers, J. C. M. and van der Werf, J. H. J. 2007. Strategies, limitations and opportunitites for marker-assisted selection in livestock. In Guimaraes, E. P., Ruane, J., Scherf, B. D., Sonnino, A. and Dargie, J. D. (eds), *Marker-Assisted Selection: Current Status and Future Perspectives in Crops, Livestock, Foresty and Fish*. Rome: FAO.

Falconer, D. S. and Mackay, T. F. C. 1996. *Introduction to Quantitative Genetics*, Harlow, Pearson Prentice Hall.

Fernando, R. L. and Grossman, M. 1989. Marker assisted selection using best linear unbiased prediction. *Genetics Selection Evolution*, 21, 467–77.

Gonzalez-Recio, O., Gianola, D., Rosa, G. J. M., Weigel, K. A. and Kranis, A. 2009. Genome-assisted prediction of a quantitative trait measured in parents and progeny: application to food conversion rate in chickens. *Genetics Selection Evolution*, 41.

Hocking, P. M. 2005. Review of QTL mapping results in chickens. *Worlds Poultry Science Journal*, 61, 215–26.

Hocking, P. M. 2014. Unexpected consequences of genetic selection in broilers and turkeys: problems and solutions. *British Poultry Science*, 55, 1–12.

Koning, D.-J. D. and Hocking, P. M. 2007. Marker-assisted selection in poultry. In Guimaraes, E. P., Ruane, J., Scherf, B. D., Sonnino, A. and Dargie, J. D. (eds), *Marker-Assisted Selection: Current Status and Future Perspectives in Crops, Livestok, Forestry and Fish*. Rome: FAO.

Kranis, A., Gheyas, A. A., Boschiero, C., Turner, F., Yu, L., Smith, S., Talbot, R., Pirani, A., Brew, F., Kaiser, P., Hocking, P. M., Fife, M., Salmon, N., Fulton, J., Strom, T. M., Haberer, G., Weigend, S., Preisinger, R., Gholami, M., Qanbari, S., Simianer, H., Watson, K. A., Woolliams, J. A. and Burt, D. W. 2013. Development of a high density 600K SNP genotyping array for chicken. *BMC Genomics*, 14.

McKay, J. C. 1998. A poultry breeder's approach to avian neoplasia. *Avian Pathology*, 27, S74–7.

Meuwissen, T. H. E., Hayes, B. J. and Goddard, M. E. 2001. Prediction of total genetic value using genome-wide dense marker maps. *Genetics*, 157, 1819–29.

Meuwissen, T. H. E., Hayes, B. J. and Goddard, M. E. 2016. Genomic selection: A paradigm shift in animal breeding. *Animal Frontiers*, 6, 6–14.

Sewalem, A., Morrice, D. M., Law, A. S., Windsor, D., Haley, C. S., Ikeobi, C. O. N., Burt, D. W. and Hocking, P. M. 2002. Mapping of Quantitative Trait Loci (QTL) for body weight at 3, 6 and 9 weeks of age in a broiler layer cross. *Poultry Science*, 81, 1775–81.

van Arendonk, J. A. M., Tier, B. and Kinghorn, B. P. 1994. Use of multiple genetic-markers in prediction of breeding values. *Genetics*, 137, 319–29.

Wolc, A., Kranis, A., Arango, J., Settar, P., Fulton, J. E., O'sullivan, N. P., Avendano, A., Watson, K. A., Hickey, J. M., de los Campos, G., Fernando, R. L., Garrick, D. J. and Dekkers, J. C. M. 2016. Implementation of genomic selection in the poultry industry. *Animal Frontiers*, 6, 23–31.

Wolc, A., Zhao, H. H., Arango, J., Settar, P., Fulton, J. E., O'sullivan, N. P., Preisinger, R., Stricker, C., Habier, D., Fernando, R. L., Garrick, D. J., Lamont, S. J. and Dekkers, J. C. M. 2015. Response and inbreeding from a genomic selection experiment in layer chickens. *Genetics Selection Evolution*, 47, 59.

Wong, G. K. S., Liu, B., Wang, J., Zhang, Y., Yang, X., Zhang, Z. J., Meng, Q. S., Zhou, J., Li, D. W., Zhang, J. J., Ni, P. X., Li, S. G., Ran, L. H., Li, H., Zhang, J. G., Li, R. Q., Li, S. T., Zheng, H. K., Lin, W., Li, G. Y., Wang, X. L., Zhao, W. M., Li, J., Ye, C., Dai, M. T., Ruan, J., Zhou, Y., Li, Y. Z., He, X. M., Zhang, Y. Z., Huang, X. G., Tong, W., Chen, J., Ye, J., Chen, C., Wei, N., Li, G. Q., Dong, L., Lan, F. D., Sun, Y. Q., Zhang, Z. P., Yang, Z., Yu, Y. P., Huang, Y. Q., He, D. D., Xi, Y., Wei, D., Qi, Q. H., Li, W. J., Shi, J. P., Wang, M. H., Xie, F., Wang, J. J., Zhang, X. W., Wang, P., Zhao, Y. Q., Li, N., Yang, N., Dong, W., Hu, S. N., Zeng, C. Q., Zheng, W. M., Hao, B. L., Hillier, L. W., Yang, S. P., Warren, W. C., Wilson, R. K., Brandstrom, M., Ellegren, H., Crooijmans, R., Van der poel, J. J., Bovenhuis, H., Groenen, M. A. M., Ovcharenko, I., Gordon, L., Stubbs, L., Lucas, S., Glavina, T., Aerts, A., Kaiser, P., Rothwell, L., Young, J. R.,

Rogers, S., Walker, B. A., Van hateren, A., Kaufman, J., Bumstead, N., Lamont, S. J., Zhou, H. J., Hocking, P. M., Morrice, D., De koning, D. J., Law, A., Bartley, N., Burt, D. W., Hunt, H., Cheng, H. H., Gunnarsson, U., Wahlberg, P., Andersson, L., et al. 2004. A genetic variation map for chicken with 2.8 million single-nucleotide polymorphisms. *Nature*, 432, 717–22.

Part 2

Animal nutrition

Chapter 4

The cellular basis of feed efficiency in poultry muscle: mitochondria and nucleic acid metabolism

Walter Bottje and Byung-Whi Kong, University of Arkansas, USA

1 Introduction

Continued improvement in animal agriculture efficiency is critical for maintaining sustainable poultry and livestock production. With feed representing 60–70% of the cost of raising an animal to market weight, feed efficiency remains a very important genetic trait in animal agriculture. As such, the efficiency of how that energy is produced and utilized ultimately contributes to cellular utilization efficiency – cellular efficiency. Across the spectrum of animal species, skeletal muscle accounts for approximately 50% of body mass, and when considered as an entire organ, skeletal muscle contributes between 25% and 40% of basal metabolic rate (Brand, 1990a; Zurlo et al., 1990; Rolfe et al., 1999). A quarter of total basal metabolic rate in an animal may be assigned to one component of mitochondrial function – proton leak (Rolfe and Brand, 1996; Rolfe and Brand, 1997). A clear link between breast muscle mitochondrial function and feed efficiency has been reported (Bottje et al., 2002; Bottje and Carstens, 2009). Thus, mitochondrial function in skeletal muscle can play a substantive role in overall efficiency.

An excellent review of avian muscle physiology, structure, growth and development in avian species is provided by Velleman and McFarland (2015). Skeletal muscle growth in poultry post-hatch is due to hypertrophy (increased size of existing cells) (Smith, 1963). In skeletal muscle, 'work' can be viewed as mechanics associated with movement

http://dx.doi.org/10.19103/AS.2016.0011.05

and exercise, but in agriculture animals, 'work' can also be considered as anabolic processes, such as protein synthesis, and assembly of myofibrils and skeletal fibres. There is a continual need for energy input as proteins are not static structures, but rather are constantly undergoing synthesis and degradation. There are numerous energy-consuming processes that are needed to maintain homeostasis; for example, cell membrane potential and transport of materials (Na_+/K^+ ATPase), construction-related processes (muscular and cytoarchitecture), repair processes (e.g. ability to 'fix' damaged structures – digestion of proteins in proteosomes, resynthesis of proteins from component amino acids) and so on. How efficiently muscle cells can carry out these processes at the cellular level has a major effect on overall production efficiency in an animal.

2 Mitochondrial processes: an overview

2.1 Introduction

An overview of mitochondrial processes is provided in Fig. 1. The first description of mitochondria is credited to a pioneering cytologist (Kölliker) over 150 years ago (Lehninger, 1965). Mitochondria are descendants of bacteria that developed a commensal relationship with a host cell several millennia ago and as such, they are the only organelle outside the nucleus with its own discrete pool of mitochondrial DNA (mtDNA). MtDNA is a circular molecule of over 16 000 base pairs that contains roughly 37 genes that code for 2 ribosomal RNAs, 22 transfer RNAs and 13 mitochondrial-encoded proteins of the respiratory chain that, along with over 70 other nuclear-encoded proteins, make up the respiratory (electron transport) chain (Anderson et al., 1981). Nuclear-encoded proteins that are synthesized in the cytosol are transported into the mitochondria by outer (TOM) and inner (TIM) membrane translocase proteins.

Mitochondria are part of an active network within the cell that constantly undergoes complex fission and fusion processes, which enables them to communicate and '... form into local and widespread mitochondrial syncytia within cells' (Hoppins et al., 2007). Fission and fusion are under the control of highly conserved dynamin-related proteins that are large GTPase proteins that regulate membrane-associated cellular processes (Praefcke and McMahon, 2004). Mitochondrial fusion is necessary to distribute mtDNA to mitochondria throughout the cell and helps maintain functionally competent mitochondria within a cell. In contrast, mitochondrial fission insures that competent mitochondria are distributed throughout the cell and is important in mitochondrial biogenesis.

2.2 Oxidative phosphorylation

Mitochondria generate 90% of cellular energy by oxidative phosphorylation. The mitochondrial electron transport chain (ETC), first reported by Kennedy and Lehninger (Kennedy and Lehninger, 1949), consists of five multi-protein enzyme complexes (CI–CV) and two mobile electron (e-) carriers, ubiquinone (Q) and cytochrome *c* (cyt *c*) (Figure 1). The five enzyme complexes are Complex I (NADH: ubiquinone oxidoreductase), Complex II (succinate-ubiquinone reductase), Complex III (ubiquinol: cytochrome *c* oxidoreductase), Complex IV (cytochrome *c* oxidase) and Complex V (F_1F_0 ATP synthase). Electrons (e-) enter the ETC at CI (for NADH-linked substrates), or CII ($FADH_2$-linked substrate).

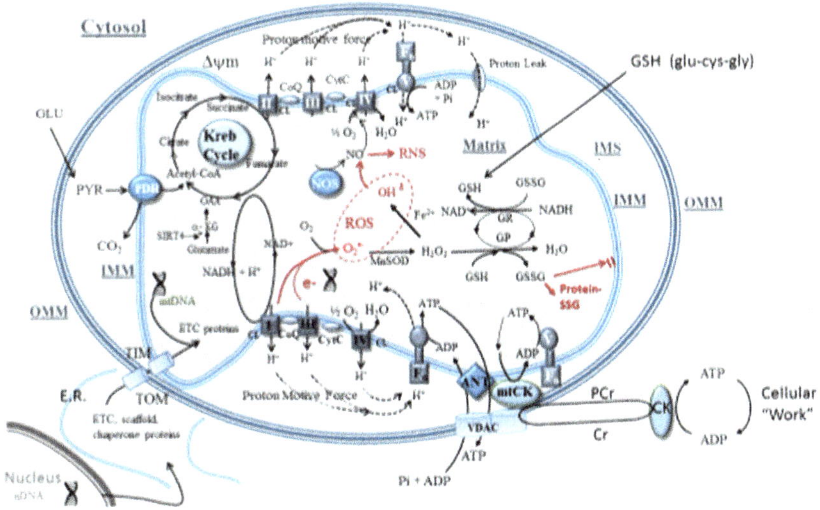

Figure 1 Mitochondrial metabolism-physiology overview.

The mitochondrial ETC is comprised of nuclear (n)- and mitochondrially (mt)-encoded proteins. Mitochondria have both an outer (OMM) and inner (IMM) mitochondrial membrane (where the ETC (I, II, III, IV, V) is located). Nuclear-encoded proteins synthesized in the cytosol are transported into the mitochondria by outer (TOM) and inner (TIM) membrane translocase proteins. Electrons enter the ETC at CII via succinate (FADH$_2$-linked substrates, top) or by NADH-linked energy substrates at CI (bottom) and then passed to CIII by coenzyme Q (CoQ) and to CIV by cytochrome C (cyt c) and finally to oxygen (that terminal e-acceptor) that is reduced to water. Electron movement along the respiratory chain is accompanied by proton (H+) pumping into the intramembranous space (IMS) to develop a proton-motive force that drives ATP synthesis when protons flow through the ATP synthase (Complex V). A mitochondrial nitric oxide synthase produces nitric oxide (NO) that can compete with oxygen for the active site on cytochrome c oxidase. Protons in the IMS may also move through the membrane at sites other than the ATP synthase in a process called proton leak. Proton leak dissipates the proton-motive force without synthesis of ATP and attenuates formation of ROS.

ATP produced by oxidative phosphorylation is transported out of the mitochondria for use by the cell through ANT on the IMM and the VDAC on the OMM. Mitochondrial CK (MTCK) present between the IMM and OMM helps transfer high-energy phosphate bonds from ATP to form PCr. PCr in the cytosol can be used as an 'energy reservoir' that transfers phosphate groups to ADP to regenerate cytosolic ATP to continue to carry out cellular or biological activities (cellular 'work'). The transfer of phosphate from PCr to ADP is catalysed by cytosolic CK.

Electrons (e-) that leak from the respiratory chain can react with oxygen to form superoxide (O_2^{-}) that is converted to hydrogen peroxide (H_2O_2) by manganese superoxide dismutase (MnSOD). In the presence of free metal ions, H_2O_2 can be converted to the highly reactive hydroxyl radical (OH·). These ROS ($O_2^{-\cdot}$, H_2O_2, OH^-) can oxidize (e.g. proteins, lipids and DNA) or react with NO to produce reactive nitrogen species. GSH is an antioxidant that is imported from the cytosol into the mitochondria. The active thiol in GSH is used to reduce H_2O_2 or lipid peroxides to water or lipid alcohols with the concomitant formation of oxidized GSH (GSSG) that can be recycled to GSH by GR that utilizes reducing equivalents from NADH. Unlike cells, mitochondria cannot export GSSG and elevations in GSSG in mitochondria can lead to protein disulphides (protein-SSG) formation. This can be particularly detrimental to ETC activity due to the presence of reactive thiol groups in these proteins.

Cardiolipin (tetra-acyl-diphosphatidyl-glycerol) is found in high amounts and is essential in membranes that conduct coupled (oxidative) phosphorylation (Hoch, 1992). Cardiolipin is a unique 'double' phosphoglyceride that has four long-chain fatty acids compared to two side chains in other phospholipids. Full activity of all respiratory chain complexes (I to V) requires the interaction of each complex with cardiolipin.

This figure was adapted from Wallace (1999) and modified from one used in *Sturkie's Avian Physiology* (Bottje, 2015), Chapter 4 (p. 40).

Coenzyme Q (CoQ, ubiquinone) carries e- from CI and CII to CIII while cyt *c* shuttles e- from CIII to CIV. Electron movement along the ETC is accompanied by proton pumping into the intermembranous space that creates a proton-motive force. Proton-motive force, consisting of a membrane potential ($\Delta\psi_{\mu}$) and a pH gradient, provides energy for ATP synthesis as protons flow back into the matrix through (CV, ATP synthase).

2.3 Endogenous oxidative stress

Mitochondria are a major site of endogenous oxidative stress from reactive oxygen species (ROS) production. In this process, O_2 is converted to ROS by univalent reduction of O_2 to superoxide ($O_2^{\bullet-}$) from e- leak (Boveris and Chance, 1973; Chance et al., 1979; Turrens and Boveris, 1980). Superoxide dismutase converts $O_2^{\bullet-}$ to H_2O_2. While H_2O_2 is not very reactive, it can be converted to the highly reactive hydroxyl radical ($^{\bullet}OH$) in the presence of Fe^{2+} and Cu^{2+}. Due to its lipid solubility, H_2O_2 can cross membranes and oxidize proteins, DNA, lipids and carbohydrates throughout the cell (Yu, 1994). Glutathione peroxidase and glutathione reductase (GR) combat mitochondrial oxidative stress. GPx converts peroxides to water or lipid alcohols whereas GR uses NADPH to reduce oxidized glutathione (GSSG) to reduced glutathione (GSH). GR is vital for mitochondrial function as mitochondria are unable to export GSSG (unlike cells) that can damage proteins by thiolation.

2.4 Proton leak

Protons can cross the mitochondrial membrane at sites other than the ATP synthase that short circuits ATP synthesis in a process termed proton leak (Fig. 1) (Brand, 1990a; Brand, 1990b). Proton leak represents an inefficiency, but attenuates ROS production. As proton leak represents up to 50% of oxygen use in a perfused muscle (Brand, 1990a; Rolfe et al., 1999), it could contribute as much as 25% of total basal metabolic rate of an animal (Rolfe and Brand, 1996; Rolfe and Brand, 1997). A self-limiting feedback on mitochondrial ROS production by superoxide through mitochondrial uncoupling was shown by Skulachev (1996, 1997).

2.5 Mitochondrial ATP/phosphocreatine transfer system

ATP is provided to the cytosol either directly as ATP or indirectly as phosphocreatine (PCr) by a protein transfer system consisting of adenine nucleotide translocase (ANT), voltage-dependent activated channel (VDAC), and cytosolic creatine kinase (CK) and mitochondrial CK (MTCK). Cytosolic CK catalyses phosphate transfer from PCr to ADP and is crucial for energy metabolism in tissues with high and/or fluctuating energy demand (e.g. skeletal, heart, smooth muscles and brain). Jacobus and Lehninger (1973) first proposed a mechanism in which the MTCK transfers phosphate groups from ATP (from oxidative phosphorylation) to form PCr, which is a high-energy phosphate reservoir for rapid regeneration of cytosolic ATP. Brain (B) and muscle (M) isoforms of CK exist as dimers in the cytosol (either BB, MM or MB), whereas mitochondrial CK is found as an octamer present between the inner and outer mitochondrial membranes [see reviews by Brdiczka et al. (2006) and Schlattner et al. (2006)]. The transfer of PO_4 from mitochondrial ATP to form PCr in the cytosol is carried out by ANT, MTCK and VDAC. The ANT-CKMT-VDAC system links mitochondrial ATP generation to a PCr reservoir that can be drawn upon when cytosolic ATP levels fall in response to energy demand (cellular work).

2.6 Mitochondrial biogenesis

Mitochondrial biogenesis is stimulated in response to increased energy demand. Paul and Sperling (1952) observed that there were more mitochondria in breast muscle in pigeons that are more active than the relatively sedentary commercial chicken. Because of its role in mitochondrial biogenesis, PGC-1α has been termed the master regulator of mitochondrial protein synthesis and biogenesis (Nisoli et al., 2003; 2004). AMP-activated protein kinase (AMPK) is critical for a) sensing energy (AMP/ATP) status and stimulating mitochondrial biogenesis (Zhou et al., 2001; Carling, 2004; Hardie, 2004b; Hardie and Sakamoto, 2006; Hardie et al., 2006), b) regulating animal food intake and overall energy balance (Minokoshi et al., 2004), and c) stimulating antioxidant protection (Choi et al. 2001; Columbo and Moncada, 2009). Once activated, AMPK phosphorylates several proteins involved in carbohydrate, lipid and protein metabolism (Kemp et al., 2003; Hardie, 2004a; Hardie, 2005; Hardie, 2007). In general, AMPK reduces ATP-utilizing (anabolic) pathways (e.g. fatty acid synthesis) and increases ATP-generating (catabolic) pathways (e.g. fatty acid oxidation and glycolysis). AMPK is required for stimulating glucose uptake and glycolysis in skeletal muscle cells and astrocytes (Zhou et al., 2001; Almeida et al., 2004). AMPK also upregulates PGC-1α expression and mitochondrial biogenesis (Ojuka, 2004). In conjunction with thyroid hormone receptor activation, PGC-1α also upregulates ANT (Masatoshi et al., 2005).

3 Evidence of mitochondrial role in high feed efficiency

Much of the content in this chapter is based on a series studies that reported an inextricable link between mitochondrial function and feed efficiency (Bottje et al., 2002; Bottje and Carstens, 2009) that were followed by proteogenomic studies of feed efficiency in muscle (Kong et al., 2011; Kong et al., 2016). These studies used a paradigm in which tissues were obtained from a single pedigree male broiler line – highly selected for feed efficiency, growth and muscle development – that were fed the same diet, housed under identical conditions and individually phenotyped for feed efficiency. In these studies, a group of 300 males were placed in floor pens and raised to 6 weeks of age. From this group, 100 males were placed in cages and individually phenotyped for FE and those with the highest or lowest FE (n = 6–8) were selected. As ROS play an important role in signal transduction, increased mitochondrial ROS was consistently observed in low FE mitochondria (Bottje et al., 2002; Bottje and Carstens, 2009). Global gene and protein expression analyses conducted in breast muscle (see Kong et al., 2011; Bottje et al., 2012; Bottje et al., 2014; 2016; Kong et al., 2016) have helped paint a proteogenomic picture of FE at the cellular level that is presented below.

The paradigm used in the pedigree broiler male line studies differs from those in which animals were divergently selected for FE (e.g. Aggrey et al., 2010; Lee et al., 2015) or were derived from an outbred commercial broiler line (Zhou et al., 2015). These studies were conducted in breast muscle that has a predominance of type IIa fibres as opposed to muscle containing a mixture of type I and type IIa fibres that are more typical of skeletal muscle. In the proteomics study (Kong et al., 2016), out of 232 proteins that were recognized in the study, 150 were upregulated (significantly or numerically) and 71 downregulated in the high FE phenotype compared to the low FE phenotype (Kong et al., 2016). Similarly, in a global gene expression study (RNAseq), 475 genes encoding mitochondrial proteins were upregulated and 224 were downregulated (Bottje et al., 2016). In both cases, the binomial statistical analysis indicated a highly significant skew ($P < 0.0001$) in the high FE phenotype. This is a clear indication that mitochondrial expression in general was enhanced in the high compared to low FE breast muscle.

3.1 Mitochondrial function, biochemistry and oxidative stress in feed efficiency

In the initial study, it was determined that mitochondrial ETC coupling was lower in breast and leg muscle mitochondria with low FE compared to high FE (Bottje et al., 2002). The lower coupling in low FE was associated with site-specific defects in electron transport, increased mitochondrial ROS production and increased protein oxidation in muscle mitochondria, heart, liver and duodenal tissue (Bottje and Carstens, 2009). Decreased respiratory chain complex activities in low FE mitochondria were hypothesized to be due to ROS-mediated protein oxidation. A study in skeletal muscle obtained from male lambs exhibiting high RFI (low FE) also reported a decrease in activity of all respiratory complexes (Sharifabadi et al., 2012).

Increased mitochondrial ETC protein expression (Kong et al., 2016) and respiratory chain complex activities (Bottje and Carstens, 2009; Sharifabadi et al., 2012) contrast with recent reports in: A) mature female Large White pigs divergently selected for high or low RFI (Vincent et al., 2015), and B) the top 5 and bottom 5 animals of a group of 238 castrated Yorkshire purebred boars individually phenotyped for RFI (Jing et al., 2015). Using a cDNA microarray, Vincent et al. (2015) reported that mRNA expression of genes encoding for five electron transport proteins and isocitrate dehydrogenase were downregulated in the low RFI pig. This reduction in mitochondrial expression is in line with the red to white muscle fibre transition that is well documented in domestic pigs. There are at least three possible explanations for differences between the present study and the study by Vincent et al. (2015): 1) species, 2) gender and 3) protein expression does not always follow gene expression differences. With respect to point 3, projection of the DE genes in the cDNA microarray dataset from Kong et al. (2011) onto the oxidative phosphorylation canonical pathway using the overlay function in IPA revealed upregulation/activation of Complex I but downregulation/inhibition of Complex III and IV. Vincent et al. (2015) did indicate that there was lower oxidative stress in the low RFI pigs, which concurs with lower oxidative stress that is characteristic of the high FE pedigree male broilers (Bottje and Carstens, 2009). Western analysis by Iqbal et al (2004) revealed that 5 proteins in Complex III were upregulated in the muscle of the low FE phenotype but there were no differences in expression of 6 other proteins in Complex I between the high and low FE phenotypes.

Proton leak attenuates mitochondrial ROS production by a self-limiting feedback mechanism (Skulachev, 1996a; 1997; Brand et al., 2004). In pedigree male broilers, uncoupling lowered ROS production in duodenal mitochondria obtained from broilers with low FE but not in mitochondria obtained from broilers with high FE (Ojano-Dirain et al., 2007). In that study, basal ROS production was higher in low FE mitochondria and therefore uncoupling would be expected to have a greater effect on mitochondrial ROS production. Bottje et al. (2009) reported that proton leak in isolated high FE muscle mitochondria was consistently less than, equal to, but never more than that observed in low FE muscle mitochondria. Thus, these results suggest that there are subtle differences in membrane characteristics (e.g. lipids and integral membrane proteins) that contribute to higher mitochondrial ROS in the low FE in this animal model (Bottje et al., 2002; Bottje and Carstens, 2009).

3.2 Enhanced energy production and transfer system in high feed efficiency

In cells with fluctuating energy demand (e.g. skeletal muscle), a transport system across the inner and outer mitochondrial membranes links oxidative phosphorylation to cytosolic phosphorylated creatine (PCr) that serves as a phosphate reservoir for rapid repletion of cytosolic ATP (Jacobus and Lehninger, 1973). Crucial proteins of this system include several CK isoforms found in the cytosol and mitochondria. In a recent proteomic studies (Kong et al., 2016), several components of this system were upregulated in high FE compared to low FE breast muscle (Fig. 1, Table 1; from Bottje et al., 2017). These components include ANT, VDAC2, the brain isoform of CK (CK-B), one form of mitochondrial CK (CKMT1A) and several proteins of the ETC. The muscle isoform of CK (CK-M) was downregulated in the high FE phenotype and there was no difference in CKMT2 expression between the high and low FE phenotypes.

Increased expression of ANT, VDAC and mitochondrial CK would connect ATP production from oxidative phosphorylation to the formation of PCr that can be used to replenish cytosolic ATP as a result of phosphorylation reactions and ATP use occurring with biological work functions. An example of an essential ATP-utilizing reaction is indicated by Na^+/K^+ ATPase. In the high FE muscle, increased expression of adenosine kinase (AdK or AK) would increase the AMP:ATP ratio (an indicator of increased energy demand). The

Table 1 Protein expression associated with energy production and conveyance in breast muscle of pedigree broiler males (from Bottje et al., 2017). Probability (*P* value) and fold differences (high vs low) are presented for upregulated (positive number) and downregulated (negative number) for proteins in the high compared to low FE pedigree male broiler phenotype ($n = 4$ per group). Mitochondrial creatine kinase (CKMT1A and CKMT2) expression data were obtained from the original dataset but not recognized by bioinformatics analysis (Kong et al., 2016).

Symbol	Protein name (significance)	*P* Value	Fold difference
CKMT1A	Mitochondrial creatine kinase	$P = 0.05$	7.1
CKMT2	Mitochondrial creatine kinase 2	$P = 0.59$	1.1
IDE	Insulin-degrading enzyme	$P = 0.004$	11.7
ANT	Solute carrier family 25[1]	$P = 0.002$	10.3
CAV1	Caveolin 1	$P = 0.009$	9.4
CK-B	Creatine kinase, brain	$P = 0.01$	8.9
VDAC1	Voltage-dependent anion channel 1 (Ca++ transport)	$P = 0.003$	2.8
VDAC2	Voltage-dependent anion channel 2 (ATP transport)	$P = 0.03$	2.4
CK-M	Creatine kinase (muscle)	$P = 0.04$	−1.4

[1] Mitochondrial carrier, adenine nucleotide translocator, ANT.

increase in AMP:ATP would be detected by AMPK (5′ AMP-activated protein kinase) that would induce peroxisome proliferator-activated receptor coactivator 1-alpha (PGC-1α) and increase mitochondrial activity and energy production through mitochondrial biogenesis. Due to its role in stimulating transcription factors that enhance mitochondrial biogenesis, PGC-1α has been called the master regulator of mitochondrial biogenesis (Nisoli et al., 2003; 2004). The predicted activation of tri-iodothyronine (T3) (Kong et al., 2016) would also enhance PGC-1α activity. The insulin signalling pathway also would be enhanced in the high FE phenotype with the predicted increase in activities of insulin receptor (INSR) and insulin-like growth factor receptor 1 (IGF1R). In this pathway, caveolin 1 (CAV1) and insulin-degrading enzyme (IDE) help internalize insulin after binding to the receptor, and in transmitting the signal to the nucleus, respectively. The predicted activation of progesterone (PG) would enhance mitochondrial activity and complement the activity of the insulin signalling pathway.

AMPK also plays a role in suppressing programmed cell death that is initiated by mitochondria. Apoptosis inhibition in the high FE muscle phenotype would be complemented by upregulation of anti-apoptotic protease-activating factor 1 (APIP) and protein phosphatase 2A (PP2A) (see review by Bottje and Kong 2013). The APIP protein is highly expressed in skeletal muscle where it decreases apoptotic stimuli by

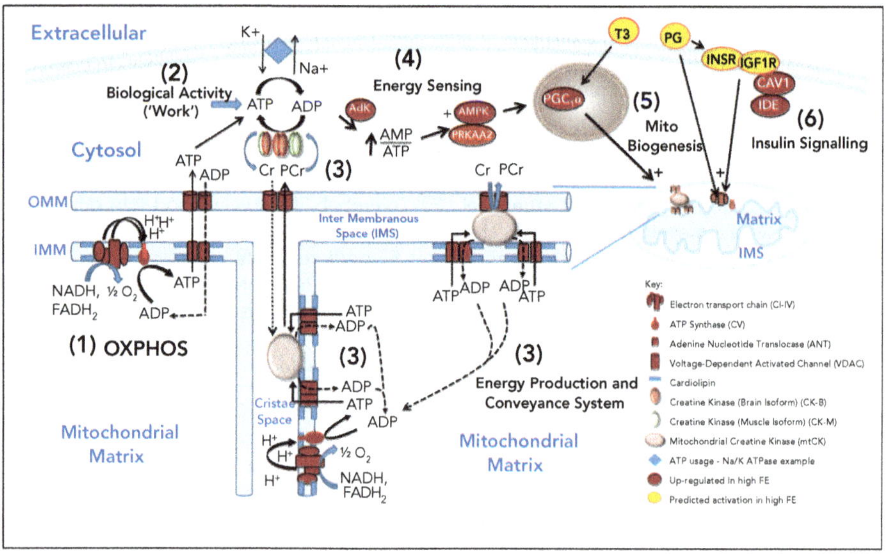

Figure 2 Enhanced energy production and transfer system in high feed efficiency muscle.

A depiction of the energy production and conveying system that is apparently enhanced in breast muscle of pedigree male broilers exhibiting a high feed efficiency phenotype is shown in Figure 1. The figure is modified from ones presented in reviews by Brdiczka et al. (2006) and Schlattner et al. (2006) using data from proteogenomic data (Kong et al., 2011; Bottje et al., 2014; Kong et al., 2016). Molecules in red and green indicate proteins or genes upregulated or downregulated, respectively, in the high FE pedigree male phenotype. Molecules in orange were predicted to be activated in the high FE pedigree male phenotype based on expression of downstream molecules (Kong et al., 2016). (Reprinted from Bottje et al. (2017).

inhibiting cytochrome *c* release and APAF-1-mediated activation of caspase-9, a critical step in apoptosis (Cho et al., 2004). Muscle tissue accounts for up to 50% of total energy consumption in the body (Brand, 1990a; Rolfe and Brand, 1996). Thus, increased expression of APIP in muscle tissue would contribute to the high FE phenotype by decreasing cell turnover.

From the findings, we hypothesize that enhanced expression of the energy production and transfer system in breast muscle of the high FE pedigree broiler male and the inhibition of apoptosis could be fundamentally important in the phenotypic expression of feed efficiency. Further investigation is warranted in which activities of mitochondrial and tissue CK activity are measured to assess whether the differences in mitochondrial expression of proteins or enzymatic activity of CK in this study could be used as an indicator feed efficiency in poultry and livestock.

4 Enhanced nucleotide metabolism in high feed efficiency

Evidence of enhanced nucleotide, purine and pyrimidine synthesis in high FE breast muscle has been reported (Bottje and Kong, 2013; Kong et al., 2016) (Table 2, Fig. 3). Figure 3 is based on established metabolic reactions provided in *Principals of Biochemistry* (Lehninger et al., 1993) using expression analysis presented in Table 2. In global gene and protein expression studies, there is evidence that nucleotide, purine and pyrimidine synthesis would be enhanced in high FE muscle. The rate-limiting step in NADH and

Table 2 Upregulated proteins and genes associated with nucleotide metabolism shown in Figure 3

Abbreviation	Protein–gene name	Fold difference	Source
AK	Adenylic kinase	1.20	1
AMPD1	Adenosine monophosphate deaminase 1	1.68	2
APRT	Adenine phosphoribosyltransferase	7.28	2
GLRX3	Glutaredoxin	5.66	2
MTHFDIL	Methylenetetrahydrofolate dehydrogenase (NADP+ dependent) 1-like	6.24	2
NAMPT	Nicotinamide phosphoribosyltransferase	1.74	3
NAMS(NMNAT1)	Nicotinamide mononucleotide pyrophosphorylase	1.67	3
PRPS 1 and 2	Phosphoribosyl pyrophosphate synthetase 1 and 2	4.60 1.80	2 3
PPA2	Pyrophosphatase (inorganic) 2	8.85	2
RRM2	Ribonucleotide reductase	1.40	4

Sources: [1]Bottje et al., 2014, [2]Kong et al., 2016, [3]Bottje et al., 2012, [4]Kong et al., 2011 (unpublished microarray data deposited in GEO).

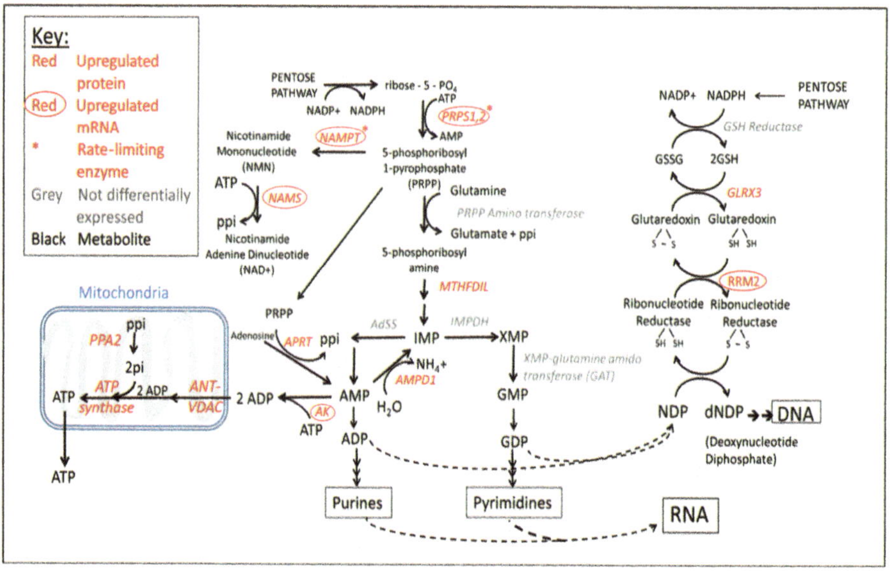

Figure 3 Enhanced nucleotide, purine and pyrimidine metabolism in high feed efficiency breast muscle (see text for details). Abbreviations in the figure are as follows: AK, adenylyl kinase; ANT, adenine nucleotide transferase; AMPD1, adenosine monophosphate deaminase 1; APRT, adenine phosphoribosyltransferase; MTHFDIL, methylenetetrahydrofolate dehydrogenase (NADP+ dependent) 1-like; NAMS (or NMPT), nicotinamide mononucleotide phosphoryltransferase; NAMPT, nicotinamide phosphoribosyltransferase; PPA2, pyrophosphatase (inorganic 2); PRPS1,2, phosphoribosyl pyrophosphate 1 and 2; VDAC, voltage-dependent activated channel. Differential expression of proteins and genes are provided in Tables 1 and 2.

nucleotide synthesis is the enzyme PRPS that catalyses the transfer of pyrophosphate from ATP to ribose-5-PO_4 to form PRPP (5-phosphoribosyl 1-pyrophosphate). The next step in purine–pyrimidine biosynthesis is catalysed by MTHF1L (methylenetetrahydrofolate dehydrogenase (NADP+ dependent 1-like)) that catalyses the transfer of a formyl group to PRPP. After four additional steps, inositate (IMP) is formed, which is a precursor of both adenosine monophosphate (AMP) and xanthylate (XMP), that in turn is converted to guanylate monophosphate (GMP) by XMP-glutamine amido transferase. AMP can also be synthesized in the purine–pyrimidine salvage pathway that is carried out with the action of adenine phosphoribosyltransferase (APRT) to form AMPT and pyrophosphate (ppi). AMP in turn is converted to IMP by AMP deaminase 1 (AMPD1) with subsequent conversion to GMP. A connection for enhanced deoxynucleotide synthesis in DNA formation (deoxynucleic acid) in high FE is indicated by the upregulation of glutaredoxin and ribonucleotide synthetase in the high FE muscle. In mitochondria, upregulation of pyrophosphatase 2 (PPA2) that catalyses the metabolism of pyrophosphate (ppi) to phosphate (pi) would be used as a substrate, along with ADP (transported into the mitochondria through VDAC and ANT channels), by ATP synthase for synthesis of ATP. Curbo et al. (2006) reported that PPA2 is located specifically in mitochondria and is found in the highest levels in heart, skeletal muscle and kidney.

5 Muscle cytoarchitecture and feed efficiency

For this section of this chapter, readers are referred once again to the review of muscle growth and development in avian muscle for detail on specific genes and proteins mentioned below (Velleman and McFarland, 2015). A large number of cytoskeletal architecture/muscle fibre genes were downregulated in the high FE pedigree broiler male (Kong et al., 2011; Bottje and Kong, 2013) that is opposite to gene expression reported in the high vs low FE commercial broilers (Zhou et al., 2015). Downregulation of cytoskeletal-muscle fibre gene expression in the pedigree male broiler model is counterintuitive in animals being selected for rapid growth and feed efficiency in the commercial broiler genetic pipeline. In addition, four isoforms of platelet-derived growth factor (PGDF) were also downregulated in the high FE pedigree male (Bottje and Kong, 2013) that could have a negative impact on muscle development (see Velleman and McFarland, 2015). It was hypothesized that lower expression of cytoarchitecture genes in the high FE muscle may indicate that less energy is expended in cytoskeletal organization and maintenance compared to low FE muscle. Support for this hypothesis was provided by the downregulation of CSRP3 (cysteine and glycine-rich protein 3) (Kong et al., 2011). CSRP3 functions as an essential nuclear regulator of myogenic differentiation, organization and in maintenance of contractile machinery in skeletal and heart muscle (Arber et al., 1994, 1997; Kong et al., 1997; Louis et al., 1997) and is found in both the cytosol and the nucleus (Louis et al., 1997). Another explanation might stem from the fact that differences in protein expression do not always coincide with differences in mRNA expression and that there are a large number of post-translational modifications that can occur to produce a fully functional protein. In this regard, closer examination of the entire proteomic dataset reported by Kong et al. (2016) indicated that of approximately 40 recognized proteins, only 11 were significantly downregulated in the high FE phenotype – the rest were not differentially expressed between the high and low FE groups, and none were significantly upregulated in the high FE phenotype.

The downregulation of one protein, myostatin, may counterbalance the seemingly incongruent downregulation of certain muscle fibre/cytoarchitecture genes in high FE muscle (Bottje and Kong, 2013). The downregulation of myostatin was confirmed by targeted gene and protein expression analyses conducted on the same set of tissue samples (Lassiter et al., 2015). Myostatin is a member of a family of transforming growth factor beta (TGF-β) proteins that have dramatic influence on muscle development. Mutations of myostatin in cattle or knockout of myostatin in mice resulted in extreme (double muscling) (McPherron and Lee, 1997; McPherron et al., 1997) and high levels of myostatin are associated with muscle atrophy (Srinivasan et al., 2004). Of particular interest is a similarity between differentially expressed genes in the Piedmontese- and Waygu-Hereford (PxH and WxH) cattle crosses (Hudson et al., 2009) and those in the low and high FE broiler study (Bottje et al., 2002; Kong et al., 2011) that appear to imply similar mechanisms of enhanced efficiency. The most differentially expressed gene in the cattle study was CSRP3, which was downregulated in the PxH (that are faster growing and more feed efficient) compared to the slower growing, less-efficient WxH cross and CSRP3 was the second most differentially expressed gene in breast muscle and was downregulated in the low compared to the high FE broiler phenotype.

A possible shortcoming of this chapter is that it is somewhat narrow in coverage of other aspects of muscle metabolism that would contribute to cellular efficiency. For example, NFE2L2 is an important regulator of antioxidant response in cells. NFE2L2 was predicted to be activated in breast muscle obtained from high FE commercial broilers (Zhou et al., 2015). This prediction was based on differential expression of downstream target molecules in the global gene expression (RNAseq) dataset.[1] Similar to the report by Zhou et al. (2015), NFE2L2 was also predicted to be activated in the proteomics dataset of the pedigree broiler male (Kong et al., 2016). However, there were very few similarities in the set of downstream target molecules used in the prediction of NFE2L2 activation between these two studies. Zhou et al. (2015) hypothesized that the activation of NFE2L2 (and subsequent coordination of the antioxidant response) was attributed to the increased expression of inflammatory myokines important in muscle remodelling and hypertrophy in the high FE commercial broiler. This suggests a very different cellular scenario in the commercial broiler compared to the pedigree broiler male in which there is higher oxidative stress in the low FE pedigree male phenotype being caused by increased mitochondrial ROS production (Bottje et al., 2002; Bottje and Carstens, 2009) that was associated with lower expression of muscle fiber/cytoarchitecture-related genes (Kong et al., 2011).

6 Summary

In this chapter, evidence from global gene and protein expression studies (proteogenomics) is presented, indicating that in breast muscle obtained from pedigree male broilers exhibiting high or low feed efficiency phenotypes, there is evidence of:

1. Enhanced mitochondrial function
2. Enhanced mitochondrial energy (ATP) production, transfer and cellular energy reserve (PCr)
3. Enhanced capability for synthesis and metabolism of purine and pyrimidine nucleotides that could lead to enhanced ability to construct RNA and DNA in muscle.

While writing this chapter, examination of the global gene and protein expression datasets revealed a large number of genes/proteins involved in ribosomal synthesis and protein translation that were upregulated in the high FE pedigree broiler male. Enhanced components for protein construction and scaffolding were hypothesized to be occurring in muscle in the high FE phenotype based on cDNA microarray results (Bottje et al., 2014), and these components now have additional traction in the cross-fertilization afforded by the proteogenomic data obtained from the same groups of muscle samples. This suggests that in the high FE pedigree male broilers, mechanisms may be in place in the muscles of the high FE phenotype that would enhance nucleotide synthesis, along with RNA and DNA building blocks and transcription, but may also possess a framework of enhanced protein translation compared to the low FE phenotype.

7 Where to look for further information

Additional information can be found at the website for European Commission (ECO) Community Research and Development Information System (CORDIS) for the ECO-FCE

1. This comparison was carried out by Ingenuity Pathway Analysis (IPA, Qiagen) software that compares expression of downstream target molecules with expression reported in the literature from an extensive literature-based database.

project: A whole-systems approach to optimising feed efficiency and reducing the ecological footprint of monogastrics) at http://cordis.europa.eu/result/rcn/153714_en.html. This 4 year project was funded by the European Union and helped coordinate research efforts on feed efficiency in poultry and livestock on more than a dozen research sites across Europe. The following, taken from directly from the website, regarding a description and objectives of this project:

Project Context and Objectives:

> 'Sustainable intensification' is a widely-used term in agricultural circles at the moment. It is clear that intensive production systems will play a key role in feeding a growing global population, but they must also be sustainable. This means being environmentally-friendly and allowing a reasonable return on investment for producers. Rising costs of energy and feed, combined with low product prices, make this latter element a continual challenge for producers. The pig and broiler chicken industries are key contributors to the European economy, and one of the main ways in which sustainability can be achieved is through improving feed conversion efficiency (FCE).
>
> This is the focus of the ECO-FCE project. Using state-of-the-art 'omics' technologies integrated with novel systems biology research, this project will gain a greater understanding of factors underpinning variation in FCE between monogastric animals. It will also identify management routes through which FCE can be improved, and nutrient and greenhouse gas emissions reduced. At the same time, the impact of FCE on product quality and on animal health and welfare is also being monitored. The project will culminate in the production of industry-ready models and tools to assist stakeholders in understanding, measuring and managing the impact of management decisions on FCE and environmental impact.

The overarching objectives of ECO-FCE are:

1. To improve European and global food security by optimising efficiency of feed use in the pig and broiler chicken industries
2. To synergistically reduce the ecological footprint of these industries to provide sustainable eco-systems and negate climate change

Copies of final reports can be obtained from one of the program directors, Dr. Elizabeth MacGowan, https://www.afbini.gov.uk/

8 Acknowledgements

The information presented in this chapter would not have existed without the support from USDA-NIFA (USDA-NIFA #2013-01953), Arkansas Biosciences Institute (Little Rock, AR) and the Arkansas Agricultural Experiment Station (Univ. of Arkansas, Fayetteville). I (WB) would like to dedicate this writing to my father (Will Gay Bottje), who through his music and life always has encouraged thoughtful creativity.

9 References

Aggrey, S. E., A. B. Karnuah, B. Sebastian, and N. B. Anthony. 2010. Genetic properties of feed efficiency parameters in meat-type chickens. *Genet. Sel. Evol.* 42:25. doi:10.1186/1297-9686-42-25.

Almeida, A., S. Moncada and J. P. Bolanos. 2004. Nitric oxide switches on glycolysis through the AMP protein kinase and 6-phosphofructo-2-kinase pathway. *Nat. Cell Biol.* 6:45-51.

Anderson, S., A. T. Bankier, B. G. Barrell, M. L. Debruijn, A. R. Coulson, J. Drouin, I. C. Eperon, D. P. Nierlich, B. A. Roe, F. Sanger, P. H. Schreier, A. J. H. Smith, R. Staden and I. G. Young. 1981. Sequence and organization of the human mitochondrial genome. *Nature* 290:457–65.

Arber, S., J. J. Hunter, J. J. Ross, M. Hongo, G. Sansig, J. Borg, J-C. Perriard, K. R. Chien and P. Caroni. 1997. MLP-deficient mice exhibit a disruption of cardiac cytoarchitectural organization, dilated cardiomyopathy, and heart failure. *Cell* 88:393–493.

Arber, S., G. Halder and P. Caroni. 1994. Muscle LIM protein, a novel essential regulator of myogenesis, promotes myogenic differentiation. *Cell* 79:221–31.

Bottje, W., M. Iqbal, Z. X. Tang, D. Cawthon, R. Okimoto, T. Wing and M. Cooper. 2002. Association of mitochondrial function with feed efficiency within a single genetic line of male broilers. *Poult. Sci.* 81:546–55.

Bottje, W., and G. E. Carstens. 2009. Association of mitochondrial function and feed efficiency in poultry and livestock species. *J. Anim. Sci.* 87:E48–63.

Bottje, W. G., M. D. Brand, C. Ojano-Dirain, K. Lassiter, M. Toyomizu and T. Wing. 2009. Mitochondrial proton leak kinetics and relationship with feed efficiency within a single genetic line of male broilers. *Poult. Sci.* 88:1683–93.

Bottje, W. G., B. W. Kong, J. J. Song, J. Y. Lee, B. M. Hargis, K. Lassiter, T. Wing and J. Hardiman. 2012. Gene expression in breast muscle associated feed efficiency in a single male broiler line using a chicken 44k microarray II. Differentially expressed focus genes. *Poult. Sci.* 91:2576–87.

Bottje, W., and B. W. Kong. 2013. Feed efficiency: Mitochondrial function to global gene expression. *J. Anim. Sci.* 91:1582–93. doi:doi:10.2527/jas.2012-5787.

Bottje, W. G., B. Kong, J. Y. Lee, T. Washington, J. I. Baum, S. Dridi, T. Wing and J. Hardiman. 2014. Potential roles of mTOR and protein degradation pathways in the phenotypic expression of feed efficiency in broilers. *J. Biochem. Physiol.* 3:1–8.

Bottje, W. G. 2015. Mitochondrial physiology. In *Sturkie's Avian Physiology* (6th Ed). C. G. Scanes, ed. Elsevier-Academic Press, San Diego, CA 92101-4495, pp. 39–51.

Bottje, W. G., K. Lassiter, S. Dridi, N. Hudson and B-W. Kong. 2017. Enhanced expression of proteins involved in energy production and transfer in breast muscle of pedigree male broilers exhibiting high feed efficiency. *Poult. Sci.* (in press).

Boveris, A., and B. Chance. 1973. The mitochondrial generation of hydrogen peroxide. General properties and effect of hyperbaric oxygen. *Biochem. J.* 134(3):707–16.

Brand, M. D. 1990a. The contribution of the leak of protons across the mitochondrial inner membrane to standard metabolic rate. *J. Theor. Biol.* 145:267–86.

Brand, M. D. 1990b. The proton leak across the mitochondrial inner membrane. *Biochem. Biophys. Acta* 1018:128–33.

Brand, M. D., C. Affourtit, T. C. K. Esteves, A. J. Kambert, S. Miwa, J. L. Pakay and N. Parker. 2004. Mitochondrial superoxide production, biological effects, and activation of uncoupling proteins. *Free Rad. Biol. Med.* 37:755–67.

Brdiczka, D. G., D. B. Zorov and S. Sheu. 2006. Mitochondrial contact sites: Their role in energy metabolism and apoptosis. *Biochem. Biophys. Acta* 1762:148–63.

Carling, D. 2004. The AMP-activated protein kinase cascade – A unifying system for energy control. *Trends Biochem. Sci.* 29:18–24.

Chance, B., H. Sies and A. Boveris. 1979. Hydroperoxide metabolism in mammalian organs. *Physiol. Rev.* 59:527–605.

Cho, D. H., Y. M. Hong, H. J. Lee, H. N. Woo, J. O. Pyo, T. W. Mak and Y. K. Jung. 2004. Induced inhibition of ischemic/hypoxic injury by APIP, a novel Apaf-1-interacting protein. *J. Biol. Chem.* 279:39942–50. doi:10.1074/jbc.M405747200.

Colombo, S. L., and S. Moncada. 2009. AMPKa1 regulates the antioxidant status of vascular endothelial cells. *Biochem. J.* 421:163–9.

Curbo, S., C. Lagier-Tourenne, R. Carrozzo, L. Palenzuela, S. Lucioli, M. Hirano, F. Santorelli, J. Arenas, A. Karlsson and M. Johansson. 2006. Human mitochondrial pyrophoshatase: cDNA cloning and analysis of the gene in patients with mtDNA deletion syndromes. *Genomics* 87:410–16.

Hardie, D. G. 2007. AMP- activated/SNF1 protein kinases: conserved guardians of cellular energy. *Nature Rev. Mol. Cell Biol.* 8:774–85.

Hardie, D. G., S. A. Hawley and J. W. Scott. 2006. AMP-activated protein kinase-development of the energy sensor concept. *J. Physiol.* 574:7–15.

Hardie, D. G., and K Sakamoto. 2006. AMPK: a key sensor of fuel and energy status in skeletal muscle. *Physiology* 21:48–60.

Hardie, D. G. 2005. New roles for the LKB1-AMPK pathway. *Current Opin. Cell. Biol.* 17:167–73.

Hardie, D. G. 2004a. AMP- activated protein kinase: a key system mediating metabolic responses to exercise. *Med. Sci. Sports Exer.* 36:28–34.

Hardie, D. G. 2004b. The AMP-activated protein kinase pathway-new players upstream, and downstream. *J. Cell Sci.* 117:5479–87.

Hudson, N. J., A. Reverter and B. P. Dalrymple. 2009. A differential wiring analysis of expression data correctly identifies the gene containing the causal mutation. *PLoS Comput. Biol.* 5:1–15. doi:10:1371.pcbi.1000382.

Iqbal, M., N. R. Pumford, Z. X Tang, K. Lassiter, T. Wing, M. Cooper and W. Bottje. 2004. Low feed efficient broilers within a single genetic line exhibit higher oxidative stress and protein expression in breast muscle with lower mitochondrial complex activity. *Poult. Sci.* 83:474–84.

Jacobus, W. E., and A. L. Lehninger. 1973. Creatine kinase of rat heart mitochondria: coupling of creatine phosphorylation to electron transport. *J. Biol. Chem.* 248:4803–10.

Jing, L., Y. Hou, Y. Miao, X. Li, J. Cao, J. M. Brameld, T. Parr and S. Zhao. 2015. Transcriptome analysis of mRNA and miRNA in skeletal muscle indicates an important network for differential residual feed intake in pigs. *Scientific Reports* 5:11953. doi:10.1038/srep11953.

Kemp, B. E., D. Stapleton, D. J. Campbell, Z. P. Chen, S. Murthy, M. Walter, A. Gupta, J. J. Adams, F. Katsis, B. VanDenderen, I. G. Jennings, T. Iseli, B. J. Michell and A. L. Witters. 2003. AMP-activated protein kinase, super metabolic regulator. *Biochem. Soc. Trans.* 31:162–8.

Kennedy, E. P., and A. L. Lehninger. 1949. Oxidation of fatty acids and tricarboxylic acid cycle intermediates by isolated rat liver mitochondria. *J. Biol. Chem.* 179:957–63.

Kong, B., J. Song, J. Y. Lee, B. M. Hargis, T. Wing, K. Lassiter and W. G. Bottje. 2011. Gene expression in breast muscle associated feed efficiency in a single male broiler line using a chicken 44k microarray. I. Top differentially expressed genes. *Poult. Sci.* 90:2535–47.

Kong, Y., M. Flick, A. J. Kudla and S. F. Konieczny. 1997. Muscle LIM protein promotes myogenesis by enhancing the activity of MyoD. *Mol. Cell Biol.* 17:4750–60.

Kong, B., K. Lassiter, A. Piekarski-Welsher, S. Dridi, A. Reverter-Gomez, N. J. Hudson and W. G. Bottje. 2016. Proteomics of breast muscle tissue associated with the phenotypic expression of feed efficiency within a pedigree male broiler line: I. Highlight on mitochondria. *PLoS ONE* 11:e0155679. doi:10.1371/journal.pone.0155679.

Lassiter, K., B. Kong, S. Dridi and W. G. Bottje. 2015. Genes involved in muscle development, muscle differentiation, and insulin signaling may play a role in the phenotypic expression of feed efficiency seen in a single genetic line of male broilers. *Poult. Sci.* 95(1):197.

Lee, J., A. B. Karnuah, R. Rekaya, N. A. Anthony and S. E. Aggrey. 2015. Transcriptomic analysis to elucidate the molecular mechanisms that underlie feed efficiency in meat-type chickens. *Mol. Genet. Genomics* 10:10.1007/s00438-015-1025-7.

Lehninger, A. L., D. L. Nelson and M. M. Cox, eds. 1993. *Principles of Biochemistry*. 2nd ed. Worth Publishers, New York, NY.

Lehninger, A. L. 1965. *The Mitochondrion*. W. A. Benjamin, Inc., New York, NY.

Louis, H. A., D. Pino, K. L. Schmeichel, P. Pomies and M. C. Beckerle. 1997. Comparison of three members of the cystine-rich protein family reveals functional conservation and divergent patterns of gene expression. *J. Biol. Chem.* 272:27484–91.

Masatoshi, U., K. Watanabe, K. Sato, Y. Akiba and M. Toyomizu. 2005. Possible role for avPGC-1alpha in the control of expression of fiber type, along with avUCP and avANT mRNA's in skeletal muscles of cold-exposed chickens. *FEBS Lett.* 579:11–17.

McPherron, A. C., A. M. Lawler and S. J. Lee. 1997. Regulation of skeletal muscle mass in mice by a new TGF-beta superfamily member. *Nature* 387:83–90.

McPherron, A. C., and S. Lee. 1997. Double muscling in cattle due to mutations in the myostatin gene. *Proc. Natl. Acad. Sci.* U. S. A. 94:12457–61.

Minokoshi, Y., T. Alquier, N. Furukawa, Y. B. Kim, A. Lee, B. Xue, J. Mu, F. Foufelle, P. Ferre and M. J. Birnbaum. 2004. AMP-Kinase regulates food intake by responding to hormonal and nutrient signals in the hypothalamus. *Nature* 428:569–74.

Nisoli, E., E. Clementi, S. Moncada and M. O. Carruba. 2004. Mitochondrial biogenesis as a cellular signaling framework. *Biochem. Pharm.* 67:1–15.

Nisoli, E., E. Clementi, C. Paolucci, V. Cozzi, C. Tonello, C. Sciorati, R. Bracale, A. Valerio, M. Francolini, S. Moncada and M. O. Carruba. 2003. Mitochondrial biogenesis in mammals: the role of endogenous nitric oxide. *Science* 299:896–9.

Ojano-Dirain, C., N. B. Tinsley, T. Wing, M. Cooper and W. G. Bottje. 2007. Membrane potential and hydrogen peroxide production in duodenal mitochondria in broilers chicken (*Gallus gallus*) with low and high feed efficiency. *Comp. Biochem. Physiol.* 147:934–41.

Ojuka, E. O. 2004. Role of calcium and AMP kinase in the regulation of mitochondrial biogenesis and GLUT4 levels in muscle. *Proc. Nutr. Soc.* 63:275–8.

Paul, M. H., and E. D. Sperling. 1952. Cyclophorus system-XXIII. Correlation of cyclophorase activity and mitochondrial density in striated muscle. *Proc. Soc. Exp. Biol. Med.* 79:352–4.

Praefcke, G. J., and H. T. McMahon. 2004. The dynamin superfamily: universal membrane tubulation and fission molecules. *Nat. Rev. Mol. Cell Biol.* 5:133–47.

Rolfe, D. F. S., J. M. B. Newman, J. A. Buckingham, M. G. Clark and M. D. Brand. 1999. Contribution of mitochondrial proton leak to respiration rate in working skeletal muscle and liver and to SMR. *Am. J. Physiol.* 276:C692–9.

Rolfe, D. F. S., and M. D. Brand. 1997. The physiological significance of mitochondrial proton leak in animal cells and tissues. *Biosci. Rep.* 17:9–16.

Rolfe, D. F., and M. D. Brand. 1996. Contribution of mitochondrial proton leak to skeletal muscle respiration and to standard metabolic rate. *Am. J. Physiol.* 271:1380–9.

Rolfe, D. F. S., J. M. Newman, J. A. Buckingham, M. G. Clark and M. D. Brand. 1999. Contribution of mitochondria proton leak to respiration rate in working skeletal muscle and liver and to SMR. *Am. J. Physiol.* 276:C692–9.

Schlattner, U., M. Tokarska-Schlattner and T. Wallimann. 2006. Mitochondrial creatinine kinase in health and disease. *Biochem. Biophys. Acta* 1762:164–80.

Sharifabadi, H. R., M. J. Zamir, E. Rowghani and W. G. Bottje. 2012. Relationship between the activity of mitochondrial respiratory chain complexes and feed efficiency in fat-tailed ghezel lambs. *J. Anim. Sci.* 90:1807–15.

Skulachev, V. P. 1996. Role of uncoupled and non-coupled oxidations in maintenance of safely low levels of oxygen and its one-electron reductants. *Q. Rev. Biophys.* 29:169–202.

Skulachev, V. P. 1997. Membrane linked systems preventing superoxide formation. *Biosci. Rep.* 17:347–66.

Smith, J. H. 1963. Relation of body size to muscle size and number in the chicken. *Poult. Sci.* 42:283–90.

Srinivasan, D., M. Dodig, M. Muc Sean, S. C. Kalhan and A. J. McCullough. 2004. Skeletal muscle atrophy is associated with an increased expression of myostatin and impaired satellite cell function in the portacaval anastomosis rat. *Am. J. Physiol. Gastrointest. Liver Physiol.* 287:G1124–30.

Turrens, J. F., and A. Boveris. 1980. Generation of superoxide anion by the NADH dehydrogenase of bovine heart mitochondria. *Biochem. J.* 191:421–7.

Velleman, S. G., and S. D. McFarland. 2015. Skeletal Muscle. In *Sturkie's Avian Physiology* (6th Ed). C. G. Scanes, ed. Elesvier Academic Press, San Diego, CA 92101-4495, pp. 379–402.

Vincent, A., I. Louveau, G. Gondret, C. Trefeu, H. Gilbert and L. Lefaucheur. 2015. Divergent selection for residual feed intake affects the transcriptomic and proteomic profiles of pig skeletal muscle. *J. Animal. Sci.* 93:2745–58.

Wallace, D. 1999. Mitochondrial diseases in man and mouse. *Science* 283:1482–8.

Yu, B. P. 1994. Cellular defenses against damage from reactive oxygen species. *Physiol. Rev.* 74:139–62.

Zhou, G., R. Myers, Y. Li, Y. Chen, X. Shen, J. Fenyk-Melody, M. Wu, J. Ventre, T. Doebber, M. Fujii, M. F. Hirshman, L. J. Goodyear and D. E. Moller. 2001. Role of AMP-activated protein kinase in mechanism of metformin action. *J. Clin. Invest.* 108:1167–74.

Zhou, N., W. R. Lee and B. Abasht. 2015. Messenger RNA sequencing and pathway analysis provide novel insights into the biological basis of chickens' feed efficiency. *BMC Genomics* 16:195–215.

Zurlo, F., K. Larson, C. Bogardus and E. Ravussin. 1990. Skeletal muscle metabolism is a major determinant of resting energy expenditure. *J. Clin. Invest.* 86:1423–7.

Chapter 5

Understanding feed and water intake in poultry

Sami Dridi, University of Arkansas, USA

1 Introduction

The food animal sector contributes 40% of the global value of agricultural output and supports the livelihoods and food security of billions of people. Owing to its high nutritive value without religious taboo, chicken production for meat and eggs has seen the largest increase during the past decade with 87 and 85 million metric tons ready-to-cook equivalent as an average annual global production and consumption of broiler meat, respectively, in 2015[1].

According to the Food and Agriculture Organization of the United Nations, the world's population is projected to grow by one-third, reaching between 9 and 10 billion people by 2050 which will result in an estimated 73% increase in meat and egg consumption[2]. However, sustainably meeting this global nutritional needs will be challenging because of the increasing pressure on the availability of natural resource (water, land, energy) combined with the potential adverse impacts of global climate change on agriculture productivity. Further challenges are the emerging increase of metabolic disorders in poultry such as muscle myopathies, including white striping and woody breast, leg disorders and the increased risk of rapid transmission of animal disease and zoonoses. Therefore, there is a critical need for both applied and fundamental intensive research to ensure sustainable agriculture growth and address the global challenge to food security.

http://dx.doi.org/10.19103/AS.2016.0011.34

These research efforts should not be limited to enhance animal agricultural productivity only, but should also address how to adapt to the significant changes in the global environment impacting on animal agriculture and vice versa, how to efficiently use our natural resource and conserve water and energy, how to improve animal health and well-being, how to improve equitable distribution of animal products in the future and how to improve the bidirectional communication between those engaged in animal agriculture and the consumers, to mention few.

Feed efficiency (FE) and water efficiency (WE) are two vital economic and agricultural traits. Although seminal work has been done in genetic selection for FE in poultry, research related to WE is scarce. Genetic selection for phenotypic FE has tremendously improved livestock productivity over the past 50 years[3]; however, the selection methods have been applied without knowledge of the fundamental molecular mechanisms – changes that might be induced by the selection. Associated with these successes, there have been a number of undesirable changes in the regulation of energy homeostasis and probably water balance. Modern broilers are hyperphagic and voraciously consume approximately 4 kg of feed to achieve a 100-fold increase in body weight at market age (56 days)[4, 5]. FE and WE encompass complex mechanisms regulating feed and water intake, energy expenditure, water retention and excretion, and intermediary metabolism related to nutrient and water utilization and partition. This chapter will summarize recent advances in our understanding of molecular pathways involved in the regulation of energy and water homeostasis and to facilitate the discussion by asking new questions which may help in developing mechanism-based strategies to improve both FE and WE.

2 Preliminaries to the discussion of feed intake regulation in poultry

Before discussing food intake regulation, one might ask the following questions: 1) Do birds have body weight set point (stable body weight) and 2) Is feed intake in birds tightly controlled or random and unplanned? Based on the metabolic set point theory of homeostasis, the body has a tendency to maintain a particular weight range, and it adjusts the complex regulatory network accordingly[6–9]. In the 1990s, Plavnik and Hurwitz showed that chickens were able to return to their initial body weight after feed restriction followed by *ad libitum* re-feeding[10]. Lepkovsky, on the other hand, reported that force-fed birds were able to return to their initial body weight when they had free access to food[11]. Together these results indicated that the 1990 broilers have a body weight set point which might be altered with the extensive genetic selection. During that period, several researchers reported that as dietary energy level increased, birds were able to satisfy their energy needs by decreasing feed intake[12, 13]. This indicated that birds adjust their feed intake according to their metabolizable energy requirements which has been widely used in formulating poultry diets[14]. However, this conclusion has been revised and re-evaluated as broilers have been shown to decrease or increase their feed intake when dietary protein content in the diet was increased or reduced, respectively[15, 16]. Finally, switching birds from a high-fibre to a low-fibre diet, or vice versa, results in compensatory changes in feed intake[17, 18]. Combined, these results not only indicate that birds control their feed intake, but also demonstrate that the mechanisms controlling appetite and satiety can be altered by genetic selection, nutrient requirement of the bird and/or diet composition.

Feed intake can also be modified by gustatory, olfactory, tactile or colour stimuli, and readers can refer to several elegant papers[19–26].

At organismal and molecular levels, a series of highly integrated regulatory mechanisms, yet not completely defined at least in avian species, exists for the control of energy intake at both peripheral and central sites. Each of these sites is discussed in the following sections.

3 Central regulation: classical neuropeptides, genetic selection and hypothalamic neuropeptides

3.1 Classical neuropeptides

Within the central nervous system (CNS), the hypothalamic satiety (arcuate nucleus, ARC and ventromedial nucleus, VMH) and hunger (paraventricular nucleus, PVH and lateral hypothalamus, LHA) centres play key roles in the regulation of feed intake[27, 28]. Significant progress has been made in the neuronal regulation of food intake in mammals and it is likely, with some differences, the similar systems appear to exist in avian species. Two separate populations of neuronal cell types are located in the ARC, one synthesizes the powerfully orexigenic peptides (neuropeptide Y, NPY and agouti-related peptide, AgRP), while the other produces the anorexigenic pro-opiomelanocortin (POMC) and cocaine and amphetamine regulated transcript (CART)[29]. Both neurons have similar electrical activity (resting membrane and fire action potentials) profiles and project to multiple hypothalamic and extra-hypothalamic sites where they communicate with second-order neurons involved in the regulation of energy balance[30, 31]. POMC neurons release α-melanocyte stimulating hormone (α-MSH), whereas NPY/AgRP neurons release NPY, AgRP and GABA peptides at their target nuclei in brain and spinal cord. These peptides interact with several other neuropeptides throughout the CNS. For instance, NPY, α-MSH (agonist) and AgRP (antagonist) interact with the central melanocortin receptor (MCR) pathways. NPY directly inhibits MC4R-expressing PVH neurons via NPY1R[32]. A series of studies demonstrated that MC4R expressed by PVH/amygdala neurons suppress food intake whereas MC4R expressed by autonomic preganglionic neurons within the dorsal vagal complex of brainstem and the intermedio-lateral cell column (IML) of spinal cord increase energy expenditure indicating a divergence of central melanocortin signalling[33, 34]. The effects of MC4R on IML and DMV are cyclic adenosine monophosphate (cAMP) and protein kinase A (PKA)-dependent, respectively. NPY suppresses the neuronal activity of POMC and GABA in ARC, orexin/hypocretin in LHA, and leptin receptor-expressing neurons in VMH[35]. The inhibitory effects of NPY were mediated through activation of G-protein-gated inwardly rectifying K^+ (GIRK) channels or inhibition of voltage-gated Ca^{2+} channel via postsynaptic NPY1R, NPY2R and NPY5R[36–38]. Previous electrophysiological data suggest that AgRP neurons send direct GABAergic projections to POMC neurons[39] and speculated that POMC neurons are one of the key downstream mediators of AgRP[40]. However, recently, Sternson laboratory showed that POMC neurons play a minimal role in mediating AgRP action on acute feeding, although they receive indeed direct GABAergic projections from AgRP neurons[31]. Whether POMC neurons play a major role in mediating AgRP action on long-term regulation of food intake and energy expenditure remains to be answered. Claret et al.[41] showed that adenosine monophosphate-activated protein kinase (AMPK) is essential for energy homeostasis regulation by POMC and AgRP neurons.

It is important to note that in addition to the aforementioned well-defined neuropeptides, there is a growing list of feeding-related peptides that differentially regulate energy balance in mammals and they are elaborated in several elegant reports[35, 42–44]. As in mammals, lesioning the medial hypothalamus of avian species increases feed intake, whereas lesioning the LHA decreases feed intake[45], indicating that these feeding-related anatomical sites are probably conserved between species. At neuronal and molecular levels, significant progress has also been made in the regulation of feed intake in birds and has been widely and elegantly reviewed[46–51]. While some chicken neuropeptides have effects similar to that described in mammals, several other peptides have opposite or no effects. Chicken NPY was first cloned and characterized by Blomqvist et al.[52] and has been shown to be, as in mammals, a potent stimulator of appetite[53]. Although with different affinity, α- and β-MSH both bind MC4R and decrease feed and water intake in chickens[54, 55]. Gamma MSH, the selective MC3R agonist, also reduced feed intake in chickens, but required a higher dose than α- or β-MSH[56]. Tachibana and colleagues[57, 58] showed that AgRP attenuated α-MSH-induced anorexia and CART reduced NPY-induced feeding in chickens. The similarity between birds and mammals on the regulation and the effects of NPY/AgRP and POMC/CART on feed intake is matched by a comparable neuroanatomical distribution of these peptide particularly in the infundibular nucleus, the equivalent of mammalian ARC, on one hand, and the highly conserved sequence (only one single residue different between the mammalian and chicken NPY, for example) and probably a similar structural conformations[52, 59–61].

Intriguingly, despite the generally conserved nature of the peptide-signalling molecules in birds and mammals, there are differences in the function of some peptides. For instance, orexins/hypocretins, galanin, melanin concentrating hormone (MCH) and motilin are potent orexigenic agents in mammals but are without any apparent effect on feed intake in chicks[62, 63]. Peptide YY and pancreatic polypeptide are potent anorexigenic in mammals, but they are orexigneic in birds[53, 64]. The exhaustive list of these peptides and their roles in the regulation of feed intake in both mammals and birds has been reviewed in a number of elegant papers[45, 47–50].

Although a significant progress has been made, our understanding of the molecular mechanisms employed by these peptides as well as their network integration in the regulation of feed intake in avian species remains quite limited. We should also note that the bulk of data about the effect of these neuropeptides on feed intake in birds mainly originated from the use of mammalian peptides by intracerebroventricular (ICV) or peripheral injection and in most cases, the sequence of the avian peptide orthologue and its distribution in the brain and in the hypothalamic nuclei was missed. Also, we generally limited the definition of feed intake to the quantity of food ingested based on the difference between the quantity of food distributed and food remaining in the feeders. It would be therefore important to include in future studies additional parameters of feeding behaviour such as time-spent eating and feeding frequency. Similarly, although it is strenuous to measure, it would be worthy to determine the effects of these peptides on water intake because, in my perspective, water is an economically important trait, especially under the climate change and drought conditions.

3.2 Genetic selection and hypothalamic neuropeptides

Several avian genetic lines have been used to determine the effect of feeding-related peptides on feed intake. Comparing layer (selected for egg production and small body weight) to broiler chickens (selected for high growth rate), Honda et al.[65] showed that central administration of α-MSH suppressed feed intake in both strains; however, β-MSH reduced feed intake in layers but not in broilers. NPY administration has been reported to increase feed intake in both layers and broilers[66]. ICV injection of norepinephrine and epinephrine increased feed intake in broilers but not in layers; however, serotonin decreased feed intake in layers but not in broilers[48]. These results indicated that genetic selection might alter the responsiveness and sensitivity to feeding-related peptides. In 1962, Paul Siegel selected two chicken lines at 8 weeks of age for either high (HWS) or low (LWS) body weight[67], resulting in anorexic and obese birds[68]. Pharmacological and physiological studies showed that the LWS birds are more sensitive to anorexigenic neuropeptides than HWS birds[56, 69, 70]. In 1981, Alain Bordas[71] selected two Rhode Island Red chicken lines for low (R^-) or high (R^+) residual feed intake resulting in dramatic increase (~40–70%) of feed consumption in the R^+ compared to R^- line. By analysing the expression profile of hypothalamic neuropeptides, our laboratory found that several key feeding-related peptides are differentially expressed between the R^+ and R^-lines[72]. Of particular interest, the orexigenic neuropeptide NPY and AgRP are highly expressed in R^+ compared to R^- chickens which may explain the hyperphagic state of the R^+ line. Interestingly, the hypothalamic expression of the suppressor of cytokine signalling 3 (SOCS3), ghrelin and the receptors for ghrelin and adiponectin were also differentially expressed between the two lines[72]. Similar analyses have been performed in two quail lines divergently selected for high (HFE) or low (LFE) feed efficiency. The FE phenotype seemed to be achieved by reduced feed intake in female and increased body weight in male quails[73]. At molecular levels, the hypothalamic expression of classical feeding-related neuropeptides was different between the two lines. Interestingly, we identified several new pathways such as AMPK, mechanistic target of rapamycin (mTOR), signal transducer and activator of transcription (STAT), acetyl-CoA carboxylase alpha (ACCα) and orexin that are differentially expressed between the LFE and HFE quails[73]. These new pathways will be discussed in more detail in the next sections.

Although these genetic materials provide very unique information and useful insights, further investigations using the existing and/or additional chicken models are warranted. For example, it would be interesting to determine the effects of some neuropeptides on feed intake in fat and lean line chickens[74]. Furthermore, in addition to FE, selecting birds for WE and performing in depth comparison at molecular levels between old and modern broilers would be worthwhile goals. These studies could help in identifying molecular signatures that may be used to adapt and prepare modern broilers for suboptimal conditions.

4 Central regulation: new central molecular pathways

New nutrient, energy and metabolite sensing pathways have been identified in the mammalian hypothalamus. One such example involve the serine threonine kinase AMPK, an evolutionary conserved signalling molecule that has emerged as one of the most important energy monitor for many organisms, from yeast to mammals[75]. It is the master energy sensor and 'fuel gauge' that not only senses but also controls and responds to altered energy status in most eukaryotic cells[76]. Several studies showed that AMPK is localized in the mammalian hypothalamus and it controls food intake[77]. Pharmacological or hormonal activation or inhibition of central AMPK activity increased and decreased food intake, respectively[77–79]. To directly address the effect of long-term manipulation of hypothalamic AMPK, Claret and colleagues[41] generated mice lacking AMPKα2 in POMC- and AgRP-expressing neurons. POMC-AMPKα2 knockout mice developed obesity due to dysregulation of food intake and energy expenditure; however, AgRP-AMPKα2 knockout mice developed a lean phenotype. Changes in the expression of well-known (an)orexigenic neuropeptides have been reported with modulation of AMPK activity. Overexpressing dominant negative (DN)-AMPK in mediobasal hypothalamus suppressed NPY and AgRP mRNA levels in ARC, whereas overexpressing active mutated (CA)-AMPK enhanced the fasting-induced increase in NPY and AgRP expression in ARC as well as MCH in LHA[79]. There are few reports about avian hypothalamic AMPK and its function in the regulation of feed intake. Denbow's group showed that ICV administration of AMPK activator, 5-aminoimidazole-4-carboxamide ribonucleotide (AICAR), caused dose-dependent decrease in feed intake in LWS but not in HWS chickens; however, AMPK

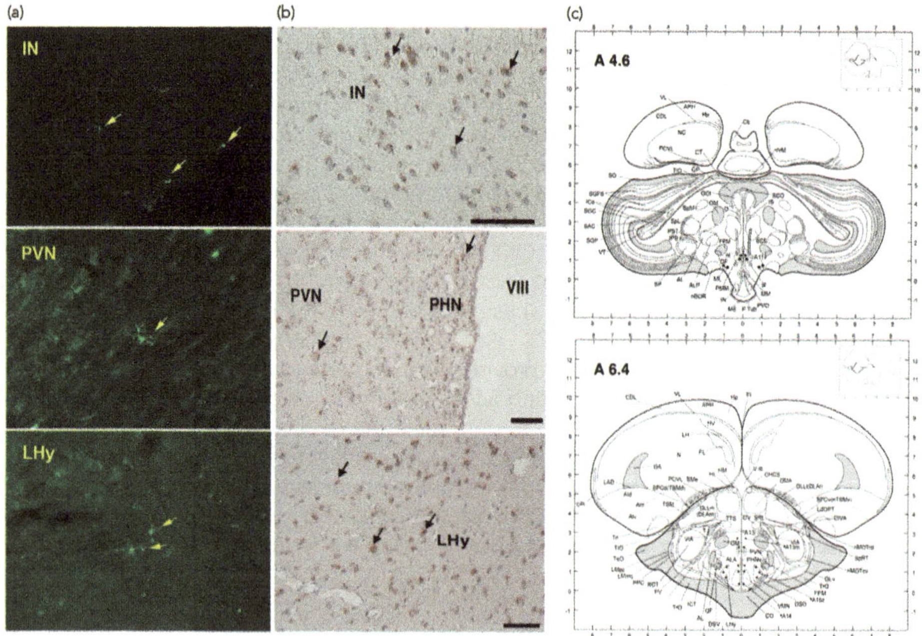

Figure 1 Localization and expression of AMPK in chicken hypothalamus. (a) Immunoflorescence, (b) immunocytochemistry adapted from Proszkowiec-Weglarz et al.[81] and (c) two transverse sections of chick brain showing key hypothalamic nuclei involved in the regulation of feeding behaviour from avian brain atlas (http://avianbrain.org). IN, infundibular nucleus; PVN, paraventricular nucleus; and LHy, lateral hypothalamus. The immunocytochemistry figure was adapted with the permission of the corresponding author.

inhibitor, compound C, increased feed intake in HWS but not in LWS lines[80]. McMurtry's group has identified and characterized AMPK subunit genes as well as two upstream genes: liver kinase B1 (LKB1), also known as serine/threonine kinase 11 (STK11), and calcium/calmodulin-dependent protein kinase kinase (CaMKK) in chicken, and showed that they are expressed in the hypothalamus[81, 82]. In agreement with McMurtry's finding[81], our unpublished data showed that AMPKα1/2 is expressed in the feeding-related IN, PVN and LHA nuclei in chicken (Fig. 1). Although these results indicated that functional AMPK pathway exist in birds, the precise role and the mode of AMPK action in regulating feed intake and energy expenditure in avian species remain to be fully elucidated.

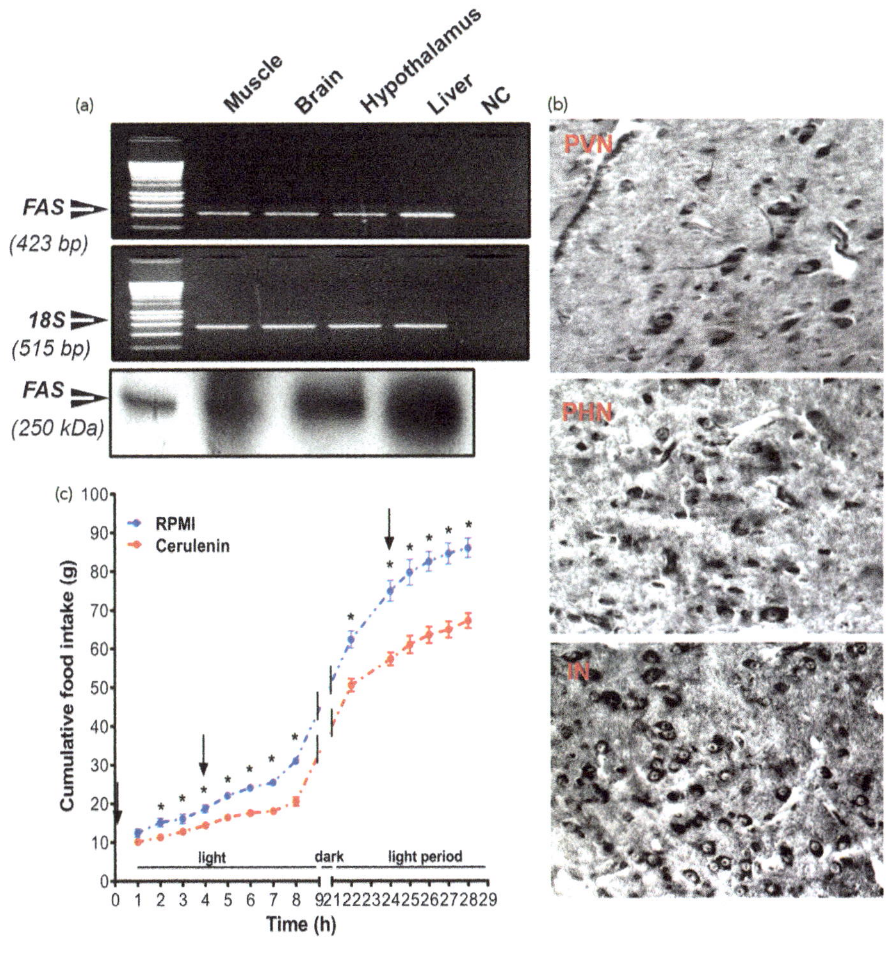

Figure 2 Expression and localization of fatty acid synthase (FAS) in chicken hypothalamus and the effect of fatty acid synthesis inhibition on feed intake. (a) PCR and Western blot, (b) immunocytochemistry and (c) effect of cerulenin, inhibitor of fatty acid synthesis, on cumulative feed intake in chickens obtained from Dridi et al.[88]

The downstream pathways of AMPK in the hypothalamus may involve the ACC (acetyl-CoA carboxylase)-malonyl-CoA-carnitine palmitoyltransferase 1c (CPT-1c) and the mTOR pathways. Some studies in mammals showed that activation of AMPK inhibited ACC activity, which in turn led to decreased intracellular malonyl-CoA levels and consequently resulted in increased CPT-1c and fatty acid oxidation. In support of this mechanism, degradation of malonyl-CoA in the mediobasal hypothalamus of rodents results in increased food intake, whereas inhibition of hypothalamic CPT1 suppresses food intake[83, 84]. Systemic or ICV administration of fatty acid synthase (FAS) inhibitors to lean or obese mice increased hypothalamic malonyl-CoA leading to the suppression of food intake[85]. Conversely, lowering malonyl-CoA with an ACC inhibitor or by the ectopic expression of malonyl-CoA decarboxylase in the hypothalamus increased food intake. Cohen et al.[86] showed that stearoyl-CoA desaturase-1 (SCD-1), rate-limiting enzyme in the biosynthesis of monounsaturated fatty acids, plays a key role in the regulation of food intake.

Although the effect of fatty acids on feed intake in chickens was studied, the underlying molecular mechanisms are still poorly understood. An early study showed that infusion of liposyn, a commercially produced lipid solution, decreased feed intake in layer but not in broilers[87]. By using various molecular techniques, it was demonstrated that FAS, rate-limiting enzyme in fatty acid biosynthesis, is expressed in the chicken hypothalamic IN and PVN nuclei and it is decreased by fasting (Fig. 2)[88]. These results indicated that central FAS is sensitive to energy shortage. Peripheral administration of cerulenin, the natural inhibitor of FAS, decreased feed intake in chickens probably via downregulation of hypothalamic MCR4/5[88].

One of the most exciting findings in the identification of downstream targets for AMPK is mTOR[89]. Activation of AMPK results in the inhibition of mTOR signalling, and thereby inhibits anabolic pathways and enhances catabolic pathways to conserve intracellular energy during low energy status. Although a direct link between the AMPK and mTOR has not been proven yet, it is a reasonable hypothesis since mTOR pathway has been shown to be not only expressed in the hypothalamus but also co-localized with NPY/AgRP and POMC/CART neurons in the ARC[90]. In addition, central treatment with both leptin and leucine has been found to inhibit food intake through mTOR activation pathway[90]. Although the mTOR pathway was differentially expressed in the hypothalamus between high and low FE quails[73], a direct mechanistic understanding of the role of mTOR in the regulation of appetite in avian species warrants further investigations.

Recent studies have implicated central autophagy in the pathogenesis of obesity. Autophagy is an important intracellular self-digestion process, conserved from yeast to human, whereby double-membrane autophagosome sequesters organelles or cytosol portions and delivers them to lysosomes for breakdown by resident hydrolases and to provide nutrients to starving cells (Fig. 3)[91]. The first studies were from Singh and Kiosses groups showing that starvation induced autophagy in the hypothalamus[92, 93]. These studies used a conditional cre-loxP approach to specifically delete autophagy-related gene 7 (Atg7) from POMC neurons and showed that these mice displayed higher body weight due to hyperphagia[94]. By using specific knockout of Atg7 in AgRP neurons, the same group showed that these mice had higher levels of POMC and α-MSH and decreased body weight compared to wild- type mice[92]. Malhotra et al.[95] showed that mice lacking Atg12 in POMC-positive neurons exhibited increased food intake and accelerated body weight gain, however Atg5 knockout mice did not display these phenotypes. Although autophagy-related genes have been characterized in chickens[96], their roles in the regulation of feed intake are still unknown and need further investigation.

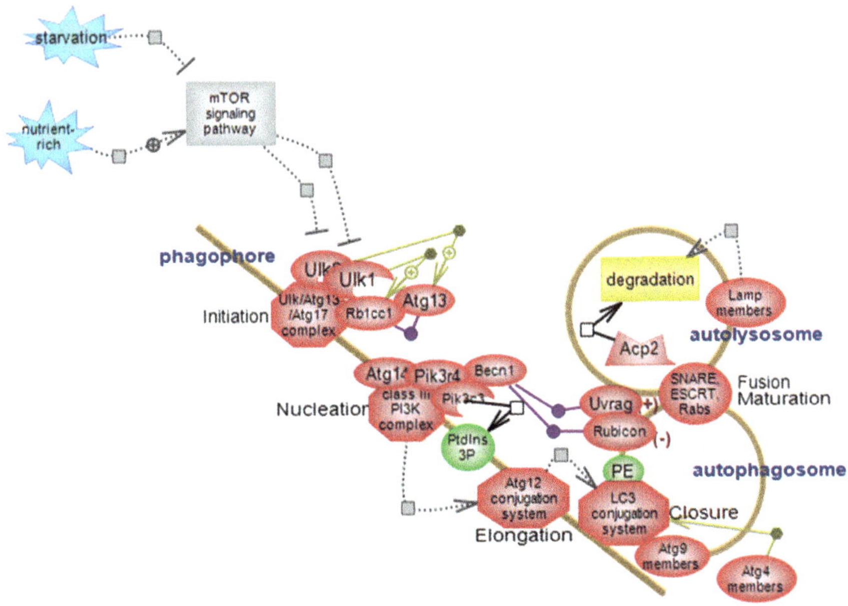

Figure 3 Autophagy steps. Initiation, nucleation, elongation and maturation are all steps of autophagy. Autophagosome formation can be initiated via mTOR inhibition or AMPK activation during starvation or nutrient limitation. This results in the activation of ULK1 which in turn phosphorylates Atg13, Atg101 and FIP200. When autophagy is activated, Beclin 1 is liberated from Bcl-2 and is associated with Vps34, Vps15 and Atg14. ULK1 phosphorylates also AMBRA, a component of the PI3K CIII complex enabling it to relocate from the cytoskeleton to the isolation membrane. The activation of Vps34 generates PI3P which catalyses the first of two types of ubiquitination-like reactions that regulates membrane elongation. First, Atg5 and Atg12 are conjugated to each other in the presence of Atg7 and Atg10. Attachment of the Atg5-Atg12-Atg16L1 complex on the isolation membrane induces the second complex to covalently conjugate PE to LC3 which facilitates in turn the closure of the isolation membrane. The complex Atg9-Atg2-Atg18 cycles between endosomes, the Golgi and the phagophore possibly carrying lipid components for membrane expansion. LC3-II is formed by LC3 conjugation to its lipid target PE and Atg4 removes LC3-II from the outer surface of newly formed autophagosome, and LC3 on the inner surface is degraded when the autophagosome fuses with lysosomes. Atg, autophagy-related genes; LC3, microtubule-associated protein light chain; PE, phosphatidylethanolamine; PI3K, phosphatidylinositol 3 kinase; PIP3, phosphatidylinositol 3-phosphate; ULK1, UNC51-like kinase 1. The figure was produced by the Pathway Studio software from Ariadne/Elsevier and is used by permission of the Rat Genome Database[200].

More recently, it has been shown that deletion of the micro RNA (miRNA) processing enzyme DICER1 in POMC- and AgRP-expressing neurons alter energy balance[97] indicating that POMC and AgRP may regulate feed intake through non-coding-RNA pathways. Canonical miRNA biogenesis involves sequential cleavage by the dsRNA specific, RNAse III enzyme Drosha and DICER1[98]. Initial nuclear processing of miRNA transcripts involves the microprocessor complex which contains Drosha and DiGeorge syndrome critical region gene 8 (DGCR8), also known as Pasha[99, 100]. The microprocessor processes pri-miRNA into pre-miRNA. The latter molecule is further cleaved by DICER1 into mature

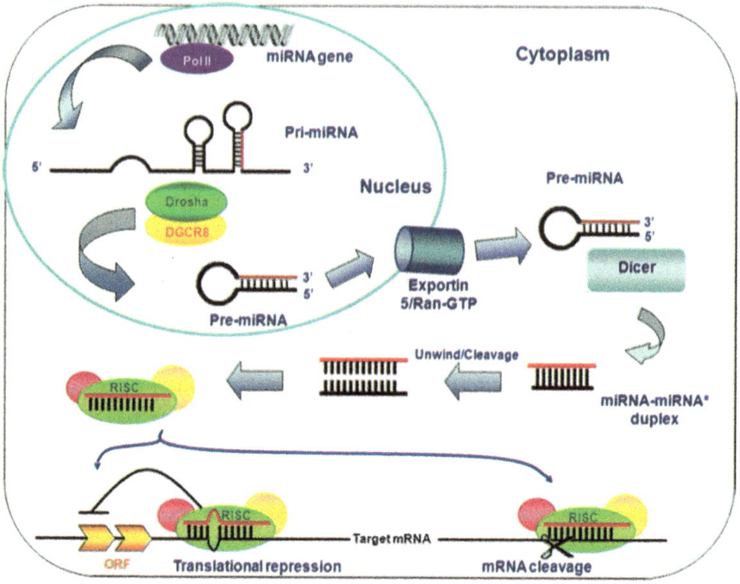

Figure 4 A general model of miRNA biogenesis. After transcription by RNA polymerase II, miRNA primary transcripts (pri-miRNAs) are cleaved by Drosha/DGCR8 in the nuclear compartment. The pre-miRNAs are then transported, via exportin 5, to the cytoplasm where they are excised by dicer1 to form mature 22-nt miRNAs. One strand is selected for stable association with Argonaute, where it serves, in coordination with RISC, as a guide to target and regulate specific mRNAs. DGCR8, DiGeorge Syndrome Critical Region gene 8; RISC, RNA-induced silencing complex. The figure was adapted from[201] with the permission of Atlas of Genetics and Cytogenetics in Oncology and Hematology.

miRNA in the cell cytoplasm (Fig. 4). One strand of the dsRNA duplex is then loaded into an Argonaute (Ago) protein and drives the recruitment of a complex of effector proteins called the RNA-induced silencing complex (RISC) that inhibits the expression of targeted transcripts[101]. More recently, Vinnikov et al.[102] have demonstrated that deletion of both alleles of the DICER1 gene in the hypothalamic ARC induce hyperphagic obesity and alter the hypothalamic expression of NPY, AgRP and MSH in mice. This effect was mediated through phosphatidylinositol-3-kinase (PI3K), protein kinase B which also known as Akt, and mTOR pathway. Injection of miRNA (miR-103) attenuates these effects, indicating that neuronal miRNA play a key role in the control of energy homeostasis in rodents. Luo et al.[103] have reported an association of single nucleotide polymorphism in the miR-1596 with residual feed intake in chickens; however, further mechanistic studies investigating the role and the mechanisms employed by miRNA and miRNA biogenesis-related enzymes to regulate feed intake in chickens are warranted.

5 Peripheral and hormonal regulation of feed intake

In response to external stimuli (feed availability, diet composition, light/dark cycle, environmental temperature, etc.) and the animal physiological status (energy stores,

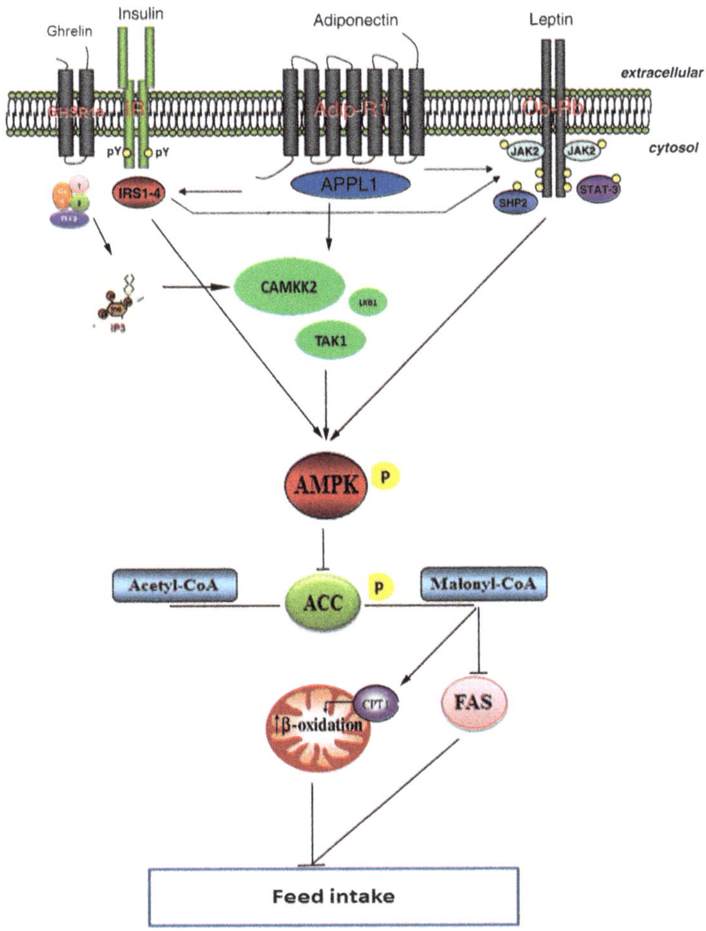

Figure 5 Intracellular AMPK downstream signalling through which peripheral hormones control feed intake in mammals. Peripheral hormones including leptin, insulin, adiponectin and ghrelin activate their receptors and alter hypothalamic expression and activation of AMPK. The anorexigenic metabolic hormones, leptin and insulin, suppress hypothalamic AMPK activity through classical JAK-STAT, mTOR and PI3K-Akt pathways. Alteration of AMPK activity increases ACC and malonyl-CoA which in turn decreases CPT-1 and fatty acid oxidation and consequently results in inhibition of feed intake. The orexigenic ghrelin activates its GHSR1a receptor and recruits PLCβ to initiate Ca^{2+} release from the endoplasmic reticulum. Released intracellular Ca^{2+} activates CaMKK and AMPK. Hypothalamic activation of AMPK inhibits ACC activity which in turn leads to decreased intracellular malonyl-CoA levels and consequently results in increased CPT-1c and fatty acid oxidation and thereby increase feed intake. The effect of adiponectin on feed intake in mammals is still non-consistent. ACC, acetyl-CoA carboxylase: Adip-R1/2, adiponectin receptor; CaMKK2, calmodulin-dependent protein kinase kinase; CPT1, carnitine palmitoyl transferase 1; FAS, fatty acid synthase; GHSR1a, growth hormone secretagogue receptor 1a; IRS1-4, insulin receptor substrate 1-4; JAK, janus kinase; LKB1, liver kinase B1; PLCβ, phospholipase C; STAT, signal transducer and activator of transcription; TAK1, TGFβ activated kinase 1.

nutrient and metabolite levels etc.), central neural circuits and peripheral target tissues regulate appetite and body weight in a coordinated and cohesive fashion involving negative feedback loops. A key component of this feedback loop is the synthesis and release of peripheral metabolic hormones. In this section, we will discuss only the interaction between major hormones (ghrelin, leptin, thyroid hormone, insulin and adiponectin) and hypothalamic AMPK in the regulation of feed intake to illustrate examples of specific signalling axes. There have been a number of reviews regarding the regulation of feed intake by CNS and peripheral tissues in poultry[45, 104].These hormones are regulated by nutritional states and convey metabolic information via their receptors to the brain. In mammals and under conditions of feed deprivation, plasma ghrelin concentrations increase, whereas circulating leptin levels decrease, indicating that these two hormones convey two opposite (negative and positive) signals to the brain. Peripheral or central administration of ghrelin induces feed intake, AMPK phosphorylation and activity in NPY/AgRP neurons in mammalian hypothalamus[105, 106]. Blocking AMPK activity with compound C inhibits ghrelin-induced food intake and ghrelin does not activate AMPK in ghrelin receptor (GHSR)-knockout mice[107]. Recent studies have demonstrated that upon binding to its receptor in the hypothalamus, ghrelin initiates Ca^{2+} influx and release in NPY neurons and induces AMPK activity[106]. Intracellular increase in Ca^{2+} has been shown to activate AMPK through its upstream regulator CaMKK[108], and upregulates NPY gene expression[109]. Knockout of CaMKK blocks the effect of ghrelin on feed intake in mice. Velasquez et al. have recently demonstrated that Sirtuin 1 (SIRT1) deacetylase activity in the hypothalamus also mediates ghrelin-induced AMPK activation and increased feed intake and SIRT1 inhibition reduced ghrelin-activated AMPK, NPY, and AgRP gene expression[110]. As P53 is a substrate of SIRT1, the same group used $P53^{-/-}$ mice and have shown that ghrelin does not induce appetite or AMPK activation[110]. The downstream signalling cascades after ghrelin-induced AMPK activation involve ACC-malonyl CoA-CPT1 and mTOR as described above[111, 112] (Fig. 5).

The ghrelin gene and its related receptor have been cloned and characterized in chicken, turkey, duck and quails[113–118]. In contrast to mammals, central injection of mammalian or chicken recombinant ghrelin inhibits feed intake in birds[119–122]. However, its peripheral effects on feed intake are not consistent. For instance, in adult Japanese quails, intraperitoneal (IP) administration of low dose (4–9 nmols $^{-1}$ BW) of mammalian ghrelin increases FI, but higher dose (24 nmols Kg^{-1} BW) suppresses feed intake[121]. Moreover, intravenous (IV) administration of ghrelin inhibits feed intake in broiler but not in layer chickens[122, 123]. Thus, this inconsistency seems to be related to the dose of ghrelin administered and/or genotype (strain) of birds used. The anorexigenic effect of ghrelin in chickens was hypothesized to be mediated through corticotropin releasing factor (CRF), but not NPY neurons as ICV injection of ghrelin increases plasma corticosterone levels leading to feed intake suppression, and this effect was inhibited when Astressin (CRF receptor antagonist) is co-injected with ghrelin[120]. Recently, Denbow's group have shown that central ghrelin differentially inhibited hypothalamic AMPKα related gene expression and phosphorylation, as well as ACC, in LWS and HWS chicken lines[124]. Together, these data indicate that ghrelin effect in chicken is opposite to that reported in mammals, and this effect as well as its related downstream-cascade (AMPK) are altered by genetic selection. Further mechanistic studies related to avian ghrelin signalling pathways and control of energy homeostasis are needed.

Leptin, on the other hand, suppresses hypothalamic AMPK activity and reduces feed intake in mammals[79, 125, 126]. This effect of leptin on AMPK seems to be independent from its canonical STAT3 pathway. Results from recent study showed that leptin suppresses AMPK activity via mTORC1 and the activation of mTORC1 occurs in a PI3K-Akt-dependent manner[112]. As for ghrelin, leptin acts on downstream components of the AMPK and increase ACC-malonyl CoA to inhibit feed intake[127, 128]. In birds, however, the leptin sequence was a matter of debate for more than 20 years. Early work led by Taouis and Ashwell groups identified chicken leptin sequences with high (>95%) homology to mammalian leptin[129, 130] and several studies showed that central or peripheral administration of recombinant leptin inhibit feed intake in birds[131–135]. These effects are mediated through chicken leptin receptor which showed approximately 60% sequence identity with the long isoform of the mammalian leptin receptor. Recently, Friedman-Einat's group has identified an original avian (chicken and duck) leptin sequence which has high guanine-cytosine content and high sequence divergent from mammalian leptin[136]. The same group has also demonstrated that the newly characterized chicken and duck leptins specifically activated chicken leptin receptor *in vitro*. Nothing is known yet about the biosynthesis, function and regulation of the newly identified chicken leptin. Although limited studies showed that leptin affect feeding-related hypothalamic neuropeptides and nitric oxide to modulate feeding behaviour in chickens[137, 138], still the underlying downstream cascades are unknown. Further studies regarding avian leptin biology and function are warranted.

Similar to leptin, insulin is an anorexigenic hormone[139]. Central administration of insulin stimulate its classical- and leptin crosstalk-signalling pathways, including JAK-STAT, Akt and FoxO1[140], and thereby suppresses AMPK activation and its downstream cascades in mammals. Insulin inhibits AMPK activation also via mTOR pathway. Intriguingly, peripheral administration of insulin increases hypothalamic AMPK activity[141]. Together, these data indicate that insulin regulates AMPK in a tissue-specific manner in mammals. Chickens are hyperglycaemic compared to mammals, with their blood glucose levels averaging three times that found in humans[142]. They do not have glucose transporter 4 (GLUT4), but they do have GLUT8 and GLUT12[143, 144], and they require insulin doses greater than four times that required in mammals to reduce glucose levels and, therefore, they are considered insulin resistant[145, 146]. As in mammals, insulin inhibits feed intake when centrally injected in chickens and this effect was mediated through alteration of both orexigenic and anorexigenic hypothalamic neuropeptides[147–149]. Similar to leptin and many other adipokines, the role of AMPK-mTOR pathways in mediating anorexic effect of insulin is still unknown in chickens.

Central injection of triiodothyronine (T_3) decreases hypothalamic AMPK pathway and reduces body weight without affecting feed intake in rodents[78]. This effect is mediated through an increased sympathetic nervous system activation, uncoupling protein 1 (UCP1) and peroxisome proliferator-activated receptor gamma (PPARγ)-coactivator 1(PGC1α) expression, and brown adipose tissue thermogenesis[78]. Coppola et al.[150] have demonstrated that thyroid hormone also acts on hypothalamic (ARC) UCP2 to regulate feeding behaviour. In chickens, T_3-supplemented diet (1 mg.Kg^{-1}) decreased feed intake and body weight[151]; however, the underlying mechanism is unknown.

Unlike leptin which circulates at picograms or nanograms per millilitre, adiponectin is the most abundant adipocytokine with high circulating levels (micrograms per millilitre)[152]. The central effect of adiponectin on feed intake in mammals is controversial. Kubota et al. (2007) showed that central injection of adiponectin increases hypothalamic AMPK phosphorylation and stimulates appetite and feed intake in mammals[153]. Blocking AMPK activity in the ARC, by using a dominant-negative AMPK, reverses these effects. Previous studies conducted by Qi et al.[154], however, reported that central administration of adiponectin causes weight loss in rodents. Results from recent study by Park et al.[155] showed that chronic central administration of adiponectin improves glucose homeostasis but did not affect hypothalamic AMPK in rodents. Such studies are currently lacking in birds, although adiponectin and its related receptors are ubiquitously expressed in chickens[156]. Moreover, the expression of adiponectin system is altered in chicken lines selected for high residual feed intake[72] and in quails selected for high feed efficiency[73]. It has been also reported that adiponectin system is associated with broiler performance traits[157], and play a key role in chicken adipogenesis and mitochondrial biogenesis[158]. Future studies are required to examine the role of adiponectin in the control of energy homeostasis in birds as well as its underlying mechanism.

6 Regulation of water homeostasis in poultry

Water quality and quantity are two key variables that significantly impact poultry productivity and sustainability. However, as for food, available global water resources are under significant challenge as population growth and climate change interact in such a way as to create regional crisis. In poultry, water is used not only for consumption, but also to cool birds during hot weather. Total water intake in the United States per year is approximately 25 billion gallons by commercial broilers[159]. This does not include breeders, layers, pullets, turkey and other avian species. According to NRC (1994), water intake of chickens increases by about 7% for each 1°C increase above 21°C[160, 161]. In addition, it has been reported that 75% of consumed water is released by the bird to the environment, which in some cases, can increase litter moisture content and induce footpad dermatitis[162]. Therefore, it is critical to consider WE when discussing poultry production and sustainability. In this portion of the chapter, we will limit the discussion to the potential molecular mechanisms involved in the control of water intake and body water balance.

Drinking plays a pivotal role in the regulation of body fluids and mainly occurs in response to thirst. This involves the stimulation of the thirst centre in the hypothalamus as a result of cellular dehydration or a decrease in the volume of extracellular fluids. The CNS integrates the information from the periphery and visceral sensory systems through the classical sensory pathways and monitors the constituents of the circulation to assess the physiological status of the animal. As in mammals, peripheral or central administration of angiotensin II (Ang II) causes an increase in water intake in chickens[163]. This indicates that Ang II regulates thirst response by two potential mechanisms: 1) Circulating Ang II may act on its hypothalamic receptor, and/or 2) Ang II produced locally in the brain may act as a neurotransmitter coordinating osmotic and hormonal information[164].

When drinking water is limited, particularly in deserts, other avenues, including reduction of water loss by evaporation or excretion, are involved to maintain body water homeostasis. Water loss is primarily by evaporation through cutaneous and respiratory routes, and excretory water loss from the kidney and the gastrointestinal tract. The loss of water by excretion can be minimized by enhanced water reabsorption in the colon, and the excretion of a small volume of concentrated urine[165]. Desert mammals can also maximize water recovery from nasal counter-current systems. Birds are also osmoregulators and have evolved three major organs to control body water homeostasis: kidney, nasal salt gland and lower gastrointestinal tract[166]. At molecular levels, at least 13 aquaporin (AQP) water-channel proteins have been discovered in mammals. Aquaporin 1 (AQP1) is highly abundant in the proximal tubule and descending thin limb and it is involved in the movement of water[167]. AQP2 is expressed in the apical plasma membrane and apical vesicles in the collecting duct principal cells and at lower abundance in connecting tubules and it is the primary target for vasopressin regulation of collecting water permeability[168]. AQP3 and AQP4 are present in the collecting duct principal cells and they have been shown to be involved in water permeability and urine concentration[169, 170]. AQP6 is an intracellular water channel, which is expressed in collecting duct-intercalated cells in cortical, outer medulla and inner medullary collecting duct[171]. Several other AQPs, including AQP7, 8 and 9, are ubiquitously expressed and their roles are not fully defined.

In the avian kidney, at least three AQPs have been characterized. In quail, AQP2 mRNA was localized in branches of the collecting duct in the medulla and cortex and it is upregulated by water deprivation suggesting its role in water retention by the kidney. AQP3 was localized in kidney's collecting tubules and ducts indicating its role in water reabsorption from collecting duct cells. AQP4 was characterized from medullary cones of quail kidney, but it was also found in several avian tissues, including hypothalamus and circumventricular; however, its physiological role is still unknown. Osmotic stress downregulated the expression of AQP1 in the endothelial cells and AQP5 in the epithelial cells in duckling salt glands[172], suggesting a decreased capillary ultrafiltration and epithelial water loss. AQP1 has been also shown to be expressed in lower gastrointestinal tract in avian species[173]. The signalling cascades and molecular mechanisms involved in mammalian AQP regulation has been elegantly detailed elsewhere (for review see[174]); however, they are not thoroughly understood in avian species.

Despite the importance of food intake as a direct and indirect source of metabolic water production, there has been little research, especially in poultry, on how birds can change their feed intake strategy and/or metabolic strategy to enhance the efficiency of their metabolic water production. We hypothesize that there is a highly integrated connection between appetite control system and osmoregulation. For instance, water-deprived desert-adapted Spinifex hopping mouse (*Notomys alexis*) for either 12 or 29 days exhibited a biphasic pattern of food consumption with an early hypophagia followed by a sustained increase in appetite for the latter phase of water restriction[175]. The initial hypophagia was driven by an increase in circulating leptin levels; however, sustained feed intake during the latter phase occurred despite the elevated plasma leptin concentrations[175]. It is possible that the hypothalamic signals controlling energy balance are outweighed by osmoregulatory needs and drives to increase substrate provision for body water homeostasis. Cline's group has investigated the effect of various peptides on water and feed intake in chickens. He has demonstrated that administration of neuropeptide FF, neuropeptide K, neuropeptide VF, visfatin, amylin, xenin, gonadotropin-inhibitory hormone, substance P, adrenomedullin, bombesin-like peptides (neuromedins and gastrin-releasing peptide), litorin, or α-MSH reduced feed intake without affecting water consumption in broilers[176–186]. However, administration of alytesin, oxyntomodulin, β-MSH, neuropeptide S, adrenocorticotropic hormone, calcitonin or stresscopin alters water intake and this effect was secondary to reduction in feed intake[54, 187–191]. Calcitonin-gene-related peptide inhibited water intake in birds independently of its effect on feed intake[192]. Denbow et al.[193] have shown that ICV injection of a $CaCl_2$ solution increased feed and water intake in broilers but not in layers, indicating that ionostatic control of food and water consumption is strain-dependent. Baghbanzadeh et al.[194] have reported that central administration of a beta-adrenergic agonist, isoproterenol, alters water intake in broilers. Although seminal work has been conducted to determine the effects of different compounds on water intake in broilers, mechanistic studies of how water homeostasis is regulated under normal and stress conditions are very limited.

Emerging studies in mammals indicated a key role of miRNAs in the control of water balance. Recent research showed that miRNAs regulate the expression of ion channels and transporters as well as neurohumoral factors involved in fluid and electrolyte balance. For instance, Mladinov et al.[195] have demonstrated that miR-192 targets the Na^+/K^+ ATPase which is the driving force of tubular transport in the kidney. Tobon et al.[196] have shown that miR-142–3p regulates the renal D1 dopamine receptor which in turn regulates diuresis and natriuresis via the Na^+/K^+ ATPase and the Na^+/H^+ exchanger. Two groups have recently shown that miR-132 and miR-212 modulate the action of AngII[197, 198]. Transgenic overexpression of miR-466a-3p was associated with disturbed ion homeostasis and altered kidney morphology[199]. More recently, we have shown that short-term water restriction alters DICER1 gene expression in the kidney of modern broilers (data not shown) indicating that miRNA biogenesis-related enzymes are involved in the control of water balance and osmoregulation in avian species. Further intensive functional studies are needed to thoroughly understand the molecular mechanisms involved in the regulation of water homeostasis in poultry.

7 Conclusion and perspectives

FE and WE are both vital economic and agricultural traits. They are closely connected, and they are adversely affected by drought conditions driven by climate changes. Although pioneering managerial, nutritional and/or genetic strategies have been conducted to improve feed efficiency, mechanism-based approaches are still needed to further improve both water and feed efficiency. New techniques involving genomics, epigenetics, proteomics, transcriptomics, mobilomics and metabolomics are now making their ways to solve the intervening puzzle between nutrients, water, genes and performances. These approaches have the potential to change the future by using personalized managerial approach based on identification, selection and optimization of nutrients and water fine-tuned with animal genetic profile. The identified molecular signatures could subsequently be used to improve water and feed efficiency via genetic selection, nutrition and/or management.

8 Where to look for further information

Interested readers are encouraged to access https://www.ncbi.nlm.nih.gov/pubmed and

Sturkie's Avian Physiology, Sixth Edition, edited by Colin G. Scanes. PP.1-995 to know more about feed and water intake in poultry.

9 References

1. Conway, A. (2015). Global poultry market grows despite avian influenza. *Poult. Trends.* 6, 6–10.
2. Alexandratos, N. B. J. (2012). World agriculture towards 2030/2050. In: FAO, editor. ESA Working paper N0 12–03. Rome.

3. Havenstein, G. B., Ferket, P. R. and Qureshi, M. A. (2003). Growth, livability, and feed conversion of 1957 versus 2001 broilers when fed representative 1957 and 2001 broiler diets. *Poult. Sci.* 82: 1500–8.
4. Scheuermann, G. N., Bilgili, S. F., Hess, J. B. and Mulvaney, D. R. (2003). Breast muscle development in commercial broiler chickens. *Poult. Sci.* 82: 1648–58.
5. Hood, R. L. (1982). The cellular basis for growth of the abdominal fat pad in broiler-type chickens. *Poult. Sci.* 61: 117–21.
6. Wirtshafter, D. and Davis, J. D. (1977). Set points, settling points, and the control of body weight. *Physiol. Behav.* 19: 75–8.
7. Harris, R. B. and Martin, R. J. (1984). Recovery of body weight from below 'set point' in mature female rats. *J. Nutr.* 114: 1143–50.
8. Hunsinger, R. N. and Wilson, M. C. (1986). Anorectics and the set point theory for regulation of body weight. *Int. J. Obes.* 10: 205–10.
9. Garn, S. M. (1992). Role of set-point theory in regulation of body weight. *FASEB J.* 6: 794.
10. Plavnik, I. and Hurwitz, S. (1991). Response of broiler chickens and turkey poults to food restriction of varied severity during early life. *Br. Poult. Sci.* 32: 343–52.
11. Lepkovsky, S. (1973). Hypothalamic-adipose tissue interrelationships. *Fed. Proc.* 32: 1705–8.
12. Plavnik, I., Wax, E., Sklan, D. and Hurwitz, S. (1997). The response of broiler chickens and turkey poults to steam-pelleted diets supplemented with fat or carbohydrates. *Poult. Sci.* 76: 1006–13.
13. Leeson, S., Caston, L. and Summers, J. D. (1996). Broiler response to diet energy. *Poult. Sci.* 75: 529–35.
14. Nutrition SoP (1994). Nutrient requirements of Poultry. In: (NRC) NRC, editor. Ninth Revised Edition ed.
15. Buyse, J., Decuypere, E., Berghman, L., Kuhn, E. R. and Vandesande, F. (1992). Effect of dietary protein content on episodic growth hormone secretion and on heat production of male broiler chickens. *Br. Poult. Sci.* 33: 1101–9.
16. Burnham, D., Emmans, G. C. and Gous, R. M. (1992). Isoleucine requirements of the chicken: the effect of excess leucine and valine on the response to isoleucine. *Br. Poult. Sci.* 33: 71–87.
17. Savory, C. J. and Gentle, M. J. (1976). Changes in food intake and gut size in Japanese quail in response to manipulation of dietary fibre content. *Br. Poult. Sci.* 17: 571–80.
18. Savory, C. J. and Gentle, M. J. (1976). Effects of dietary dilution with fibre on the food intake and gut dimensions of Japanese quail. *Br. Poult. Sci.* 17: 561–70.
19. Bang, B. G. (1960). Anatomical evidence for olfactory function in some species of birds. *Nature* 188: 547–9.
20. Bertin, A., Calandreau, L., Arnould, C. and Levy, F. (2012). The developmental stage of chicken embryos modulates the impact of in ovo olfactory stimulation on food preferences. *Chem. Senses* 37: 253–61.
21. Bonadonna, F. and Nevitt, G. A. (2004). Partner-specific odor recognition in an Antarctic seabird. *Science* 306: 835.
22. Burne, T. H. and Rogers, L. J. (1999). Changes in olfactory responsiveness by the domestic chick after early exposure to odorants. *Anim. Behav.* 58: 329–36.
23. Mabayo, R. T., Okumura, J., Hirao, A., Sugita, S., Sugahara, K. et al. (1996). The role of olfaction in oil preference in the chicken. *Physiol. Behav.* 59: 1185–8.
24. Ganchrow, D. and Ganchrow, J. R. (1985). Number and distribution of taste buds in the oral cavity of hatchling chicks. *Physiol. Behav.* 34: 889–94.
25. Henton, W. W., Smith, J. C. and Tucker, D. (1966). Odor discrimination in pigeons. *Science* 153: 1138–9.
26. Hughes, B. O. and Wood-Gush, D. G. (1971). A specific appetite for calcium in domestic chickens. *Anim. Behav.* 19: 490–9.
27. Morgane, P. J. (1961). Distinct 'feeding' and 'hunger motivating' systems in the lateral hypothalamus of the rat. *Science* 133: 887–8.
28. Elmquist, J. K., Elias, C. F. and Saper, C. B. (1999). From lesions to leptin: hypothalamic control of food intake and body weight. *Neuron* 22: 221–32.

29. Friedman, J. M. and Halaas, J. L. (1998). Leptin and the regulation of body weight in mammals. *Nature* 395: 763–70.
30. Mercer, A. J., Hentges, S. T., Meshul, C. K. and Low, M. J. (2013). Unraveling the central proopiomelanocortin neural circuits. *Front Neurosci.* 7: 19.
31. Atasoy, D., Betley, J. N., Su, H. H. and Sternson, S. M. (2012). Deconstruction of a neural circuit for hunger. *Nature* 488: 172–7.
32. Ghamari-Langroudi, M., Srisai, D. and Cone, R. D. (2011). Multinodal regulation of the arcuate/paraventricular nucleus circuit by leptin. *Proc. Natl. Acad. Sci. U.S.A.* 108: 355–60.
33. Balthasar, N., Dalgaard, L. T., Lee, C. E., Yu, J., Funahashi, H. et al. (2005). Divergence of melanocortin pathways in the control of food intake and energy expenditure. *Cell* 123: 493–505.
34. Rossi, J., Balthasar, N., Olson, D., Scott, M., Berglund, E. et al. (2011). Melanocortin-4 receptors expressed by cholinergic neurons regulate energy balance and glucose homeostasis. *Cell Metab.* 13: 195–204.
35. Sohn, J. W., Elmquist, J. K. and Williams, K. W. (2013). Neuronal circuits that regulate feeding behavior and metabolism. *Trends Neurosci.* 36: 504–12.
36. Roseberry, A. G., Liu, H., Jackson, A. C., Cai, X. and Friedman, J. M. (2004). Neuropeptide Y-mediated inhibition of proopiomelanocortin neurons in the arcuate nucleus shows enhanced desensitization in ob/ob mice. *Neuron* 41: 711–22.
37. Fu, L. Y., Acuna-Goycolea, C. and van den Pol, A. N. (2004). Neuropeptide Y inhibits hypocretin/orexin neurons by multiple presynaptic and postsynaptic mechanisms: tonic depression of the hypothalamic arousal system. *J. Neurosci.* 24: 8741–51.
38. Acuna-Goycolea, C., Tamamaki, N., Yanagawa, Y., Obata, K. and van den Pol, A. N. (2005). Mechanisms of neuropeptide Y, peptide YY, and pancreatic polypeptide inhibition of identified green fluorescent protein-expressing GABA neurons in the hypothalamic neuroendocrine arcuate nucleus. *J. Neurosci.* 25: 7406–19.
39. Cowley, M. A., Smart, J. L., Rubinstein, M., Cerdan, M. G., Diano, S. et al. (2001). Leptin activates anorexigenic POMC neurons through a neural network in the arcuate nucleus. *Nature* 411: 480–4.
40. Cone, R. D. (2005). Anatomy and regulation of the central melanocortin system. *Nat. Neurosci.* 8: 571–8.
41. Claret, M., Smith, M. A., Batterham, R. L., Selman, C., Choudhury, A. I. et al. (2007). AMPK is essential for energy homeostasis regulation and glucose sensing by POMC and AgRP neurons. *J. Clin. Invest.* 117: 2325–36.
42. Vettor, R., Fabris, R., Pagano, C. and Federspil, G. (2002). Neuroendocrine regulation of eating behavior. *J. Endocrinol. Invest.* 25: 836–54.
43. Woods, S. C., Seeley, R. J., Porte, D., Jr. and Schwartz, M. W. (1998). Signals that regulate food intake and energy homeostasis. *Science* 280: 1378–83.
44. Berthoud, H. R. (2002). Multiple neural systems controlling food intake and body weight. *Neurosci. Biobehav. Rev.* 26: 393–428.
45. Kuenzel, W. J., Beck, M. M. and Teruyama, R. (1999). Neural sites and pathways regulating food intake in birds: a comparative analysis to mammalian systems. *J. Exp. Zool.* 283: 348–64.
46. Kuenzel, W. J. (1994). Central neuroanatomical systems involved in the regulation of food intake in birds and mammals. *J. Nutr.* 124: 1355S–70S.
47. Richards, M. P. and Proszkowiec-Weglarz, M. (2007). Mechanisms regulating feed intake, energy expenditure, and body weight in poultry. *Poult. Sci.* 86: 1478–90.
48. Denbow, D. M. (1985). Food intake control in birds. *Neurosci. Biobehav. Rev.* 9: 223–32.
49. Cline, M. A. and Furuse, M. (2012). Neuropeptide regulation of food intake in chicks. In Morrison, J. L. (ed.), *Food Intake: Regulation, Assessing and Controlling*. NY: Nova Science Publishers, Inc., pp. 1–34.
50. Denbow, M. C. M. (2014). Food intake regulation. In Scanes, C. G. (ed.), *Sturkie's Avian Physiology*. Sixth Edition ed: Elsevier, pp. 469–79.
51. Wayne, J. K. (2014). The avian subpallium and autonomic nervous system. In Scanes, C. G. (ed.), *Sturkie's Avian Physiology*. Sixth Edition ed: Elsevier, pp. 135–55.

52. Blomqvist, A. G., Soderberg, C., Lundell, I., Milner, R. J. and Larhammar, D. (1992). Strong evolutionary conservation of neuropeptide Y: sequences of chicken, goldfish, and Torpedo marmorata DNA clones. *Proc. Natl. Acad. Sci. U.S.A.* 89: 2350–4.
53. Kuenzel, W. J., Douglass, L. W. and Davison, B. A. (1987). Robust feeding following central administration of neuropeptide Y or peptide YY in chicks, Gallus domesticus. *Peptides* 8: 823–8.
54. Smith, M. L., Prall, B., Nandar, W. and Cline, M. A. (2008). Beta-melanocyte-stimulating hormone potently reduces appetite via the hypothalamus in chicks. *J. Neuroendocrinol.* 20: 220–6.
55. Kawakami, S., Bungo, T., Ando, R., Ohgushi, A., Shimojo, M. et al. (2000). Central administration of alpha-melanocyte stimulating hormone inhibits fasting- and neuropeptide Y-induced feeding in neonatal chicks. *Eur. J. Pharmacol.* 398: 361–4.
56. Smith, M. L., Prall, B. C., Siegel, P. B. and Cline, M. A. (2011). The threshold of insulin-induced hypophagia is lower in chicks selected for low rather than high juvenile body weight. *Behav. Brain Res.* 216: 719–22.
57. Tachibana, T., Sugahara, K., Ohgushi, A., Ando, R., Kawakami, S. et al. (2001). Intracerebroventricular injection of agouti-related protein attenuates the anorexigenic effect of alpha-melanocyte stimulating hormone in neonatal chicks. *Neurosci. Lett.* 305: 131–4.
58. Tachibana, T., Takagi, T., Tomonaga, S., Ohgushi, A., Ando, R. et al. (2003). Central administration of cocaine- and amphetamine-regulated transcript inhibits food intake in chicks. *Neurosci. Lett.* 337: 131–4.
59. Larhammar, D. (1996). Evolution of neuropeptide Y, peptide YY and pancreatic polypeptide. *Regul. Pept.* 62: 1–11.
60. Wang, X., Day, J. R. and Vasilatos-Younken, R. (2001). The distribution of neuropeptide Y gene expression in the chicken brain. *Mol. Cell Endocrinol.* 174: 129–36.
61. Boswell, T., Li, Q. and Takeuchi, S. (2002). Neurons expressing neuropeptide Y mRNA in the infundibular hypothalamus of Japanese quail are activated by fasting and co-express agouti-related protein mRNA. *Brain Res. Mol. Brain Res.* 100: 31–42.
62. Furuse, M., Ando, R., Bungo, T., Ao, R., Shimojo, M. et al. (1999). Intracerebroventricular injection of orexins does not stimulate food intake in neonatal chicks. *Br. Poult. Sci.* 40: 698–700.
63. Ando, R., Bungo, T., Kanwakami, S., Shimojo, M., Masuda, Y. et al. (2000). Intracerebroventricular injection of mammalian motilin, melanin-concentrating hormone or galanin does not stimulate food intake in neonatal chicks. *Br. Poult. Sci.* 41: 508–11.
64. Ando, R., Kawakami, S. I., Bungo, T., Ohgushi, A., Takagi, T. et al. (2001). Feeding responses to several neuropeptide Y receptor agonists in the neonatal chick. *Eur. J. Pharmacol.* 427: 53–9.
65. Honda, K., Saneyasu, T., Hasegawa, S. and Kamisoyama, H. (2012). A comparative study of the central effects of melanocortin peptides on food intake in broiler and layer chicks. *Peptides* 37: 13–17.
66. Saneyasu, T., Honda, K., Kamisoyama, H., Ikura, A., Nakayama, Y. et al. (2011). Neuropeptide Y effect on food intake in broiler and layer chicks. *Comp. Biochem. Physiol. A Mol. Integr. Physiol.* 159: 422–6.
67. Siegel, P. B. (1962). A double selection experiment for body weight and breast angle at eight weeks of age in chickens. *Genetics* 47: 1313–19.
68. Zelenka, D. J., Dunnington, E. A., Cherry, J. A. and Siegel, P. B. (1988). Anorexia and sexual maturity in female white rock chickens. I. Increasing the feed intake. *Behav. Genet.* 18: 383–7.
69. Cline, M. A., Kuo, A. Y., Smith, M. L., Nandar, W., Prall, B. C. et al. (2009). Differential feed intake responses to central corticotrophin releasing factor in lines of chickens divergently selected for low or high body weight. *Comp. Biochem. Physiol. A Mol. Integr. Physiol.* 152: 130–4.
70. Cline, M. A., Nandar, W., Bowden, C., Calchary, W., Smith, M. L. et al. (2010). The threshold of amylin-induced anorexia is lower in chicks selected for low compared to high juvenile body weight. *Behav. Brain Res.* 208: 650–4.
71. Bordas, A. and Merat, P. (1981). Genetic variation and phenotypic correlations of food consumption of laying hens corrected for body weight and production. *Br. Poult. Sci.* 22: 25–33.
72. Sintubin, P., Greene, E., Collin, A., Bordas, A., Zerjal, T. et al. (2014). Expression profile of hypothalamic neuropeptides in chicken lines selected for high or low residual feed intake. *Neuropeptides* 48: 213–20.

73. Blankenship, K., Gilley, A., Piekarski, A., Orlowski, S., Greene, E. et al. (2015). Differential expression of feeding-related hypothalamic neuropeptides in the first generation of quails divergently selected for low or high feed efficiency. *Neuropeptides* 38: 31–40
74. Leclercq, B. (1983). The influence of dietary protein content on the performance of genetically lean or fat growing chickens. *Br. Poult. Sci.* 24: 581–7.
75. Carling, D., Thornton, C., Woods, A. and Sanders, M. J. (2012). AMP-activated protein kinase: new regulation, new roles? *Biochem. J.* 445: 11–27.
76. Stein, S. C., Woods, A., Jones, N. A., Davison, M. D. and Carling, D. (2000). The regulation of AMP-activated protein kinase by phosphorylation. *Biochem. J.* 345 Pt 3: 437–43.
77. Andersson, U., Filipsson, K., Abbott, C. R., Woods, A., Smith, K. et al. (2004). AMP-activated protein kinase plays a role in the control of food intake. *J. Biol. Chem.* 279: 12005–8.
78. Lopez, M., Varela, L., Vazquez, M. J., Rodriguez-Cuenca, S., Gonzalez, C. R. et al. (2010). Hypothalamic AMPK and fatty acid metabolism mediate thyroid regulation of energy balance. *Nat. Med.* 16: 1001–8.
79. Minokoshi, Y., Alquier, T., Furukawa, N., Kim, Y. B., Lee, A. et al. (2004). AMP-kinase regulates food intake by responding to hormonal and nutrient signals in the hypothalamus. *Nature* 428: 569–74.
80. Xu, P., Siegel, P. B. and Denbow, D. M. (2011). AICAR and Compound C regulate food intake independently of AMP-activated protein kinase in lines of chickens selected for high or low body weight. *Comp. Biochem. Physiol. A Mol. Integr. Physiol.* 159: 401–12.
81. Proszkowiec-Weglarz, M., Richards, M. P., Ramachandran, R. and McMurtry, J. P. (2006). Characterization of the AMP-activated protein kinase pathway in chickens. *Comp. Biochem. Physiol. B Biochem. Mol. Biol.* 143: 92–106.
82. Proszkowiec-Weglarz, M., Richards, M. P. and McMurtry, J. P. (2006). Molecular cloning, genomic organization, and expression of three chicken 5′-AMP-activated protein kinase gamma subunit genes. *Poult. Sci.* 85: 2031–41.
83. Hu, Z., Cha, SH., Chohnan, S. and Lane, M. D. (2003). Hypothalamic malonyl-CoA as a mediator of feeding behavior. *Proc. Natl. Acad. Sci. U.S.A.* 100: 12624–9.
84. Obici, S., Feng, Z., Arduini, A., Conti, R. and Rossetti, L. (2003). Inhibition of hypothalamic carnitine palmitoyltransferase-1 decreases food intake and glucose production. *Nat. Med.* 9: 756–61.
85. Loftus, T. M., Jaworsky, D. E., Frehywot, G. L., Townsend, C. A., Ronnett, G. V. et al. (2000). Reduced food intake and body weight in mice treated with fatty acid synthase inhibitors. *Science* 288: 2379–81.
86. Cohen, P., Miyazaki, M., Socci, N. D., Hagge-Greenberg, A., Liedtke, W. et al. (2002). Role for stearoyl-CoA desaturase-1 in leptin-mediated weight loss. *Science* 297: 240–3.
87. Lacy, M. P., Van Krey, H. P., Skewes, P. A., Denbow, D. M. (1986) Food intake in the domestic fowl: effect of intrahepatic lipid and amino acid infusions. *Physiol. Behav.* 36: 533–8.
88. Dridi, S., Ververken, C., Hillgartner, F. B., Arckens, L., Van der Gucht, E. et al. (2006). FAS inhibitor cerulenin reduces food intake and melanocortin receptor gene expression without modulating the other (an)orexigenic neuropeptides in chickens. *Am. J. Physiol. Regul. Integr. Comp. Physiol.* 291: R138–47.
89. Bolster, D. R., Crozier, S. J., Kimball, S. R. and Jefferson, L. S. (2002). AMP-activated protein kinase suppresses protein synthesis in rat skeletal muscle through down-regulated mammalian target of rapamycin (mTOR) signaling. *J. Biol. Chem.* 277: 23977–80.
90. Cota, D., Proulx, K., Smith, K. A., Kozma, S. C., Thomas, G. et al. (2006). Hypothalamic mTOR signaling regulates food intake. *Science* 312: 927–30.
91. Mizushima, N., Levine, B., Cuervo, A. M. and Klionsky, D. J. (2008). Autophagy fights disease through cellular self-digestion. *Nature* 451: 1069–75.
92. Kaushik, S., Rodriguez-Navarro, J. A., Arias, E., Kiffin, R., Sahu, S. et al. (2011). Autophagy in hypothalamic AgRP neurons regulates food intake and energy balance. *Cell Metab.* 14: 173–83.
93. Alirezaei, M., Kemball, C. C., Flynn, C. T., Wood, M. R., Whitton, J. L. et al. (2010). Short-term fasting induces profound neuronal autophagy. *Autophagy* 6: 702–10.

94. Coupe, B., Ishii, Y., Dietrich, M. O., Komatsu, M., Horvath, T. L. et al. (2012). Loss of autophagy in pro-opiomelanocortin neurons perturbs axon growth and causes metabolic dysregulation. *Cell Metab.* 15: 247–55.
95. Malhotra, R., Warne, J. P., Salas, E., Xu, A. W. and Debnath, J. (2015). Loss of Atg12, but not Atg5, in pro-opiomelanocortin neurons exacerbates diet-induced obesity. *Autophagy* 11: 145–54.
96. Piekarski, A., Khaldi, S., Greene, E., Lassiter, K., Mason, J. G. et al. (2014). Tissue distribution, gender- and genotype-dependent expression of autophagy-related genes in avian species. *PLoS ONE* 9: e112449.
97. Schneeberger, M., Altirriba, J., Garcia, A., Esteban, Y., Castano, C. et al. (2012). Deletion of miRNA processing enzyme Dicer in POMC-expressing cells leads to pituitary dysfunction, neurodegeneration and development of obesity. *Mol. Metab.* 2: 74–85.
98. Murchison, E. P. and Hannon, G. J. (2004). miRNAs on the move: miRNA biogenesis and the RNAi machinery. *Curr. Opin. Cell Biol.* 16: 223–9.
99. Gregory, R. I., Yan, K. P., Amuthan, G., Chendrimada, T., Doratotaj, B. et al. (2004). The Microprocessor complex mediates the genesis of microRNAs. *Nature* 432: 235–40.
100. Lee, Y., Ahn, C., Han, J., Choi, H., Kim, J. et al. (2003). The nuclear RNase III Drosha initiates microRNA processing. *Nature* 425: 415–19.
101. Zamore, P. D., Tuschl, T., Sharp, P. A. and Bartel, D. P. (2000). RNAi: double-stranded RNA directs the ATP-dependent cleavage of mRNA at 21 to 23 nucleotide intervals. *Cell* 101: 25–33.
102. Vinnikov, I. A., Hajdukiewicz, K., Reymann, J., Beneke, J., Czajkowski, R. et al. (2014). Hypothalamic miR-103 protects from hyperphagic obesity in mice. *J. Neurosci.* 34: 10659–74.
103. Luo, C., Sun, L., Ma, J., Wang, J., Qu, H. et al. (2015). Association of single nucleotide polymorphisms in the microRNA miR-1596 locus with residual feed intake in chickens. *Anim. Genet.* 46: 265–71.
104. Denbow, D. M. (1994). Peripheral regulation of food intake in poultry. *J. Nutr.* 124: 1349S–54S.
105. Andrews, Z. B. (2011). Central mechanisms involved in the orexigenic actions of ghrelin. *Peptides* 32: 2248–55.
106. Kohno, D., Sone, H., Minokoshi, Y. and Yada, T. (2008). Ghrelin raises [Ca2+]i via AMPK in hypothalamic arcuate nucleus NPY neurons. *Biochem. Biophys. Res. Commun.* 366: 388–92.
107. Lopez, M., Lage, R., Saha, A. K., Perez-Tilve, D., Vazquez, M. J. et al. (2008). Hypothalamic fatty acid metabolism mediates the orexigenic action of ghrelin. *Cell Metab.* 7: 389–99.
108. Hawley, S. A., Pan, D. A., Mustard, K. J., Ross, L., Bain, J. et al. (2005). Calmodulin-dependent protein kinase kinase-beta is an alternative upstream kinase for AMP-activated protein kinase. *Cell Metab.* 2: 9–19.
109. Anderson, K. A., Ribar, T. J., Lin, F., Noeldner, P. K., Green, M. F. et al. (2008). Hypothalamic CaMKK2 contributes to the regulation of energy balance. *Cell Metab.* 7: 377–88.
110. Velasquez, D. A., Martinez, G., Romero, A., Vazquez, M. J., Boit, K. D. et al. (2011). The central Sirtuin 1/p53 pathway is essential for the orexigenic action of ghrelin. *Diabetes* 60: 1177–85.
111. Andrews, Z. B., Liu, Z. W., Walllingford, N., Erion, D. M., Borok, E. et al. (2008). UCP2 mediates ghrelin's action on NPY/AgRP neurons by lowering free radicals. *Nature* 454: 846–51.
112. Watterson, K. R., Bestow, D., Gallagher, J., Hamilton, D. L., Ashford, F. B. et al. (2013). Anorexigenic and orexigenic hormone modulation of mammalian target of rapamycin complex 1 activity and the regulation of hypothalamic agouti-related protein mRNA expression. *Neurosignals* 21: 28–41.
113. Ahmed, S. and Harvey, S. (2002). Ghrelin: a hypothalamic GH-releasing factor in domestic fowl (Gallus domesticus). *J. Endocrinol.* 172: 117–25.
114. Kaiya, H., Van Der Geyten, S., Kojima, M., Hosoda, H., Kitajima, Y. et al. (2002). Chicken ghrelin: purification, cDNA cloning, and biological activity. *Endocrinology* 143: 3454–63.
115. Tanaka, M., Miyazaki, T., Yamamoto, I., Nakai, N., Ohta, Y. et al. (2003). Molecular characterization of chicken growth hormone secretagogue receptor gene. *Gen. Comp. Endocrinol.* 134: 198–202.
116. Richards, M. P., Poch, S. M. and McMurtry, J. P. (2006). Characterization of turkey and chicken ghrelin genes, and regulation of ghrelin and ghrelin receptor mRNA levels in broiler chickens. *Gen. Comp. Endocrinol.* 145: 298–310.

117. Nie, Q., Fang, M., Xie, L., Peng, X., Xu, H. et al. (2009). Molecular characterization of the ghrelin and ghrelin receptor genes and effects on fat deposition in chicken and duck. *J. Biomed. Biotechnol.* 2009: 567120.
118. Kitazawa, T., Maeda, Y. and Kaiya, H. (2009). Molecular cloning of growth hormone secretagogue-receptor and effect of quail ghrelin on gastrointestinal motility in Japanese quail. *Regul. Pept.* 158: 132–42.
119. Furuse, M., Tachibana, T., Ohgushi, A., Ando, R., Yoshimatsu, T. et al. (2001). Intracerebroventricular injection of ghrelin and growth hormone releasing factor inhibits food intake in neonatal chicks. *Neurosci. Lett.* 301: 123–6.
120. Saito, E. S., Kaiya, H., Tachibana, T., Tomonaga, S., Denbow, D. M. et al. (2005). Inhibitory effect of ghrelin on food intake is mediated by the corticotropin-releasing factor system in neonatal chicks. *Regul. Pept.* 125: 201–8.
121. Shousha, S., Nakahara, K., Kojima, M., Miyazato, M., Hosoda, H. et al. (2005). Different effects of peripheral and central ghrelin on regulation of food intake in the Japanese quail. *Gen. Comp. Endocrinol.* 141: 178–83.
122. Geelissen, S. M., Swennen, Q., Geyten, S. V., Kuhn, E. R., Kaiya, H. et al. (2006). Peripheral ghrelin reduces food intake and respiratory quotient in chicken. *Domest Anim. Endocrinol.* 30: 108–16.
123. Kaiya, H., Saito, E. S., Tachibana, T., Furuse, M. and Kangawa, K. (2007). Changes in ghrelin levels of plasma and proventriculus and ghrelin mRNA of proventriculus in fasted and refed layer chicks. *Domest Anim. Endocrinol.* 32: 247–59.
124. Xu, P., Siegel, P. B. and Denbow, D. M. (2011). Genetic selection for body weight in chickens has altered responses of the brain's AMPK system to food intake regulation effect of ghrelin, but not obestatin. *Behav. Brain Res.* 221: 216–26.
125. Halaas, J. L., Gajiwala, K. S., Maffei, M., Cohen, S. L., Chait, B. T. et al. (1995) .Weight-reducing effects of the plasma protein encoded by the obese gene. *Science* 269: 543–6.
126. Cohen, S. L., Halaas, J. L., Friedman, J. M., Chait, B. T., Bennett, L. et al. (1996). Human leptin characterization. *Nature* 382: 589.
127. Gao, S., Kinzig, K. P., Aja, S., Scott, K. A., Keung, W. et al. (2007). Leptin activates hypothalamic acetyl-CoA carboxylase to inhibit food intake. *Proc. Natl. Acad. Sci. U.S.A.* 104: 17358–63.
128. Wolfgang, M. J., Cha, S. H., Sidhaye, A., Chohnan, S., Cline, G. et al. (2007). Regulation of hypothalamic malonyl-CoA by central glucose and leptin. *Proc. Natl. Acad. Sci. U.S.A.* 104: 19285–90.
129. Taouis, M., Chen, J. W., Daviaud, C., Dupont, J., Derouet, M. et al. (1998). Cloning the chicken leptin gene. *Gene* 208: 239–42.
130. Ashwell, C. M., Czerwinski, S. M., Brocht, D. M. and McMurtry, J. P. (1999). Hormonal regulation of leptin expression in broiler chickens. *Am. J. Physiol.* 276: R226–32.
131. Raver, N., Taouis, M., Dridi, S., Derouet, M., Simon, J. et al. (1998). Large-scale preparation of biologically active recombinant chicken obese protein (leptin). *Protein Expr. Purif.* 14: 403–8.
132. Dridi, S., Raver, N., Gussakovsky, E. E., Derouet, M., Picard, M. et al. (2000). Biological activities of recombinant chicken leptin C4S analog compared with unmodified leptins. *Am. J. Physiol. Endocrinol. Metab.* 279: E116–23.
133. Kuo, A. Y., Cline, M. A., Werner, E., Siegel, P. B. and Denbow, D. M. (2005). Leptin effects on food and water intake in lines of chickens selected for high or low body weight. *Physiol. Behav.* 84: 459–64.
134. Lohmus, M., Sundstrom, L. F., El Halawani, M. and Silverin, B. (2003). Leptin depresses food intake in great tits (Parus major). *Gen. Comp. Endocrinol.* 131: 57–61.
135. Lohmus, M., Sundstrom, L. F. and Silverin, B. (2006). Chronic administration of leptin in Asian Blue Quail. *J. Exp. Zool. A Comp. Exp. Biol.* 305: 13–22.
136. Seroussi, E., Cinnamon, Y., Yosefi, S., Genin, O., Smith, J. G. et al. (2016). Identification of the Long-Sought Leptin in Chicken and Duck: Expression Pattern of the Highly GC-Rich Avian leptin Fits an Autocrine/Paracrine Rather Than Endocrine Function. *Endocrinology* 157: 737–51.
137. Yang, S. J. and Denbow, D. M. (2007). Interaction of leptin and nitric oxide on food intake in broilers and Leghorns. *Physiol. Behav.* 92: 651–7.

138. Dridi, S., Swennen, Q., Decuypere, E. and Buyse, J. (2005). Mode of leptin action in chicken hypothalamus. *Brain Res.* 1047: 214–23.
139. Woods, S. C., Lotter, E. C., McKay, L. D. and Porte, D., Jr. (1979). Chronic intracerebroventricular infusion of insulin reduces food intake and body weight of baboons. *Nature* 282: 503–5.
140. Fruhbeck, G. (2006). Intracellular signalling pathways activated by leptin. *Biochem. J.* 393: 7–20.
141. Han, S. M., Namkoong, C., Jang, P. G., Park, I. S., Hong, S. W. et al. (2005). Hypothalamic AMP-activated protein kinase mediates counter-regulatory responses to hypoglycaemia in rats. *Diabetologia* 48: 2170–8.
142. Simon, J. and Rosselin, G. (1979). Effect of intermittent feeding on glucose-insulin relationship in the chicken. *J. Nutr.* 109: 631–41.
143. Seki, Y., Sato, K., Kono, T., Abe, H. and Akiba, Y. (2003). Broiler chickens (Ross strain) lack insulin-responsive glucose transporter GLUT4 and have GLUT8 cDNA. *Gen. Comp. Endocrinol.* 133: 80–7.
144. Coudert, E., Pascal, G., Dupont, J., Simon, J., Cailleau-Audouin, E. et al. (2015). Phylogenesis and Biological Characterization of a New Glucose Transporter in the Chicken (Gallus gallus), GLUT12. *PLoS ONE* 10: e0139517.
145. Akiba, Y., Chida, Y., Takahashi, T., Ohtomo, Y., Sato, K. et al. (1999). Persistent hypoglycemia induced by continuous insulin infusion in broiler chickens. *Br. Poult. Sci.* 40: 701–5.
146. Shiraishi, J., Yanagita, K., Fukumori, R., Sugino, T., Fujita, M. et al. (2011). Comparisons of insulin related parameters in commercial-type chicks: evidence for insulin resistance in broiler chicks. *Physiol. Behav.* 103: 233–9.
147. Honda, K., Kamisoyama, H., Saneyasu, T., Sugahara, K. and Hasegawa, S. (2007). Central administration of insulin suppresses food intake in chicks. *Neurosci. Lett.* 423: 153–7.
148. Shiraishi, J., Yanagita, K., Fujita, M. and Bungo, T. (2008). Central insulin suppresses feeding behavior via melanocortins in chicks. *Domest Anim. Endocrinol.* 34: 223–8.
149. Shiraishi, J., Yanagita, K., Fujita, M. and Bungo, T. (2008). Micro-opioid receptor agonist diminishes POMC gene expression and anorexia by central insulin in neonatal chicks. *Neurosci. Lett.* 439: 227–9.
150. Coppola, A., Liu, Z. W., Andrews, Z. B., Paradis, E., Roy, M. C., et al. (2007). A central thermogenic-like mechanism in feeding regulation: an interplay between arcuate nucleus T3 and UCP2. *Cell Metab.* 5: 21–33.
151. Rosebrough, R. W., Mitchell, A. D. and McMurtry, J. P. (1996). Carry-over effects of dietary crude protein and triiodothyronine (T3) in broiler chickens. *Br. J. Nutr.* 75: 573–81.
152. Kadowaki, T. and Yamauchi, T. (2005). Adiponectin and adiponectin receptors. *Endocr. Rev.* 26: 439–51.
153. Kubota, N., Yano, W., Kubota, T., Yamauchi, T., Itoh, S. et al. (2007). Adiponectin stimulates AMP-activated protein kinase in the hypothalamus and increases food intake. *Cell Metab.* 6: 55–68.
154. Qi, Y., Takahashi, N., Hileman, S. M., Patel, H. R., Berg, A. H. et al. (2004). Adiponectin acts in the brain to decrease body weight. *Nat. Med.* 10: 524–9.
155. Park, S., Kim, D. S., Kwon, D. Y. and Yang, H. J. (2011). Long-term central infusion of adiponectin improves energy and glucose homeostasis by decreasing fat storage and suppressing hepatic gluconeogenesis without changing food intake. *J. Neuroendocrinol.* 23: 687–98.
156. Maddineni, S., Metzger, S., Ocon, O., Hendricks, G. and Ramachandran, R. (2005). Adiponectin gene is expressed in multiple tissues in the chicken: food deprivation influences adiponectin messenger ribonucleic acid expression. *Endocrinology* 146: 4250–6.
157. da Cruz, V. A., Schenkel, F. S., Savegnago, R. P., Grupioni, N. V., Stafuzza, N. B. et al. (2015). Association of Apolipoprotein B and Adiponectin Receptor 1 Genes with Carcass, Bone Integrity and Performance Traits in a Paternal Broiler Line. *PLoS ONE* 10: e0136824.
158. Gan, L., Yan, J., Liu, Z., Feng, M. and Sun, C. (2015). Adiponectin prevents reduction of lipid-induced mitochondrial biogenesis via AMPK/ACC2 pathway in chicken adipocyte. *J. Cell Biochem.* 116: 1090–1100.

159. Vilsack, T. and Clark, C. (2014). 2012 Census of agriculture: United States Summary and State data. In: USDA, editor: USDA NASS.
160. Diarra, S. (2014). Feeding management of poultry in high environmental temperatures. *Int. J. Pout. Sci.* 13: 657–61.
161. NRC (1994). Nutrient requirement of poultry. In: NAS, editor. Washington, DC: National Academy Press.
162. Ziaei, N. G., Edwards, S. A., Blanchard, P. J., Ward, J. and Feuerstein, D. (2007). Effect of gender on factors affecting excreta dry matter content of broiler chickens. *J. Appl. Poult. Res.* 16: 226–33.
163. Snapir, N., Robinzon, B. and Godschalk, M. (1976). The drinking response of the chicken to peripheral and central administration of angiotensin II. *Pharmacol. Biochem. Behav.* 5: 5–10.
164. Ferguson, A. V. (2014). Circumventricular organs: integrators of circulating signals controlling hydration, energy balance, and immune function. In De Luca, L. A., Menani, J. V., Johnson, A. K. (eds), Chapter 2, *Neurobiology of Body Fluid Homeostasis: Transduction and Integration*. CRC, Roca Raton, pp. 23–36.
165. Diaz, G. B., Ojeda, R. A. and Dacar, M. (2001). Water conservation in the South American desert mouse opossum, Thylamys pusilla (Didelphimorphia, Didelphidae). *Comp. Biochem. Physiol. A Mol. Integr. Physiol.* 130: 323–30.
166. Braum, E. (2014). Osmoregulatory systems of birds. In Scanes, C. G. (ed.), *Sturkie's Avian Physiology*: Elseveir, pp. 285–99.
167. Nielsen, S., Smith, B. L., Christensen, E. I., Knepper, M. A. and Agre, P. (1993). CHIP28 water channels are localized in constitutively water-permeable segments of the nephron. *J. Cell Biol.* 120: 371–83.
168. Kishore, B. K., Mandon, B., Oza, N. B., DiGiovanni, S. R., Coleman, R. A. et al. (1996). Rat renal arcade segment expresses vasopressin-regulated water channel and vasopressin V2 receptor. *J. Clin. Invest.* 97: 2763–71.
169. Ma, T., Song, Y., Yang, B., Gillespie, A., Carlson, E. J. et al. (2000). Nephrogenic diabetes insipidus in mice lacking aquaporin-3 water channels. *Proc. Natl. Acad. Sci. U. S. A.* 97: 4386–91.
170. Ma, T., Yang, B., Gillespie, A., Carlson, E. J., Epstein, C. J. et al. (1997). Generation and phenotype of a transgenic knockout mouse lacking the mercurial-insensitive water channel aquaporin-4. *J. Clin. Invest.* 100: 957–62.
171. Yasui, M., Kwon, T. H., Knepper, M. A., Nielsen, S. and Agre, P. (1999). Aquaporin-6: An intracellular vesicle water channel protein in renal epithelia. *Proc. Natl. Acad. Sci. U. S. A.* 96: 5808–13.
172. Muller, C., Sendler, M. and Hildebrandt, J. P. (2006). Downregulation of aquaporins 1 and 5 in nasal gland by osmotic stress in ducklings, Anas platyrhynchos: implications for the production of hypertonic fluid. *J. Exp. Biol.* 209: 4067–76.
173. Casotti, G., Waldron, T., Misquith, G., Powers, D. and Slusher, L. (2007). Expression and localization of an aquaporin-1 homologue in the avian kidney and lower intestinal tract. *Comp. Biochem. Physiol. A Mol. Integr. Physiol.* 147: 355–62.
174. Nielsen, S., Frokiaer, J., Marples, D., Kwon, T. H., Agre, P. et al. (2002). Aquaporins in the kidney: from molecules to medicine. *Physiol. Rev.* 82: 205–44.
175. Takei, Y., Bartolo, R. C., Fujihara, H., Ueta, Y. and Donald, J. A. (2012). Water deprivation induces appetite and alters metabolic strategy in Notomys alexis: unique mechanisms for water production in the desert. *Proc. Biol. Sci.* 279: 2599–608.
176. Cline, M. A., Nandar, W. and Rogers, J. O. (2007). Central neuropeptide FF reduces feed consumption and affects hypothalamic chemistry in chicks. *Neuropeptides* 41: 433–9.
177. Prall, B. C. and Cline, M. A. (2008). Anorexigenic effects of central neuropeptide K are associated with hypothalamic changes in juvenile Gallus gallus. *Gen. Comp. Endocrinol.* 159: 130–5.
178. Nandar, W., Milligan, J. M. and Cline, M. A. (2008). Mechanisms of xenin-induced anorectic response in chicks (Gallus gallus). *Gen. Comp. Endocrinol.* 157: 58–62.

179. Cline, M. A., Bowden, C. N., Calchary, W. A. and Layne, J. E. (2008). Short-term anorexigenic effects of central neuropeptide VF are associated with hypothalamic changes in chicks. *J. Neuroendocrinol.* 20: 971–7.
180. Cline, M. A., Nandar, W., Prall, B. C., Bowden, C. N. and Denbow, D. M. (2008). Central visfatin causes orexigenic effects in chicks. *Behav. Brain Res.* 186: 293–7.
181. Cline, M. A., Nandar, W., Smith, M. L., Pittman, B. H., Kelly, M. et al. (2008). Amylin causes anorexigenic effects via the hypothalamus and brain stem in chicks. *Regul. Pept.* 146: 140–6.
182. McConn, B. R., Yi, J., Gilbert, E. R., Siegel, P. B., Chowdhury, V. S. et al. (2016). Stimulation of food intake after central administration of gonadotropin-inhibitory hormone is similar in genetically selected low and high body weight lines of chickens. *Gen. Comp. Endocrinol.* 232: 96–100.
183. Mace, J. D., Tachibana, T., Wang, G., Newmyer, B. A., Guilliams, E. et al. (2014). Substance P is associated with hypothalamic paraventricular nucleus activation that coincides with increased urotensin 2 mRNA in chicks. *Neuropeptides* 48: 305–11.
184. Tachibana, T., Matsuda, K., Khan, S. I., Ueda, H. and Cline, M. A. (2010). Feeding and drinking response following central administrations of bombesin-like peptides in chicks. *Comp. Biochem. Physiol. A Mol. Integr. Physiol.* 156: 394–9.
185. Cline, M. A., Cofield, S. A. and Tachibana, T. (2010). Central litorin injection is associated with primary anorexigenic effects that coincide with activation of the magnocellular division of the paraventricular nucleus. *Neuropeptides* 44: 247–52.
186. Cline, M. A., Nandar, W., Bowden, C., Hein, P. P., Denbow, D. M. et al. (2008). Differential feeding responses to central alpha-melanocyte stimulating hormone in genetically low and high body weight selected lines of chickens. *Life Sci.* 83: 208–13.
187. Cline, M. A., Fouse, D. N. and Prall, B. C. (2008). Central and peripheral alytesin cause short-term anorexigenic effects in neonatal chicks. *Neuropeptides* 42: 283–91.
188. Cline, M. A., Bowden, C. N., Nandar, W. and Rogers, J. O. (2008). Central oxyntomodulin causes anorexigenic effects associated with the hypothalamus and alimentary canal in chicks (Gallus gallus). *Comp. Biochem. Physiol. A Mol. Integr. Physiol.* 149: 405–10.
189. Shipp, S. L., Yi, J., Dridi, S., Gilbert, E. R. and Cline, M. A. (2015). The central anorexigenic mechanism of adrenocorticotropic hormone involves the caudal hypothalamus in chicks. *Neuropeptides* 53: 29–35.
190. Layne, J. E., Hunt, K. E., True, T., Gill, R. S., Combos, V. E. et al. (2009). Central calcitonin exerts anorectic effects via the hypothalamus in chicks. *Pharmacol. Biochem. Behav.* 92: 433–8.
191. Cline, M. A., Prall, B. C., Rogers, J. O. and Tachibana, T. (2009). Satiety induced by central stresscopin is mediated by corticotrophin-releasing factor receptors and hypothalamic changes in chicks. *Pharmacol. Biochem. Behav.* 92: 663–9.
192. Cline, M. A., Calchary, W. A. and Nandar, W. (2009) Effect of calcitonin gene-related peptide (CGRP) on avian appetite-related processes. *Behav. Brain Res.* 196: 242–7.
193. Denbow, D. M. and Van Krey, H. P. (1987). Ionostatic control of food intake in the domestic fowl. *Poult. Sci.* 66: 1229–35.
194. Baghbanzadeh, A., Hajinezhad, M. R., Shohreh, B. and Maleklou, R. (2010). Intralateral hypothalamic area injection of isoproterenol and propranolol affects food and water intake in broilers. *J. Comp. Physiol. A Neuroethol. Sens. Neural. Behav. Physiol.* 196: 221–6.
195. Mladinov, D., Liu, Y., Mattson, D. L. and Liang, M. (2013). MicroRNAs contribute to the maintenance of cell-type-specific physiological characteristics: miR-192 targets Na+/K+-ATPase beta1. *Nucleic. Acids Res.* 41: 1273–83.
196. Tobon, K. E., Chang, D. and Kuzhikandathil, E. V. (2012). MicroRNA 142–3p mediates post-transcriptional regulation of D1 dopamine receptor expression. *PLoS ONE* 7: e49288.
197. Jin, W., Reddy, M. A., Chen, Z., Putta, S., Lanting, L. et al. (2012). Small RNA sequencing reveals microRNAs that modulate angiotensin II effects in vascular smooth muscle cells. *J. Biol. Chem.* 287: 15672–83.

198. Eskildsen, T. V., Jeppesen, P. L., Schneider, M., Nossent, A. Y., Sandberg, M. B. et al. (2013). Angiotensin II regulates microRNA-132/-212 in hypertensive rats and humans. *Int. J. Mol. Sci.* 14: 11190–11207.
199. Luo, Y., Liu, Y., Liu, M., Wei, J., Zhang, Y. et al. (2014). Sfmbt2 10th intron-hosted miR-466(a/e)-3p are important epigenetic regulators of Nfat5 signaling, osmoregulation and urine concentration in mice. *Biochim. Biophys. Acta* 1839: 97–106.
200. Shimoyama, M., De Pons, J., Hayman, G. T., Laulederkind, S. J., Liu, W. et al. (2015). The Rat Genome Database 2015: genomic, phenotypic and environmental variations and disease. *Nucleic. Acids Res.* 43: D743–50.
201. Calore, F. and Fabbri, M. (2012). MicroRNAs and Cancer. *Atlas of Genetics and Cytogenetics in Oncology and Hematology* 16, 51–69.

Chapter 6

Advances and future directions in poultry feeding: an overview

Velmurugu Ravindran and Mohammad R. Abdollahi, Massey University, New Zealand

1 Introduction

The poultry industry has advanced remarkably over the past 50 years. In particular, poultry meat production has been the most successful than any in the animal industry. Production standards of broilers and layers have continually improved over this period, with contemporary male broilers currently reaching a live weight of 2.5 kg at 33–35 days of age, and white egg layers capable of producing 330 eggs in 52 weeks of lay. Over this period, the body weight of broilers at 42 days has increased by 25–50 g per year and the feed conversion ratio to 2 kg body weight has improved 2–3 points annually. As shown by Havenstein et al. (2003), genetic selection brought about by breeding companies is responsible for 85–90% of the improvements in broiler growth, and advances in nutritional management have provided only 10–15% of the changes. When these researchers compared the performance of the 1957 broiler strain to the 2001 broiler strain, which were fed their representative diets, the birds from the 2001 genetic strain were 4.96 times heavier than those from the 1957 strain and averaged 8% more breast meat yield. The necessity to achieve and sustain these improvements in genetic potential was the driving force behind recent advances in poultry nutrition, and there had been continuous refinement in the nutrition and feeding practices of commercial poultry.

Although broilers are highly efficient among farm animals in converting feed to food products, they still excrete significant amounts of unutilised nutrients. For example, broilers lose almost 25–30% of ingested dry matter, 20–25% of gross energy, 30–50% of nitrogen and 45–55% of phosphorus intake in the manure. Thus, there is considerable room to improve the conversion of feed-to-meat efficiency. Much of this inefficiency results from nutrient over-formulation and inherent limitations in the digestion and utilisation of nutrients.

http://dx.doi.org/10.19103/AS.2016.0011.06

The overall aim is poultry feeding should be precision nutrition that will meet not only the economic expectations of the industry but also of the consumers. In the industry, there is a widespread tendency to over-formulate diets when doubt exists on the content or availability of critical nutrients (especially amino acids, calcium and phosphorus), mixing and delivery capabilities of nutrients to the bird or if the nutrient requirement was uncertain. This practice is no longer acceptable because this is not only wasteful but also excess nutrients are excreted in the manure and are ultimately a source of pollution. By 'fine tuning' the diets that more closely match the requirements of the bird, the efficiency of nutrient utilisation can be optimised. The intention of this chapter is not to review the available literature on the progress of poultry nutrition research, but instead to provide a critical overview of the current status and future directions in poultry feeding. Only the key advances in nutrition are discussed, and the focus will be particularly on five main categories: (i) understanding of nutrient metabolism and nutrient requirements; (ii) quantification of the supply and availability of nutrients in raw materials; (iii) formulation of least-cost diets that bring nutrient requirements and nutrient supply together in an effective manner; (iv) contribution of biotechnological advances to nutrition, especially the feed additives and (v) progress in feed processing. The overall target of nutritional practices is feeding at lower costs and maximising economic efficiency.

2 Advances in poultry feeding

A major challenge in defining the nutrient needs is the fact that they are influenced by a number of factors and are subject to constant changes. Nutrient requirements are influenced by two main factors, namely bird-related factors (genetics, sex and type and stage of production) and external factors (thermal environment, stress and husbandry conditions). Precision in defining requirements involves accuracy at both these levels. Requirements of major nutrients for various classes of poultry are now available, and these developments are made possible largely because of increasing uniformity of genotypes, housing and husbandry practices in the poultry industry.

Historically, the industry has utilised the nutrient requirements recommended in the publication by National Research Council (NRC). The most recent publication on poultry was in 1994 (now 22 years old), which is a long period, given the genetic advances that have been made in both broilers and layers during this period. Currently, the recommendations suggested by commercial breeding companies provide guidelines that more closely match the requirements of modern bird strains than those recommended by NRC (1994).

Of all the dietary components, essential amino acids and energy are the most expensive and critical. Defining the requirements for the ten essential amino acids poses considerable degree of difficulty, but it has been made easier by the acceptance of ideal protein concept. Like other nutrients, the requirements for amino acids are influenced by various factors, including genetics, sex, physiological status, environment and health status. However, most changes in amino acid requirements do not lead to changes in the relative proportion of the different amino acids. Thus, the actual changes in amino acid requirements can be expressed in relation to a balanced protein or 'ideal protein'. The ideal protein concept uses lysine as the reference amino acid, and the requirements for other essential amino acids are then set as a percentage (or ratio) of the lysine requirement. The ideal protein balance for meat chickens at three growth phases is shown in Table 1. The advantage of this system is that once the lysine requirements under a

Table 1 Ideal amino acid ratios of meat chickens at three growth periods

Amino acid	1–21 days	22–42 days	43–56 days
Lysine[1]	100	100	100
Arginine	105	108	108
Histidine	35	35	35
Isoleucine	67	69	69
Leucine	109	109	109
Methionine + cysteine	72	72	72
Phenylalanine + tyrosine	105	105	105
Threonine	67	68.5	68.5
Tryptophan	16	17	17
Valine	77	80	80

[1]Recommended digestible lysine requirements for meat chickens during 1–21 days, 22–42 days and 43–56 days are 1.070, 0.865 and 0.745%, respectively (Baker, 1996).

variety of conditions are determined, the needs of all other essential amino acids can be calculated. This approach has now become an accepted practice in the industry to set the amino acid specifications in feed formulations.

The requirements for energy and amino acids may also be affected by the feed form (pellet vs. mash). Increasing the diet density through the pelleting process has been reported to markedly increase productive energy. Metabolisable energy contributions of 0.63 (Skinner-Noble et al., 2005) to 0.78 (McKinney and Teeter, 2004) MJ/kg diet from pellets compared to those from mash diets have been reported. Broilers fed with pellets have lower heat increment and utilise more of the feed energy for productive purposes than those fed with mash (Latshaw and Moritz, 2009). The ability of pelleted diets to provide extra productive energy can be favourably used, as a non-nutritional factor, by the broiler industry to reduce dietary energy content.

Because of the extra productive energy available for growth due to pelleting, amino acid requirements, particularly of lysine, may need to increase to maintain the balance between dietary energy and amino acid concentrations (Greenwood et al., 2004). If nutritional requirements of broilers per unit of feed in a pelleted diet are higher than those in a mash diet, application of recommendations of amino acid requirements determined through assays based on purified, semi-purified, dose-response or mash diets to an industry situation where good pellet quality exists may result in suboptimal broiler performance due to underestimation of the bird's amino acid needs (Abdollahi et al., 2013).

2.1 Defining nutrient requirements, composition and ingredient quality

The principal role of feed ingredients is to provide the nutrients that can be digested and utilised for maintenance and productive functions by the bird. Over the years, enormous volume of data has been generated and compiled on the gross nutrient composition of raw materials. The variability that is inherent to each raw material is also recognised and such variability places pressure on precise feed formulations. Data on variation (or matrices) are available for the major feed ingredients and applied in feed formulation packages to achieve better precision. A related development is the availability of rapid tests, such as the near-infrared reflectance analysis, to predict gross nutrient composition and to access the variability in ingredient supplies on an on-going basis.

However, not all of the nutrients in ingredients are available for maintenance and production purposes, and a portion of nutrients is excreted undigested or unutilised. With the development and advances in feed evaluation techniques, sufficient data have been accumulated on the digestibility of nutrients, especially of amino acids and phosphorus. In the case of amino acids, a recent development had been the wider use of digestible amino acid concentrations, rather than total amino acid concentrations, in feed formulations (Ravindran et al., 1998, 2005; Bryden et al., 2009). The use of digestible amino acids is particularly relevant to situations where diet formulations consist of a range of poorly digestible ingredients. Formulating diets based on digestible amino acids makes it possible to increase the range and inclusion levels of alternative ingredients in poultry diets. In effect, this approach improves the precision of formulation, may lower feed cost and ensures more predictable bird performance.

The use of an appropriate energy system is another critical issue because of the importance of energy to bird performance and diet cost. Despite its limitations, apparent metabolisable energy (AME) has been the system of choice for describing available energy. Net energy (NE) system, a refinement of the AME concept, has received attention from time to time. In theory, NE will more closely describe the energy available in an ingredient for a bird's metabolic functions and is more predictive of animal performance (Pirgozliev and Rose, 1999). NE is nowadays routinely used in dairy cattle and pig-feeding systems in many parts of the world, but no real progress has been made in determining the NE of raw materials for poultry. It is, however, difficult to assay, costly and time-consuming, and has limited use in the routine screening of ingredients. There is some recent interest on NE determination of different diet combinations (Swick et al., 2013; Carré et al., 2014), but to be acceptable to the commercial industry, formulations based on NE values should demonstrate an economic advantage over the current system, which is yet to be proven.

It must be noted that the current evaluation of energy and nutrient utilisation in ingredients for broiler chickens are conducted using older birds (typically 21–35 days of age) and the determined estimates are applied to broilers and layers of all ages. The effect of age of broilers and of bird type (broilers vs. layers) on the AME and amino acid digestibility needs to be tested in well-planned studies (Huang et al., 2005, 2006; Adedokun et al., 2009). In particular, the limited capacity of chicks during first week of life after the hatch to digest nutrients is now well documented (Noy and Sklan, 1995; Batal and Parsons, 2002; Thomas et al., 2008; Tancharoenrat et al., 2013). The application of a single value to all growth phases of broilers or of layers is clearly fraught with serious flaw and highlights the need for age-dependent estimates for use in feed formulations.

As discussed later, another major constraint in the endeavour to better define nutrient quality is the lack of rapid tests to determine amino acid digestibility or AME of major ingredients.

2.2 Better feed formulation

Once the nutritional needs are defined, the next step is to match these needs using combinations of ingredients and supplements. The object of formulation is to derive a balanced diet that will provide appropriate quantities of available nutrients at least cost. Given the range of possible ingredients and nutrients involved, a large number of arithmetical calculations are needed to produce a least-cost diet.

Over the years, feed formulation has evolved from a simple balancing of few feedstuffs for a limited number of nutrients to computer-aided linear programming systems. Currently, newer systems of stochastic non-linear programme are becoming popular with the commercial availability of this formulation software. Because variability in ingredient composition is non-linear, stochastic programmes address this issue in the most cost-effective manner possible. In recent years, feed formulation systems have evolved from a technological standpoint by refocusing to meet the demands to maximise profits. New approaches, which predict the maximum profit for a given ingredient combination and price of the broiler/special cuts, are being increasingly used by the commercial industry.

2.3 Products of biotechnology in poultry feeding – Additives

Progress in biotechnology during the past two decades has offered new opportunities to enhance the productivity and efficiency of animals through improved nutrition. Biotechnology covers a vast field of applications in animal nutrition. Some of these applications are already in use (Table 2) as distinct from others whose potentialities are

Table 2 Examples of some biotechnological applications that are widely used in animal nutrition

Application	Aim(s) of developing the technology
1. New ingredients	Production of microbial proteins as new feed sources in animal feeding (e.g. single cell protein and yeast protein).
2. Designer ingredients	Nutritional enhancement (e.g. high-oil corn and high-methionine lupins) or reduction in the level of anti-nutritive components in common feed ingredients (e.g. low-phytate corn).
3. Feed additives	
a. Antimicrobials	To suppress the growth of harmful bacteria and promote the establishment of a desirable gut flora balance (e.g. antibiotics).
b. Crystalline and synthetic amino acids	To increase dietary supply of a specific amino acid and improve protein balance in diet formulations.
c. Feed enzymes	To improve availability of nutrients (energy, amino acids, phosphorus etc.) in feed ingredients by reducing the negative effects of anti-nutritive components (e.g. microbial phytases acting on phytate and xylanases acting on arabinoxylans in wheat).
4. Gut ecosystem enhancers	
a. Probiotics	To promote the establishment of a desirable gut ecosystem through the proliferation of beneficial species (e.g. direct-fed microbials).
b. Prebiotics	To competitively exclude harmful organisms and promote the establishment of a desirable gut ecosystem (e.g. mannan oligosaccharides).

known but are yet to be commercially applied because of technical limitations and public concerns (Table 3).

Among these, the advent of feed additives is of particular importance. Feed additives are products used in animal nutrition for purposes of improving the quality of feed and of food from animal origin, and to improve the performance and health of animals, for example, providing enhanced digestibility of the feed materials. In-feed antibiotics have been thus far the most effective and successful additive used by the poultry industry. One could say that in-feed antibiotics are partly responsible for the performance efficiency currently enjoyed by the industry. However, the recent mandatory or voluntary removal of in-feed antibiotics from poultry diets, spurred by reports of potential antibiotic resistance in humans, is creating a major challenge to maintain bird health, production improvements and efficiency of nutrient utilisation. As discussed later, a number of alternatives are being tested and researched, but are yet to be broadly accepted by the commercial industry.

The growth in acceptance of other feed additives in pig and poultry production over the last two decades has been an extraordinary development. Perhaps the most important additive to enter the animal feed market is exogenous feed enzymes, which have evolved from an undefined entity to a well-accepted tool to improve nutrient utilisation (Bedford and Partridge, 2010). The availability of glycanases (xylanases and glucanases) in the 1990s has effectively overcome the anti-nutritive effects of non-starch polysaccharides and enabled the increased use of viscous grains in poultry diets. Today, the use of these enzymes in wheat- and barley-based poultry diets is routine. During the past decade, the use of microbial phytase in poultry diets is on the increase, in response to concerns over phosphorus pollution from effluents from intensive animal operations (Selle and Ravindran, 2007). Most recently, combinations of carbohydrase enzymes such as xylanases, amylases and glucanases, as well as other exogenous enzymes such as proteases, are also gaining commercial relevance. Cocktails of these enzymes have been shown to be effective even in corn-based diets (Cowieson, 2010), which contain low levels of non-starch polysaccharides.

Table 3 Examples of some biotechnological applications with future potential in animal nutrition

Application	Aim(s) of developing the technology
Modification of gut microbes	To genetically modify microorganisms naturally present in the gut to enhance their capacity for defined functions or add new functions (e.g. rumen microbes to improve cellulose digestion).
Introduction of new gut microbes	To introduce new species or strains of microorganisms into the gut.
Bioactive peptides	Improved growth and efficiency (e.g. growth hormone-releasing peptides), improved gut function, immunomodulation and antibacterial properties.
Nucleotides	Stimulation of intestinal development, tissue growth and better immune response.
Antimicrobial replacers	Antimicrobial enzymes (e.g. lysozyme) and delivery of specific antibodies via spray-dried plasma and egg products.
Transgenesis	To modify nutrient metabolism and improve growth efficiency by transfer of genes.

The availability of synthetic and crystalline amino acids is another major contribution of biotechnology development, and this additive has enabled the nutritionists to more precisely meet the ideal amino acid profile and to improve the performance and yield of high-producing modern birds. Currently, three crystalline amino acids, namely methionine, lysine and threonine, are available to the industry at competitive prices. Valine and isoleucine are expected to become available in the near future and may allow further improvements in feed formulation.

With high energy prices, there is recent interest in the use of digestion aids (emulsifiers) to improve the utilisation of fats by breaking up the fat into small, finely divided globules and increase the access for lipase action. In recent years, some effective nutritional emulsifiers such as lysolecithins, compared to the older crude lecithin versions, have come into the market.

Among the other additives, mycotoxin binders need special mention. The negative effects of mycotoxins have been known for many years. But until the 1990s, the only avenue of control was the use of clays such as bentonite. During the past two decades, the availability of more effective and specific mycotoxin binders and deactivators has substantially lowered the risk of compromised bird productivity.

2.4 Feed processing

The progress in the technology of feed manufacture during the past 50 years represents another major and necessary development in improving bird performance. The technology has progressed from simple mixing of mash feed to pelleting, which involves various physical and thermal processing operations with subsequent physical and chemical alterations of feed (Schofield, 2005).

Currently, a majority of the feed used in the production of broilers is fed in pelleted or crumbled form. Offering feed to poultry in pellet or crumbled form has improved the economics of production by bettering feed efficiency and growth performance. These improvements are attributed to decreased feed wastage, higher nutrient density, reduced selective feeding, decreased time and energy spent for eating, destruction of pathogenic organisms and, more importantly, increased feed consumption (Behnke, 1996; Amerah et al., 2007).

2.5 Phase feeding

Phase feeding, a form of precise feeding, is another development during the past two decades. This is a feeding system in which dietary amino acid levels are reduced steadily over time in an attempt to reduce costs associated with excess dietary protein or amino acids. Commercial phase-feeding programmes may include several phases to step down amino acids and other nutrients for broilers and layers. The number of phases to be implemented in production cycle is dictated by both economics and the practicability.

The wider implementation of phase/precise feeding, however, is limited *inter alia* by the following issues: (i) data on ingredient variation and the reliability of matrix values need to be updated on a continuous basis; (ii) more data on digestible amino acids, at least regarding the major raw materials, are needed; (iii) information is needed on the comparative digestibility of amino acids for different classes of chickens – layers and broilers of different age groups. In particular, it is known that digestibility of various nutrients and AME during the first week is lower compared to that of older birds (Noy and Sklan, 1995;

Thomas et al., 2008; Tancharoenrat et al., 2013); (iv) information on digestible amino acid requirements for different classes of poultry is lacking and (v) finally, it is unfortunate that we do not still have objective rapid tests, which the industry can use to estimate AME or digestible amino acids, as the raw materials are received at the feed mill.

3 Future directions in poultry feeding

Future directions in poultry feeding will be driven by on-going changes in world animal agriculture and by societal issues. Sometime in the future, we may have to modify feed formulations to accommodate not only science-based needs but also societal needs. The impact made by social issues (in-feed antibiotics, environment, welfare, traceability, use of meat and bone meal, genetically modified ingredients etc.) will influence the decision making from farm level to retail distribution of poultry products (Leeson, 2007).

3.1 Sustainability

With increasing public interest over environment, the reduction of nutrient excretion in effluents from intensive animal operations has now become a major issue. Not long ago, when feeds were formulated, the main objective was how to supply the nutrients (nutrient input). Today, there is much public concern about what comes out of the bird (nutrient output). Animal agriculture, including commercial poultry sector, clearly has a problem of releasing excess nutrients into the environment and must assume ownership of its impact on environment, especially water quality.

From the nutritional point of view, the most obvious strategy is to feed the bird to match the requirement and to improve the efficiency of nutrient utilisation by the bird, which in turn will reduce nutrient load in the manure. Among the other possibilities to improve the nutrient utilisation efficiency, the use of feed additives is most promising.

3.2 Ingredient quality

Raw material quality had always been a major constraint in the feed industry. The relevant quality issues are variability, nutrient digestibility, AME and the presence of non-desirable substances such as fibre and anti-nutritional factors. Mycotoxins pose another major challenge, as presence of even low levels of mycotoxins will affect nutrient utilisation (Bryden, 2012).

As feed prices increase, control and assurance of ingredient quality will also become more important, and the industry will need to invest on quality issues. Clearly, it is not possible to improve the quality of raw materials after the delivery to the mill. It logically follows that any improvement in quality can be achieved only by working closely with the supplier. Those supplying the raw materials, however, are often not well versed in animal nutrition, which implies that ways of assuring quality ultimately depends on the nutritionists. This will be relevant for all raw materials, including mineral sources, additives and pre-mixes.

Rapid tests to predict gross nutrient composition and to assess the variability in ingredient supplies received at the feed mill are already available. The investment on rapid tests should be extended to the measurement of AME or digestible amino acids to formulate the diets precisely.

3.3 Antibiotic-free nutrition and gut health

The association between nutrition and health is well recognised, and this link must be exploited to ensure continued growth in animal production (Ravindran, 2012). Currently, poultry industry is facing a future without in-feed antibiotics, which is the most effective feed additive that was available to maintain immunity and normal balance of intestinal microbiota. The ban in the European Union and different degrees of voluntary withdrawal in other parts of the world on the use of in-feed antibiotics will put extra pressure on the gut health and general health of animals. As a result, there had been increasing focus on alternatives to in-feed antibiotics and tested products including enzymes, probiotics, prebiotics, essential oils, botanicals and organic acids. During the past decade, these products have been widely tested and this evaluation will continue in the future.

Exhaustive reviews are available on in-feed antibiotic alternatives, and there has been accumulation of publications in large volume on their influence in modifying gut microbiota profile and animal performance (Dibner and Buttlin, 2002; Patterson and Burkholder, 2003; Ricke, 2003; Dibner and Richards, 2005; Gianneanas, 2008; Yang et al., 2009). In summary, most alternatives have been shown to 'mimic' the working effects of in-feed antibiotics in terms of physiologic, immunologic, nutritive and/or microbiologic responses; however, none of the current generation of alternatives, on their own, is capable of fully replacing them. This is not to suggest that they should be discounted because they may well have a significant role to play within combination products. For example, there is some suggestion that synergistic effects may be expected if probiotics and prebiotics are administered in combination (Roberfroid, 1998).

One major weakness of reported evaluations is that most scientific data come from studies conducted under controlled experimental conditions that often are not repeated when the products are applied under commercial conditions. Furthermore, the alternative options are currently more costly than conventional in-feed antibiotics (Table 4); this is at a time when consumers are demanding lower food prices, coupled with improved quality and food safety.

Table 4 Feed cost of various alternative products, compared to antibiotics and coocidiostats[1]

	Cost (€ per ton)
In-feed antibiotics	1–2
Anticoccidials	2–3
Enzymes	2–3
Direct-fed microbials	4–7
Acidifiers	3–12
Oligosaccharides	2–15
Botanicals	3–25

[1]From Huyghebaert et al. (2011). One Euro (€) = 1.05 US Dollar (December 2015).

Another major problem faced by the commercial industry is that, within each class of alternatives, numerous products are available in the market and their efficacy is highly variable. As noted by Huyghebaert et al. (2011), the main characteristic of a good antibiotic alternative is practicality; it must consistently improve animal performance. It must also be noted that to achieve success, in addition to the alternatives, a combination of other strategies, including modifications in husbandry and nutritional management, needs to be considered to promote gut health and good gut flora (Mateos et al., 2002; Dahiya et al., 2006). These include (i) use of highly digestible pre-starter diets, (ii) use of lower dietary protein levels and better balance of amino acids, (iii) use of coarse particle size or whole-grain feeding to enhance gizzard development (Amerah et al., 2007; Singh et al. 2014), (iv) maintenance of good litter quality and (v) proper stocking density and improved climate control.

3.4 Gut integrity

Gut integrity is as important as good microbiota balance. Intestinal integrity for commercial poultry can be defined as the maintenance of intestinal health to enable the expression of the full genetic potential for growth and yield, and to fully utilise the dietary nutrients. Normal biota plays an important role in maintaining gut structure, strengthening the gut mucosal barrier and protein metabolism of the gut. In situations where the profiles are shifted by pathogenic biota (e.g. clostridium and coliforms), there is significant inflammation and damage to the mucosal layer and the barrier function. Coccidiosis is a major cause of poor gut integrity, and an effective anti-coccidial programme must be in place. Raw material quality is another contributing factor. Substances (e.g. mycotoxins) or raw materials (e.g. fibrous feeds) that can irritate the gut (Yegani and Korver, 2008) must be closely monitored.

3.5 Focus on gizzard development

Fore gut, particularly gizzard, if fully developed, can be regarded as an important barrier in preventing pathogenic bacteria from entering the distal intestinal tract (Svihus, 2011). A well-developed gizzard enhances the grinding action, generates stronger reverse peristalsis contractions within the gastrointestinal tract, increases proteolysis by pepsin and stimulates secretion of hydrochloric acid, which reduces the pH. Harmful bacteria entering the intestinal tract via the feed have a greater chance of being suppressed in a highly acidic environment. Because it is being increasingly recognised that poultry have a requirement for a certain degree of physical structure in their feed to meet their innate feeding behaviour development (Svihus, 2011), the inclusion of dietary structural components, such as coarse particles, insoluble fibre sources and whole grains, in poultry diets should be given consideration. The major motivation for inclusion of structural components in poultry diets is to stimulate gizzard development and functionality, which will favourably influence gut health and the bird's ability to better utilise nutrients. Because fine grinding is generally favoured to obtain a high pellet quality, and because it is difficult to avoid further reduction in particle size during the pelleting process, fine particle size is almost inevitable during pelleting; this results in a suboptimal gizzard development with potential negative influence on nutrient digestibility. The use of structural components, therefore, becomes even more critical in highly processed diets.

3.6 Crystalline/synthetic amino acids

Protein is a costly item in poultry diets, so maximising the efficiency of protein and amino acid utilisation is very important. Geneticists have done their part in providing current strains of poultry that are capable of producing protein gain at greater efficiencies than ever before. The challenge to the nutritionists is to sustain these improvements in genetic potential by refining the amino acid nutrition of poultry. In this context, the commercial availability of synthetic and crystalline amino acids was important and has enabled and will continue to assist the nutritionists (i) to more precisely meet the ideal amino acid profile; (ii) to use digestible amino acids, rather than total amino acids, as the basis of feed formulations (Lemme et al., 2004); (iii) to reduce dietary crude protein levels and meet the amino acid requirements more precisely. This will lead to greater efficiency of nitrogen utilisation and protein accretion, eventually lowering the nitrogen output in the manure (Pesti, 2009), and (iv) to develop phase-feeding programmes, wherein dietary amino acid levels are reduced steadily over time in an attempt to reduce costs associated with excess dietary protein or amino acids.

The use of digestible amino acids will become particularly relevant as we start using a range of non-traditional alternative ingredients that are poorly digestible. Formulating diets based on digestible amino acids makes it possible to increase the diversity and inclusion levels of non-traditional ingredients, despite the fact that they may contain less than optimal natural amino acid profiles, and are poorly digested. This will facilitate the dietary creation of 'ideal' protein.

Currently, three crystalline amino acids, namely DL-methionine, L-lysine HCl and L-threonine, are available to the industry at competitive prices. Though somewhat expensive, L-tryptophan can also be purchased in feed-grade forms. Valine, arginine and isoleucine, the next limiting amino acids in practical corn-soy diets, are expected to become available in the near future and may allow further improvements in feed formulation.

The outlook for crystalline amino acids as additives is arguably better than any time in the past. Amino acid supplements have important nutritional, economic and environmental roles to play in future poultry production systems. Concerns, however, have been raised in some quarters about the faster absorption of free amino acids compared to protein-bound amino acids, but available evidence indicates that supplements of limiting amino acids are utilised more efficiently by poultry for growth than equivalent quantities supplied as intact proteins (D'Mello, 2003).

3.7 Feed enzyme technology

In the future, there will be more pressure to extract every kcal of energy and every unit of nutrients. A combination of strategies has to be employed and exogenous feed enzymes will have a key role to play in maximising the release of nutrients. One can expect the development of new enzyme products that are effective in a range of diet formulations. There is evidence suggesting that preparations with multiple enzyme activities may provide a competitive strategy to improve nutrient utilisation in poultry diets (Cowieson et al., 2006; Selle and Ravindran, 2007). Such enzyme cocktails, rather pure single enzymes, represent the next generation of feed enzymes. This is because feed ingredients are structurally exceedingly complex. In the 'native' stage, nutrients in raw materials are not isolated entities but exist as complexes with various linkages to protein, fat, fibre and other complex carbohydrates.

Advances in enzyme technology will continue and one can expect that better forms of enzymes will be developed in the future. The 'next-generation' enzymes will be close to being 'perfect', with rapid and high specific catalytic activity (per unit of protein), good thermostability, high activity under a wide range of gut pH, resistance to proteolysis and good stability under ambient temperatures. Technologies are also being evolved to maintain enzyme activities in their dry enzyme products in order to protect them from the heat, moisture and high pressures generated during feed processing, and a number of thermostable enzymes, especially phytases, are now commercially available (Amerah et al., 2011).

3.8 Early nutrition

A good start is critical to achieve the potential of the chick. To achieve their genetic potential, the neonate must quickly adapt to efficiently digesting and utilising nutrients from relatively complex exogenous dietary sources in which energy is supplied predominantly by carbohydrates. Nutrition at this stage should be designed for (i) stimulation of feed intake, (ii) early development of the digestive system, (iii) higher nutrient digestibility, so that minimal substrates is available for the growth of gut pathogens, (iv) promotion of good gut microbiota development and (v) development of immune system.

Two nutritional strategies are relevant in this context. First, any delay to feed access after hatch must be avoided, as adverse effects of depriving immediate access on growth and feed conversion ratio have been demonstrated in a number of studies. Second, the nutritional requirements are very high during the first week of life, whereas the capacity to digest the feed, and absorb and transport nutrients is limiting. Data presented in Table 5 highlights the poor digestion of lipids during the first week. Similar deficiencies in protein digestibility and energy utilisation have been observed. The exact reasons for these findings are unclear, but indicate opportunities to manipulate and uplift the nutrient utilisation during the first week. The implications are that we need to (i) assign lower metabolisable energy and digestible amino acid values for raw materials in pre-starter diets, (ii) use highly digestible/nutrient dense diets and (iii) use additives to promote good gut flora.

Table 5 Influence of fat type and age of broilers on the total tract digestibility coefficient of fats (Tancharoenrat et al., 2013)

Fat type	Age (weeks)			
	1	2	3	4
Tallow	0.37	0.65	0.73	0.73
Soybean oil	0.59	0.90	0.97	0.96
Tallow: soybean oil	0.50	0.83	0.85	0.88
Poultry fat	0.60	0.85	0.93	0.91
Palm oil	0.60	0.81	0.84	0.84

3.9 Use of growth models

To continue to improve the production efficiency, alternative concepts of feeding need to be explored by modelling. Because biological systems involve complex interactions among a myriad of factors, it is impossible for the human brain to integrate these processes and gain insight into the system (Gous, 2014). The key to understanding complex systems is the science of simulation and construction of models. Growth models can be effective tools to (i) compare actual versus potential performance, which can indicate the extent of management or health problems in the flock, (ii) provide economic analysis of alternative feeding regimens and (iii) identify areas within the growth or production process where information is lacking or changes are needed. Several growth models to simulate feed intake and production parameters under a given husbandry condition are available and some of which are robust and useful. It must be noted, however, that the models are only as good as the data sets used to develop them and importantly they should be 'open' and 'dynamic'.

In addition to growth models, simulation models could also be used to evaluate risks and to optimise financial return using data mining such as neural networks (Mehri, 2012; Filipe et al., 2015) and data analysis by bioinformatics, meta-analysis and holo-analysis techniques (Hooge, 2004; Rosen, 2007; Cowieson and Roos, 2013).

3.10 Optimal feed processing and hygiene practices

Though the benefits of feed processing and pelleting in particular are not questionable, there are challenges associated with pelleting which need to be addressed in order to achieve optimal feed processing. The major aim of feed technology, from the nutritional point of view, should be to mould a balanced mash diet to high physical quality of pellets, which are also highly digestible. However, the need to achieve this high physical quality and to reduce potential levels of feed-borne pathogens such as *Salmonella* and *Campylobacter* for feed safety has led to the application of relatively high conditioning temperatures during conventional pelleting processes, a practice that does not favour high nutrient utilisation. The true impact of high conditioning temperatures application on nutrient utilisation of pelleted diets has been neglected due to focus on physical pellet quality and feed safety. Several strategies have been suggested to improve the physical quality of the pellets, instead of applying high conditioning temperatures (Abdollahi et al., 2013). Further research is warranted to identify and evaluate other possible approaches to manufacture high-quality pellets at low conditioning temperatures.

Whilst manufacturing high-quality pellets at low conditioning temperatures seems promising, the only remaining concern would be the need to eliminate feed-borne pathogens. Thermal treatment is currently thought to be the most practical method to achieve satisfying levels of feed safety; however, application of high temperatures during pelleting does not guarantee complete elimination of *Salmonella*, and the sanitising effects of high-temperature pelleting may be lost by dust contamination in the feed mill or during transport and delivery (Jones and Richardson, 2004). Considering the ever-increasing cost of feed ingredients and the negative impact of high pelleting temperatures on nutrient utilisation, there is an urgent need to find new approaches to improve feed hygiene which are not detrimental to feed nutrients.

3.11 Other issues of importance

There are also a number of other nutritional strategies which are important, but those will not be discussed here for want of space. These include *inter alia* water quality, *in ovo* nutrition, calcium nutrition and estimation of digestible phosphorus and calcium in raw materials. Special mention, however, is needed on the interactions between nutrition and the immune system, which have received considerable attention during the past decade (Korver, 2012). Using nutrients to alter immune function has become especially important in regions where in-feed antibiotics have been banned. It is well known that almost all nutrients are involved in the development of the immune system and immune responses. Deficient and excessive intakes can have negative consequences on immune status and susceptibility to a variety of pathogens.

Brief mention must also be made on the potential future role and applications of nutrigenomic tools (transcriptomics, proteomics and metabolomics) in animal production, which provide new strategies to investigate complex biological systems (Zdunczyk and Pareek, 2008). Data can be collected across the whole system of interest at several levels such as gene expression, protein expression and changes in metabolite concentrations. The dynamic nature of this active field of study, founded on the interaction between many basic sciences – genetics, molecular biology, cell biology, microbiology, immunology, chemistry, pharmacology and chemical engineering – is evident. This is an exciting time for biological scientists as the 'omics' era continues to evolve and may shape the nutritional strategies in the future. At this time, however, the issues of technical complexity and sources of variation and computational strategies for data integration remain to be resolved (Dunn et al., 2005). The translation of this fundamental science into everyday animal production is still a long way, but the possibilities cannot be ignored.

4 Future trends in research

Feed represents the greatest single expenditure associated with poultry production. Broilers and layers are highly efficient in converting feed to food products among farm animals, but (as noted earlier) they still excrete significant amounts of unutilised nutrients. Thus, there is considerable room to improve the conversion efficiency of feed to animal products. A good portion of this inefficiency results from incomplete digestion and/or utilisation of nutrients. For this reason, future nutritional research in poultry should focus on issues relating to identifying barriers to effective digestion and utilisation of nutrients. In this endeavour, poultry nutritionists need to combine their expertise with those specialising in other biological sciences, including immunology, microbiology, histology, molecular biology and plant science. Such collaboration across scientific areas will be necessary for the industry not only to improve production efficiency but also to address issues of food safety, environmental stewardship and bird welfare.

Future challenges in poultry feeding and key areas that present immediate opportunities are highlighted in this chapter. Overall, the challenge for the future is to 'produce more and efficiently, with less resources and in a sustainable manner'. It must be recognised, however, that because production needs and challenges are constantly changing, the industry will be in a steady state of evolution. Ultimately, the direction of future poultry production will have to be determined by the needs of the industry, particularly dealing with their most pressing problems, and scientific opportunities.

5 Where to look for further information

Church, D. C. (2010), *Livestock Feeds and Feeding*. Prentice Hall, Boston, MA.
Daghir, N. J. (1995), *Poultry Production in Hot Climates*. CAB International, Wallingford, UK.
Ensminger, M. E., Oldfield, J. E. and Heinemann, W. W. (1990), *Feeds and Nutrition*. The Ensminger Publishing Company, Clovis, CA.
Ewing, W. N. (1997), *The Feeds Directory*. Context. Packingdon, UK.
FAO. Feed Resources Information System, Animal Health and Production Division, Food and Agriculture Organization of the United Nations, Rome. http://www.fao.org/ag/aga/agap/frg/afris/default.htm.
FAO. Poultry Development Review. Food and Agriculture Organization of the United Nations, Rome. http://www.fao.org/ag/againfo/themes/en/poultry/AP_nutrition.htm.
Jensen, W., Devine, C. and Dikeman, M. (Eds, 2014), *Encyclopaedia of Meat Sciences*, 2nd edn. Elsevier Ltd, Oxford, UK.
Klein, R. (2014), *Chicken Nutrition*. Context. Ashby-de-la-Zouch, UK.
Leeson, S. and Summers, J. D. (2001), *Scott's Nutrition of the Chicken*, 4th edn. Nottingham University Press, Nottingham, UK.
Leeson, S. and Summers, J. D. (2009), *Commercial Poultry Nutrition*, 3rd edn. Nottingham University Press, Trumpton, Nottingham, UK.
Moughan, P. J., Ravindran, V. and Sorbara, J. O. B. (2014), 'Dietary protein and amino acids – consideration of the undigestible fraction', *Poult. Sci.*, 93, 2400–10.
Scanes, C. G., Brant, G. and Ensminger, M. E. (2004), *Poultry Science*, Pearson, Prentice Hall, Upper Saddle River, NJ.
Weidenborner, M. (2008), *Mycotoxins in Food Safety*, Springer, New York, NY.

6 References

Abdollahi, M. R., Ravindran, V. and Svihus, B. (2013), 'Pelleting of broiler diets: an overview with emphasis on pellet quality and nutritional value', *Anim. Feed Sci. Technol.*, 179, 1–23.
Adedokun, S. A., Utterback, P., Parsons, C. M., Adeola, O., Lilburn, M. S. and Applegate, T. A. (2009), 'Comparison of amino acid digestibility of feed ingredients in broilers, laying hens, and roosters', *Brit. Poultry Sci.*, 50, 350–8.
Amerah, A. M., Ravindran, V., Lentle, R. G. and Thomas, D. G. (2007), 'Feed particle size: implications on the digestion and performance of poultry', *Wld's. Poult. Sci. J.*, 63, 439–55.
Amerah, A. M., Gilbert, C., Simmins, P. H. and Ravindran, V. (2011), 'Influence of feed processing on the efficacy of exogenous enzymes in broiler diets', *Wld's. Poult. Sci. J.*, 67, 29–46.
Baker, D. H. (1996), 'Advances in amino acid nutrition and metabolism of swine and poultry', In *Nutrient Management of Food Animals to Enhance and Protect the Environment*, E. T. Kornegay (Ed.), Lewis Publishers, New York, NY, pp. 41–53.
Batal, A. B. and Parsons, C. M. (2002), 'Effects of age on nutrient digestibility in chicks fed different diets', *Poult. Sci.*, 81, 400–7.
Bedford, M. R. and Partridge, G. G. (Eds, 2010), *Enzymes in Farm Animal Nutrition*, CAB International, Wallingford, UK.
Behnke, K. C. (1996), 'Feed manufacturing technology: current issues and challenges', *Anim. Feed Sci. Technol.*, 62, 49–57.

Bryden, W. L. (2012), 'Mycotoxin contamination of the feed supply chain: implications for animal productivity and feed security', *Anim. Feed Sci. Technol.*, 173, 134–58.

Bryden, W. L., Li, X., Ravindran, G., Hew, L. I. and Ravindran, V. (2009), 'Ileal digestible amino acid values in feedstuffs for poultry', In *Rural Industries Research and Development Corporation*, Canberra, Australia, p. 76.

Carré, B., Lessire, M. and Juin, H. (2014), 'Prediction of the net energy value of broiler diets', *Animal.*, 8, 1395–401.

Cowieson, A. J. (2010), 'Strategic selection of exogenous enzymes for corn/soy-based poultry diets', *J. Poult. Sci.*, 47, 1–7.

Cowieson, A. J., Hruby, M. and Pierson, E. E. M. (2006), 'Evolving enzyme technology: impact on commercial poultry nutrition', *Nutr. Res. Rev.*, 11, 91–114.

Cowieson, A. J. and Roos, F. F. (2013), 'Bioefficacy of a mono-component protease in the diets of pigs and poultry: a meta-analysis of effect on ileal amino acid digestibility', *J. Appl. Anim. Nutr.*, 2, 1–8.

Dahiya, J. P., Wilke, D. C., Van Kessel, A. G. and Drew, M. D. (2006), 'Potential strategies for controlling necrotic enteritis in broiler chickens in post-antibiotic era', *Anim. Feed Sci. Technol.*, 129, 60–88.

Dibner, J. J. and Buttlin, P. (2002) 'Use of organic acids as a model to study the impact of gut flora on nutrition and metabolism', *J. Appl. Poult. Res.*, 11, 453–63.

Dibner, J. J. and Richards, J. D. (2005), 'Antibiotic growth promoters in agriculture: history and mode of action', *Poult. Sci.*, 84, 634–43.

D'Mello, J. P. F. (Ed., 2003), *Amino Acids in Animal Nutrition*, CABI Publishing, Wallingord, UK.

Dunn, W. B., Bailey, N. J. C. and Johnson, H. E. (2005), 'Measuring the metabolome: current analytical techniques', *Analyst*, 130, 606–25.

Filipe, V. P. S., Silva, M. A., Valente, B. D. and Rosa, G. J. M. (2015), 'Using multiple regression, Bayesian networks and artificial neural networks for prediction of total egg production in European quails based on earlier expressed phenotypes', *Poult. Sci.*, 94, 772–80.

Gianneanas, I. (2008), 'How to use plant extracts and phytogenics in animal diets', In *The Future of Animal Production*, E. M. Binder and G. Schatzmayr (Eds), Nottingham University Press, Nottingham, pp. 111–29.

Gous, R. M. (2014), 'Modeling as a research tool in poultry science'. *Poult. Sci.*, 93, 1–7.

Greenwood, M. W., Cramer, K. R., Clark, P. M., Behnke, K. C. and Beyer, R. S. (2004), 'Influence of feed form on dietary lysine and energy intake and utilisation of broilers from 14 to 30 days of age', *Int. J. Poult. Sci.*, 3, 189–94.

Havenstein, G. B., Ferket, P. R. and Qureshi, M. A. (2003), 'Growth, liveability and feed conversion of 1957 versus 2001 broilers when fed representative 1957 and 2001 broiler diets', *Poult. Sci.*, 82, 1500–8.

Hooge, D. M. (2004), 'Meta-analysis of broiler chicken pen trials evaluating dietary mannan oligosaccharide, 1993-2003', *Int. J. Poult. Sci.*, 3, 163–74.

Huang, K. H., Li, X., Ravindran, V. and Bryden, W. L. (2006), 'Comparison of apparent ileal amino acid digestibility in feed ingredients for broilers, layers and roosters', *Poult. Sci.*, 85, 624–35.

Huang, K. H., Ravindran, V., Li, X. and Bryden, W. L. (2005), 'Influence of age on the apparent ileal amino acid digestibility of feed ingredients for broiler chickens', *Br. Poult. Sci.*, 46, 236–45.

Huyghebaert, G., Ducatelle, R. and Immerseel, F. V. (2011), 'An update on alternatives to antimicrobial growth promoters for broilers', *Vet. J.*, 187, 182–8.

Jones, F. T. and Richardson, K. E. (2004), 'Salmonella in commercially manufactured feeds', *Poult. Sci.*, 83, 384–91.

Korver, D. R. (2012), 'Implications of changing immune function through nutrition in poultry', *Anim. Feed Sci. Technol.*, 173, 54–64.

Latshaw, J. D. and Moritz, J. S. (2009), 'The partitioning of metabolisable energy by broiler chickens', *Poult. Sci.*, 88, 98–105.

Leeson, S. (2007), 'Balancing science versus societal issues in poultry nutrition', *CAB Rev.: Perspect. Agric., Vet. Sci., Nutr. Nat. Res.*, 2, 1–5.

Lemme, A., Ravindran, V. and Bryden, W. L. (2004), 'Ileal digestibility of amino acids in feed ingredients for broilers', *Wld's. Poult. Sci J.*, 60, 421–35.

Mateos, G. G., Lazaro, R. and Gracia, M. I. (2002), 'The feasibility of nutritional modifications to replace drugs in poultry feeds', *J. Appl. Poult. Res.*, 11, 437–52.

McKinney, L. J. and Teeter, R. G. (2004), 'Predicting effective caloric value of non-nutritive factors: I. Pellet quality and II. Prediction of consequential formulation dead zones', *Poult. Sci.*, 83, 1165–74.

Mehri, M. (2012), 'Development of artificial neural network models based on experimental data of response surface methodology to establish the nutritional requirements of digestible lysine, methionine, and threonine in broiler chicks', *Poult. Sci.*, 91, 3280–5.

Noy, Y. and Sklan, D. (1995), 'Digestion and absorption in the young chick', *Poult. Sci.*, 74, 366–73.

NRC (1994), *Nutrient Requirements of Poultry*, 9th rev edn., National Research Council, National Academic Press, Washington, DC.

Patterson, J. A. and Burkholder, K. M. (2003), 'Application of prebiotics and probiotics in poultry production', *Poult. Sci.*, 82, 627–31.

Pesti, G. M. (2009), 'Impact of dietary amino acid and crude protein levels in broiler feeds on biological performance', *J. Appl. Poult. Res.*, 18, 477–86.

Pirgozliev, V. and Rose, S. P. (1999), 'Net energy systems for poultry feeds: a quantitative review', *Wld's. Poult. Sci. J.*, 55, 23–36.

Ravindran, V. (2012), 'Nutrition and pathology of non-ruminants', *Anim. Feed Sci. Technol.*, 173, 1–2.

Ravindran, V., Hew, L. I. and Bryden, W. L. (1998), 'Digestible amino acids in poultry feedstuffs', Rural Industries Research and Development Corporation, Canberra and Poultry Research Foundation, The University of Sydney, Camden, NSW, Australia.

Ravindran, V., Hew, L. I., Ravindran, G. and Bryden, W. L. (2005), 'Apparent ileal digestibility of amino acids in feed ingredients for broiler chickens', *Anim. Sci.*, 81, 85–97.

Ricke, S. C. (2003), 'Perspectives on the use of organic acids and short-chain fatty acids as antimicrobials', *Poult. Sci.*, 82, 632–9.

Roberfroid, M. B. (1998), 'Prebiotics and synbiotics: concepts and nutritional properties', *Br. J. Nutr.*, 80 (Suppl. 2), S197–202.

Rosen, G. D. (2007), 'Halo-analysis of the efficacy of Bio-Mos® in broiler nutrition', *Br. Poult. Sci.*, 48, 21–6.

Schofield, E. K. (Ed., 2005), *Feed Manufacturing Technology V*, American Feed Industry Association, Arlington, VA.

Selle, P. H. and Ravindran, V. (2007), 'Microbial phytase in poultry nutrition', *Anim. Feed Sci. Technol.*, 135, 1–41.

Singh, Y., Amerah, A. M. and Ravindran, V. (2014), 'Whole grain feeding: methodologies and effects on performance, digestive tract development and nutrient utilisation of poultry', *Anim. Feed Sci. Technol.*, 190, 1–18.

Skinner-Noble, D. O., McKinney, L. J. and Teeter, R. G. (2005), 'Predicting effective caloric value of non-nutritive factors: III. Feed form affects broiler performance by modifying behaviour patterns', *Poult. Sci.*, 84, 403–11.

Svihus, B. (2011), 'The gizzard: function, influence of diet structure and effects on nutrient availability', *Wld's. Poult. Sci. J.*, 67, 207–23.

Swick, R. A., Wu, S., Zuo, J., Rodgers, N., Barekatain, M. R. and Choct, M. (2013), 'Implications and development of a net energy system for broilers', *Anim. Prod. Sci.*, 53, 1231–7.

Tancharoenrat, P., Ravindran, V., Zaefarian, F. and Ravindran, G. (2013), 'Influence of age on the apparent metabolisable energy and total tract apparent fat digestibility of different fat sources for broiler chickens', *Anim. Feed Sci. Technol.*, 186, 186–92.

Thomas, D. V., Ravindran, V. and Ravindran, G. (2008), 'Nutrient utilisation of diets based on wheat, sorghum or maize by the newly hatched broiler chick', *Br. Poult. Sci.*, 49, 429–35.

Yang, Y., Iji, P. A. and Choct, M. (2009), 'Dietary modulation of gut microflora in broiler chickens: a review of the role of six kinds of alternatives to in-feed antibiotics', *Wld's. Poult. Sci. J.*, 65, 97–114.

Yegani, M. and Korver, D. R. (2008), 'Factors affecting intestinal health in poultry', *Poult. Sci.*, 87, 2052–63.

Zdunczyk, Z. and Pareek, S. C. (2008), 'Application of nutrigenomics tools in animal feeding and nutritional research', *J. Anim. Feed Sci.*, 17, 3–16.

Chapter 7

Advances in understanding and improving the role of amino acids in poultry nutrition

William A. Dozier III, Auburn University, USA; and Paul B. Tillman, Poultry Technical Nutrition Services, USA

1 Introduction

Genetic selection of poultry over the past several years has increased the rate and efficiency of growth of poultry (Havenstein et al., 2003a,b, 2007; Aviagen, 2015; Cobb, 2015; Hubbard, 2015). Primary breeding companies for meat birds have placed emphasis on selecting for improved feed conversion and breast meat yield (pectoralis major and minor muscles) due to feed ingredient price volatility and demand for further-processed poultry products. Over the last decade, genetic selection of broilers has increased body weight gain and breast yield, and improved feed conversion by 41 g, 0.50% and 0.02 points, respectively, on an annual basis (Aviagen, 2015). The juvenile broiler undergoes rapid growth rate during the first few days post-hatching. For example, 7-day body weight increases by approximately four-fold from hatching and weighs 3 kilograms in body weight by 6–7 weeks of age. The improvement in feed conversion of the modern broiler relates to increased growth rate per unit of feed consumed, which translates into increased white meat accretion compared with the commercial broiler used in the previous decade (Havenstein et al., 2003a,b). Therefore, while broiler feed intakes have increased, this increase cannot fully compensate for proportionately higher lean muscle growth. Thus, it does stand to reason that the modern broiler will respond to higher concentrations of dietary amino acid density to increase lean tissue mass since lean tissue is relatively high in amino acid content.

http://dx.doi.org/10.19103/AS.2016.0011.07

In addition to meat birds, egg production rates of laying hens have increased markedly over the last decade (Wu et al., 2007; Hendrix Genetics, 2010; Hy-Line International, 2016). The modern hen optimizes peak egg production rate of 97% from 20 to 60 weeks of age (Hy-Line International, 2016), whereas Wu et al. (2007) determined maximum egg production with Phase 1 Hy-Line hens to be 89%. This difference can translate to approximately 30 more eggs per hen under commercial practice from 20 to 60 weeks of age. As egg mass output has increased through genetic selection, body weight and feed intake have decreased concurrently. This has placed emphasis on dietary amino acid density to enhance egg size and egg production early in layer production. Moreover, the industry is moving towards a 90–100-week single production cycle. It is not uncommon for these high-performing laying hens to exhibit egg production rates over 90%, resulting in a higher persistency of egg production than laying hens of previous decades. Digestible amino acid requirements are not well established for the modern laying hen.

This chapter will provide an overview of research evaluating the utilization of amino acids in feed ingredients, the essential amino acid requirements/digestible ratios of poultry, and the implications of amino acid density on nitrogen balance. The majority of this research has been conducted with broiler chickens in the last several years, and as a result, this review will focus on broilers, but when applicable, differences occurring with layers and turkeys will also be addressed. Finally, the knowledge gaps in amino acid nutrition as it relates to poultry will be discussed.

2 History of crystalline amino acids

Inclusion of synthetic and crystalline amino acids in poultry diets has been a milestone in the area of poultry nutrition, allowing nutritionists within the last four decades to produce diets more economically without compromising growth rate, meat yield, or egg production rate of poultry (Kidd and Tillman, 2016). The price spread between the primary protein (amino acid) source, soybean meal (SBM), and the primary grain source, corn, along with fat price often dictates the inclusion rate of supplemental amino acids, particularly, lysine and threonine.

2.1 Synthetic methionine and L-Lysine

Least-cost formulation was developed in the 1950s to determine the least expensive combination of feedstuffs to meet the nutrient requirements of the animal through the use of linear programming techniques (Waugh, 1951). Diets were formulated on a crude protein basis to meet the needs for growth of the animals. With the use of a least-cost formulation, synthetic methionine sources (DL-methionine or methionine hydroxy analogue) were developed (Matterson et al., 1953; Blake and Wineman, 1960), which significantly reduced dietary costs during the 1970s. By the 1980s, inclusion of crystalline lysine was used to further reduce dietary cost (Georgen and Tintgnac, 1982). The use of the first and second limiting amino acids in poultry diets enabled to place more emphasis on formulating on a total amino acid basis rather than just on crude protein. However, the benefits of synthetic methionine and crystalline lysine could not be realized as maximum constraints were placed on methionine and lysine inclusion rates due to poor performance associated with reduced crude protein diets. With the demand increasing for deboning operations and fewer broilers marketed as whole birds, the broiler used starting in the late 1990s responded to higher dietary lysine than birds used in previous decades. Kerr et al. (1999) provided evidence that placing a maximum inclusion of feed-grade lysine was not necessary to optimize broiler growth performance and breast meat yield. The increase in the production of feed-grade methionine and lysine has enabled nutritionists to reduce dietary cost without compromising performance objectives. Therefore, the industry started to place more emphasis on formulating diets using digestible amino acid values during the late 1990s to early 2000.

2.2 L-Threonine

L-threonine was commercially produced in the late 1990s, but received resistance by the industry as growth performance would be negatively affected by a further reduction in crude protein if methionine, lysine and threonine were used in least-cost formulation. This was particularly true if the next limiting digestible amino acid did not have a minimum constraint set, and crude protein dropped more than 3% points. In 2000, the broiler industry did not report the use of threonine. Over the past 15 years of research with L-threonine, it has been shown that diet cost could be reduced by $4.0 to $5.0 per ton with the use of 0.5 to 1.0 kg inclusion of L-threonine. The U.S. broiler industry currently utilizes approximately 15 million metric tons of L-threonine. This increased utilization of L-threonine has coincided with formulating diets on amino acid ratios relative to digestible lysine over the last 10 years.

2.3 L-Valine

In 2009, L-valine was commercially produced for poultry and animal diets. Supplementation of commercially available L-valine into broiler diets has been shown to provide acceptable performance (Corzo et al., 2011). Due to the selling price and opportunity/shadow value of L-valine, it typically has not entered into formulations for broilers to any major degree. If it does enter into broiler feed formulas, the inclusion rate is much lower than methionine, lysine or threonine. As with all amino acids, the SBM minus grain price spread along with fat price, primarily determines the value of L-valine. During times of high fat prices, and a wide SBM minus grain price spread, L-valine will enter into more commercial formulations, particularly as the cost of production decreases efficiency of production and/or increases production capacity.

3 Amino acid digestibility of feed ingredients

Formulating on a digestible amino acid basis allows the poultry nutritionist to precisely meet the needs of the bird to optimize rate and efficiency of growth, meat yield, and egg production while minimizing nitrogen excretion as amino acids are not completely utilized in feed ingredients (Lemme et al., 2004). This method of formulation allows for the use of alternative feed ingredients that can reduce dietary cost without compromising performance objectives, particularly, with the large volume of feed-grade amino acids being utilized by the feed industry (Rostagno et al., 1995). Over the past decade, one of the largest priorities related to amino acid digestibility research has been to consistently implement similar bioassays across poultry laboratories (Ravindran et al., 1999; Lemme et al., 2004; Adedokun et al., 2011).

3.1 Amino acid digestibility assays

Digestibility refers to the proportion of consumed nutrients that are not excreted by the bird (Lemme et al., 2004). In the literature, there has been confusion about the terms 'bioavailability' and 'digestibility', as if they are interchangeable for biological activity (Lemme et al., 2004). Bioavailability assays are conducted by evaluating graded concentrations of a single nutrient among the test diets using slope-ratio assay technique (Lewis and Balyley, 1995). This method can be laborious and expensive, and only generates information on a single amino acid, and does not provide usable information for diet formulation. Conversely, digestibility assays generate amino acid digestibility coefficients on several amino acids of a feed ingredient that in turn can be used in diet formulation. Hence, this is the preferred method for establishing amino acid bioavailability of feed ingredients.

Precision feeding assays to determine apparent amino acid digestibility of feed ingredients based on excreta collections using adult roosters were developed over three decades ago (Sibbald, 1979). Two drawbacks of the excreta collection method are that amino acids are present in both urine and faeces and microbial contribution of amino acids occur in the caeca (Lemme et al., 2004). Apparent amino acid digestibility coefficients are not corrected for endogenous losses of amino acids entering into the small intestine, and not applying such a correction factor for endogenous losses can underestimate amino acid digestibility coefficients for feed ingredients. In order to avoid the microbial contamination and correct for the endogenous amino acid losses, the caecetomized precision-fed rooster assay (true digestibility assay) and the standardized ileal digestibility assay were developed (Parsons, 1985; Ravindran et al., 1999; Lemme et al., 2004). The true digestibility assay is relatively fast, inexpensive, requires little labour, avoids palatability issues with feed ingredients, and caecectomized roosters may be used for multiple assays. The disadvantages of this assay are that roosters require a surgical procedure (caecectomy) for the removal of the caeca, which compromises normal feeding behaviour (fasting and defined amount of test ingredient), and endogenous losses are determined with fasted-cockerels within the same assay.

The standardized ileal digestibility assay is conducted in growing birds having free access to feed. An inert marker (chromic oxide, titanium dioxide, insoluble ash, etc.) is mixed in the diet to aid in determining the amino acid content of both the feed and digesta, in order to calculate apparent amino acid digestibility. Digesta contents are collected at the lower part of the ileum, between the Meckel's diverticulum and proximally to the ileocaecal junction. However, researchers have been inconsistent in collecting samples at the same location and this can influence digestibility values among ingredients. Some researchers use the full ileum, whereas others may collect samples at the lower 1/3 or lower 2/3 of the small intestine. This method avoids microbial activity in the caeca and excludes amino acids from the urine. The disadvantages of this method are that it is laborious (requires several people to obtain digesta) and may have palatability issues; moreover, since the birds are sacrificed at the end of the experiment, they are utilized for only one experiment. Endogenous losses can be measured within the same experiment, or estimated values from previous research can be applied to correct the apparent amino acid digestibility coefficients for endogenous losses.

3.2 Endogenous amino acid flow

Endogenous amino acid losses consist of basal endogenous losses (independent of diet composition) and specific losses influenced by the test ingredient (Lemme et al., 2004). The endogenous amino acid flow in the small intestine originates from saliva, bile, pancreatic and gastric secretions, sloughed epithelial cells, and mucoproteins (Leeson and Summers, 2001). Aspartic acid, glutamic acid, serine and threonine have been reported to be relatively high in endogenous flow compared with other amino acids in poultry (Salter and Fulford, 1974; Ankanaporn et al., 1997). The four classical methods estimating endogenous amino acid losses include feeding birds a nitrogen-free diet, using the regression method to estimate endogenous losses at zero protein intake, feeding a highly digestible protein source, and using fasted caecetomized roosters. Three review papers have adequately addressed advantages and disadvantages of these methodologies (Ravindran et al., 1999; Lemme et al., 2004; Adedokun et al., 2011).

Adedokun et al. (2011) prepared an extensive review on factors affecting endogenous amino acid flow in poultry and presented some knowledge gaps in the literature. One area which was outlined was the differences between the basal and ingredient specific losses affected by both feed ingredients and methodology. Additional research is needed to determine the additive effects of endogenous amino acid flow among ingredients in complete diets. Another area that needs additional research is to determine the effects of birds subjected to a health challenge on endogenous amino acid flow. Epithelial cell turnover and mucin production in the small intestine are more pronounced when broilers are exposed to a disease challenge, particularly with the increased use of antibiotic-free production in the United States and Europe (Fernando and McCraw, 1973; Montagne et al., 2004; Rochell et al., 2016).

3.3 Digestible amino acid coefficients for feed ingredients

Digestible amino acid coefficients for feed ingredients are needed to formulate diets on a digestible amino acid basis. True digestibility amino acid values have been determined on feedstuffs over the past decades, but this method does not account for differences in digestibility coefficients among species and birds are not in a normal physiological state when using the intubation feeding method (Parsons, 1985). The ileal amino acid digestibility assay has been gaining popularity as the method of choice for determining amino acid coefficients for feed ingredients (Lemme et al., 2004). The amount of data available for certain ingredients is sparse and has been conducted with a single species, but more information is needed with ingredients utilizing broilers, layers and turkeys (Adedokun et al., 2014). In addition, the effects of age on digestibility coefficients have been reported to differ for chicks having an immature gastrointestinal tract vs. older birds (Batal and Parsons, 2002). Endogenous amino acid flow of young poultry can be approximately twice the amount within the first few days posthatching compared with chicks and poults at 15 or 21 d of age, leading to differences in digestibility amino acid coefficients among feed ingredients (Adedokun et al., 2007).

Future research should focus on developing a database of digestibility coefficients for chicks and poults during the first few days posthatch. This will enable nutritionists the flexibility of utilizing digestible amino acid coefficients specific for the starter period and an alternate set of digestible amino acid coefficients for feed ingredients used in formulating diets for poultry having a mature gastrointestinal tract (>16 days of age). A knowledge gap in the literature exist for examining the additivity of values from a single ingredient versus values from a complete diet as values from specific ingredients used in diet formulation do not always translate to digestible values of a complete diet. Amino acid suppliers are developing equations to predict digestible amino acid coefficients from near-infrared technology. It would be useful to assemble a comprehensive set of sample numbers to develop robust equations. There is more work to be conducted on near-infrared technology and reliable amino acid digestibility coefficients for feed formulation.

4 Digestible amino acid requirements/ratios

In recent years, digestible amino acid requirements/ratios have been evaluated with broilers more extensively than layers and turkeys. Through the improvements associated with genetic selection, poultry are responding to higher dietary amino acid densities than birds used in research of previous decades. The use of requirement prediction models, which have been verified through feeding trials, particularly if coupled with growth prediction and economic models, hold some potential as tools in help determining the optimal amino acid requirements of these modern broiler strains (Sakomura et al., 2015).

4.1 Digestible amino acid requirements/ratios of broilers

Digestible amino acid ratios are a popular strategy used, around the globe, to express amino acid requirements relative to digestible lysine (Baker et al., 2002). As such, an inaccurate dietary digestible lysine requirement can result in essential amino digestible acid ratios/requirements being either excessive or sub-marginal. Within the past decade, researchers have evaluated lysine requirements during various phases of production to better define digestible lysine requirement of the modern broiler (Dozier et al., 2008, 2009, 2010; Dozier and Payne, 2012; Mejia et al., 2012). The accelerated growth rate of broilers used in the recent research has resulted in 10–12% higher digestible lysine requirements (Dozier and Payne, 2012), when expressed as a percentage of the diet, than for those birds used in previous research (Baker et al., 2002; Garcia and Batal, 2005). The optimum requirement for feed conversion or breast meat yield may vary by 5–7% between genetic strains, and males tend to have higher requirements than females (ranging from 5 to 12%), with the larger spread between requirements occurring as broilers approach 4.0 kg in body weight. Digestible ratios of TSAA (total sulphur amino acids: methionine plus cysteine) (Mack et al., 1999; Rostagno et al., 2011, Mehri et al., 2012; Dozier and Mercier, 2013), threonine (Corzo et al., 2009; Mejia et al., 2012; Mehri et al., 2012; Dozier et al., 2015), valine (Corzo et al., 2007, 2008 and 2011; Campos et al., 2012; Tavernari et al., 2013), arginine (Campos et al., 2012; Corzo, 2012; Mejia et al., 2012; Neto et al., 2013; Wecke and Liebert, 2013), isoleucine (Corzo et al., 2004, 2008b; Tavernari et al., 2012, 2013), and tryptophan (Campos et al., 2012; Corzo, 2012; Wecke and Liebert, 2013) have been evaluated with modern broilers.

In the United States, an increasing number of broilers are being fed to 3.8–4.0 kg in body weight to accommodate the demand for further-processed products. Amino acid requirements are sparse in broilers produced beyond 50 days of age while many broilers can be marketed at 59–63 d of age. In the future, research is needed to address the amino acid requirements of these heavy broilers based on growth performance and meat yield. Muscle quality parameters such as 'white striping' and 'wooden breast' should also be considered as response criteria until these muscle quality problems are resolved in commercial production.

Another area that has gained interest in amino acid nutrition is the role of glycine in low crude protein diets for broilers during the prestarter and starter periods. Glycine supplementation to low crude protein diets has been reported to ameliorate poor growth performance of chicks fed diets from vegetable origin, inferring that glycine may have essentiality for the young chick (Corzo et al., 2005; Dean, 2006; Awad et al., 2015; Siegert et al., 2016). While some research has been conducted examining the glycine plus serine needs of young broiler chicks (Waguespack et al., 2009), the requirement has not fully been delineated, particularly as it pertains to the optimal ratio to digestible lysine. Research is sparse on the optimum glycine + serine to digestible lysine ratio and a high ratio for the young chick fed low crude protein diet could potentially change the amino acid order of limitation for chicks fed diets of vegetable origin. In addition, interactive effects of methionine, threonine and glycine warrant future investigation as methionine and threonine play a role in glycine metabolism (Corzo et al., 2009; Powell et al., 2011; Siegert et al., 2015a,b).

4.2 Digestible amino acid requirements/ratios of laying hens

Genetic selection has made significant progress on increasing egg production of laying hens. It is not uncommon for a commercial flock to maintain egg production of 92% until 60 weeks of age. Due to the persistency of lay coupled with welfare concerns with moulting, single production cycles of 90–100 weeks are being utilized with hens producing upwards of 300 eggs. In contrast to broilers, new research evaluating amino acid requirements and amino acid ratios of laying hens have not been commensurate with their improvements in performance. Lower body weight along with reduced feed consumption have occurred simultaneously with the increase in egg production of commercial layers, so amino acid requirements may differ than those previously determined values to optimize egg production and egg size. Feed intake is known to vary among modern genetic strains (white vs. brown egg layers), translating into different amino acid requirements (Bonekamp et al., 2010). Hence, genetic strains should be considered with diet formulation. Formulating diets utilizing amino acid ratios can be advantageous in reducing diet cost without compromising performance objectives (Bregendahl et al., 1998). The optimum ratios of amino acids can change with older hens to optimize egg size and egg production. When supplementing diets with crystalline amino acids (DL-methionine, L-lysine, L-tryptophan and L-threonine), ratios of the less limiting amino acids (isoleucine or valine) can be the threshold in formulation. Digestible methionine, isoleucine and tryptophan ratios may differ for the optimal response criteria throughout the production cycle.

Research is warranted to determine amino acid requirements/ratios to optimize egg size and egg production early in the laying cycle. Amino acid requirements in pre-peak diets are central to optimizing both egg size and egg production early in the laying cycle. In addition, research on amino acid requirements late in production is warranted due to the persistency in lay with these high-performing layers. Amino acid requirements/ratios could also change during a phase-feeding programme to optimize egg size or output. In addition, other factors such as egg quality and/or egg price can also alter the economically optimal amino acid level or ratio to use within the feeding programme. As with broilers, the potential use of requirement prediction models, which have been verified through feeding trials, particularly if coupled with economic profitability models, have some potential in suggesting the optimal amino acid needs of modern laying hen strains depending upon the desired output (Sakomura et al., 2015).

In the future, cage-free production could have an impact on the maintenance needs of hens, thus affecting feed intake translating into different amino acid requirements compared with hens reared in cages. Further research is warranted to define amino acid needs with egg layers reared in cages, but particularly for those hens in cage-free production facilities.

4.3 Digestible amino acid requirements/ratios of turkeys

Recent digestible amino acid requirement research with modern turkeys is lacking in the published literature (Linares et al., 2012). Several studies evaluating TSAA, lysine and threonine requirements of turkeys were conducted approximately 12–18 years ago, and these genetic strains are not representative of turkeys used in current production systems (Boling and Firman, 1998; Kidd et al., 1998; Kamyab and Firman, 2000; Moore et al., 2004). However, the turkey industry and particularly the larger integrated turkey companies, conduct 'in-house' amino acid research to develop their feeding programmes utilizing modern genetic strains. It is known that growth and meat yield responses to dietary amino acid concentrations in turkeys are highly variable, and ample replication is paramount when determining dietary amino acid requirements. In addition, considerable strain differences exist for amino acid requirements (Aviagen, 2010; Hybrid, 2013). Turkeys are able to adjust their feed intake when fed diets having sub-optimum dietary amino acid concentrations and this can contribute to the variability encountered in research studies. In contrast to broilers, it is believed that the order of the six most limiting amino acids in turkey diets consists of TSAA, lysine, threonine, valine, arginine and tryptophan. Digestible amino acid ratios are not well defined for toms versus hens throughout a 16- or 19-week production cycle. Specifically, knowledge is limited on the optimum digestible valine, isoleucine and tryptophan to lysine ratios of turkeys at various ages. Digestible valine ratios are particularly important as the demand for 'all-vegetable' diets increases to meet marketing expectations.

5 Nitrogen balance

Nitrogen runoff into water sources from livestock and poultry operations has created environmental concerns of contaminating water sources. In addition, ammonia emissions from poultry operations have generated environmental concerns and nuisance complaints (Fairchild et al., 2009). Heightened ammonia concentrations in poultry houses are known to negatively affect well-being of the bird and compromise worker health (Kristensen and Wathes, 2000; Miles et al., 2004). Feeding reduced crude protein diets has been shown as a strategy to reduce nitrogen excretion and ammonia concentrations in poultry production (Ferguson et al., 1998a,b; Bregendahl et al., 2002; Novak et al., 2006).

5.1 Nitrogen balance in broilers

A strong relationship exists between nitrogen intake and nitrogen excretion with broilers fed diets varying in crude protein (Ferguson et al., 1998a,b; Bregendahl et al., 2002). For example, a 6.3 g/chick decrease of nitrogen excretion can be obtained from a 1% point decrease in dietary crude protein with diets ranging in crude protein from 23.2 to 18.6% from 7 to 21 d of age (Bregendahl et al., 2002). Ferguson et al. (2008a,b) fed broilers diets varying in dietary CP (26.4 to 21.9% in the starter period; 21.5 to 16.5% in the grower period) during a 43 d production period. These researchers reported a 16.5% reduction in litter nitrogen content and a 31% decrease in ammonia concentrations when dietary crude protein was reduced (21.9% and 16.5% in the starter and grower periods). Decreasing nitrogen intake through dietary crude protein reduction can have a pronounced effect on nitrogen excretion and ammonia emissions, but performance objectives and meat yield responses should not be compromised at the expense of nitrogen excretion. Hence, it is important to maintain dietary amino acid ratios to avoid adverse performance. Previous research evaluating growth responses of broilers fed low protein diets may have been confounded by variable concentrations of less limiting amino acids such as, valine, isoleucine, arginine and tryptophan (Pinchasov et al., 1990; Bregendahl et al., 2002; Hernandez et al., 2012). Future research should ensure that adequate concentrations of less limiting amino acids do not confound growth and nitrogen excretion responses of broilers fed reduced crude protein diets. In addition, previous research typically has only evaluated nitrogen excretion at a given time point. Research is warranted to conduct nitrogen mass balance studies with real-time ammonia emission concentrations. These data should be expressed per unit of BW gain and total amount of saleable meat, which would indicate the most efficient diet to manipulate nitrogen balance without compromising production efficiency.

5.2 Nitrogen balance in layers

In parallel with broilers, reducing dietary crude protein has also been reported to reduce nitrogen excretion and ammonia emissions from laying hens (Novak et al., 2006; Roberts et al., 2007). Egg production and egg mass should be considered when reducing dietary amino acid density or crude protein as a strategy of dietary manipulation to reduce nitrogen output. Roberts et al. (2007) fed laying hens (45–58 weeks of age) diets formulated to contain either 17 or 16% CP. Feeding the reduced CP diet decreased nitrogen excretion (g/d) by 10%, but egg production (90.3 vs. 87.6%) and egg mass (55.6 vs. 53.8 g/d) were adversely affected, indicating the importance of balancing production efficiency with nitrogen excretion. Novak et al. (2006) provided hens from 20 to 43 and 44 to 63 weeks of age 18.9, 17.0 and 14.4 and 16.3, 14.6 and 13.8 g/hen per d of CP, respectively, and evaluated egg production and nitrogen balance measurements. Excreta nitrogen content was decreased by 19 (6.32, 5.88 and 5.12%) and 9% (4.95, 4.74 and 4.49%), respectively, when hens consumed 14.4 and 13.8 g/hen per d of CP during 20–43 and 44–63 weeks of age. However, percent egg production was adversely affected when hens were fed the low CP diets, regardless of feeding phase. In addition, feeding the low CP diet in phase 1 negatively influenced feed conversion (feed intake/egg mass).

6 Summary

Significant accomplishments have been made in understanding factors influencing amino acid digestibility coefficients for poultry (broilers, layers and turkeys) over the last several years. The use of near-infrared technology in establishing a robust database is needed to more accurately determine total amino acid composition and digestibility coefficients for diet formulation, but only after consistent approaches to ingredient digestibility bioassays are adopted. Moreover, understanding and modelling factors affecting indigestibility of different ingredients are warranted. Digestible amino acid requirements have been established for broilers more so than for layers or turkeys. Future research is needed to define amino acid requirements among various phases of broiler (especially in heavy broilers past 50 days of age), laying hens, and turkey production systems. Amino acid requirements for immune function, disease/microbial load, and physiological needs may differ from growth performance and meat yield for specific amino acids such as arginine, threonine and branch chain amino acids, which will be very relevant to the emphasis placed on antibiotic-free production. Genetic selection, welfare, antibiotic-free production, meat quality, and further processing of meat birds (bird size) are factors that will shape the direction of future research evaluating amino acid responses of poultry.

7 Where to look for further information

Adedokun, S. A., O. Adeola, C. M. Parsons, M. S. Lilburn and T. J. Applegate. 2011. Factors affecting endogenous amino acid flow in chickens and the need for consistency in methodology. *Poult. Sci.* 90:1737–48.

Adeola, O., P. C. Xue, A. J. Cowieson, and K. M. Ajuwon. 2016. Basal endogenous losses of amino acids in protein nutrition research for swine and poultry. *Anim. Feed Sci. Technol.* 221:274–83.

Baker, D. H. 2009. Advances in protein-amino acid nutrition of poultry. *Amino Acids* 37:29–41.

Bryden, W. L., X. Li, G. Ravindran, L. I. Hew and V. Ravindran. 2009. Ileal digestible amino acid values in feedstuffs for poultry. RIRDC publication No 09/071. 86 pages. https://rirdc.infoservices.com.au/downloads/09-071.pdf

CVB Feed Table 2016. Chemical composition and nutritional values of feedstuffs. CVB. Internet: www.cvbdiervoeding.nl

D'Mello, J. P. F. (ed.) 2003. *Amino Acids in Animal Nutrition*. 2nd ed. CABI Publishing, Cambridge, MA. 513 pages. ISBN 0 85199 654 X

Lemme, A., V. Ravindran and W. L. Bryden. 2004. Ileal digestibility of amino acids in feed ingredients for broilers. *World's Poult. Sci.* 60:423–37.

Ravindran, V. 2016. Feed-induced specific ileal endogenous amino acid losses: Measurement and significance in the protein nutrition of monogastric animals. *Anim. Feed Sci. Technol.* 221:304–13.

Rostagno, H. R., L. F. T. Albino, J. L. Donzele, P. C. Gomes, R. F. de Olveira, D. C. Lopes, A. S. Firiera and S. L. de T. Barreto. 2011. Brazilian Tables for Poultry and Swine. Composition of Feedstuffs and Nutritional Requirements. 3rd ed. 251 pages. Universidade Federal de Vicosa, Dept. Zootecnia, Vicosa, MG, Brazil.

Sauvant, D., J. M. Perez and G. Tran. 2004. Tables of composition and nutritional value of feed materials. 2nd ed. 304 pages. Wageningen Academic Publishers and INRA Editions. ISBN 2-7380-1170-5

Wu, Guoyao, F. W. Bazer, Z. Dai, D. Li, J. Wang and Z. Wu. 2014. Amino acid nutrition in animals: Protein synthesis and beyond. *Annu. Rev. Anim. Biosci.* 2:387–417.

8 References

Adedokun, S. A., C. M. Parsons, M. S. Lilburn, O. Adeola and T. J. Applegate. 2007. Enogenous amino acid flow in broiler chicks is affected by the age of birds and method of estimation. *Poult. Sci.* 86:2590–7.

Adedokun, S. A., O. Adeola, C. M. Parsons, M. S. Lilburn and T. J. Applegate. 2011. Factors affecting endogenous amino acid flow in chickens and the need for consistency in methodology. *Poult. Sci.* 90:1737–48.

Adedokun, S. A., P. Jaynes, M. E. Abd El-Hack, R. L. Payne and T. J. Applegate. 2014. Standarized ileal amino acid digestibility of meat and bone meal and soybean meal in laying hens and broilers. *Poult. Sci.* 93:420–8.

Angkanaporn, K., M. Choct, W. L. Bryden, E. F. Annison and G. Annison. 1997. Evaluation of homoarginine as a marker for the endogenous amino acid concentrations in poultry excreta. *Br. Poul. Sci.* 38:577–85.

Aviagen. 2015. Ross Broiler Nutrition Supplement. http://www.aviagen.com

Aviagen. 2010. Feeding Recommendations for Commercial Stock. http://www.aviagenturkeys.com

Awad, E. A., I. Zulkifi, A. F. Soleimani and T. C. Loh. 2015. Individual non-essential amino acids fortification of a low-protein diet for broilers under the hot and humid tropical climate. *Poult. Sci.* 94:2772–7.

Baker, D. H., A. B. Batal, T. M. Parr, N. R. Augspurger and C. M. Parsons. 2002. Ideal ratio (relative to lysine) of tryptophan, threonine, isoleucine, and valine for chicks during the second and third weeks posthatch. *Poult. Sci.* 81:485–94.

Batal, A. B. and C. M. Parsons. 2002. Effects of age on nutrient digestibility in chicks fed different diets. *Poult. Sci.* 81:400–7.

Blake, E. S. and R. J. Wineman. 1960. Amino acid analogues. U.S. Patent 2938053 A.

Boling, S. D. and J. D. Firman. 1998. Digestible lysine requirement of female turkeys during the starter period. *Poult. Sci.* 77:547–51.

Bonekamp, R. P. R. T., A. Lemme, P. J. A. Wijtten and J. K. W. M. Sparla. 2010. Effects of amino acids on egg number and egg mass of brown (heavy breed) and white (light breed) laying hens. *Poult. Sci.* 89:522–9.

Bregendahl, K., J. L. Sell and D. R. Zimmerman. 2002. Effect of low-protein diets on growth performance and body composition of broiler chicks. *Poult. Sci.* 81:1156–67.

Bregendahl, K., S. A. Roberts, B. Kerr and D. Hoehler. 2008. Ideal ratios of isoleucine, methionine, methionine plus cysteine, threonine, tryptophan, and valine relative to lysine for white leghorn-type laying hens of twenty-eight to thirty-four weeks of age. *Poult. Sci.* 87:744–58.

Campos, A., H. S. Rostagno, E. T. Nogueira, L. F. T. Albino, J. P. L. Pereira and R. C. Maia. 2012. Updating of the ideal protein for broilers: arginine, isoleucine, valine and tryptophan. *Revista Brasileira de Zootecnia* 4:326–32.

Cobb. 2015. Broiler Performance and Nutrition Supplement. http://www.cobb-vantress.com

Corzo, A., E. T. Moran Jr and D. Hoehler. 2004. Isoleucine need of heavy broiler males. *Arch. Geflugelk* 68:194–8.

Corzo, A., C. A. Fritts, M. T. Kidd and B. J. Kerr. 2005. Response of broiler chicks to essential and non-essential amino acid supplementation of low crude protein diets. *Anim. Feed Sci. Technol.* 118:319–27.

Corzo, A., M. T. Kidd, W. A. Dozier III and B. J. Kerr. 2009. Dietary glycine and threonine interactive effects in broilers. *J. Appl. Poult. Res.* 18:79–84.

Corzo, A., M. T. Kidd, W. A. Dozier III and S. L. Vieira. 2007. Marginality and needs of dietary valine for broilers fed certain all-vegetable diets. *J. Appl. Poult. Res.* 16:546–54.

Corzo, A., W. A. Dozier III and M. T. Kidd. 2008a. Valine nutrient recommendations for Ross × Ross 308 broilers. *Poult. Sci.* 87:335–8.

Corzo, A., W. A. Dozier III, M. T. Kidd and D. Hoehler. 2008b. Impact of dietary isoleucine on heavy-broiler production. *International J. Poult. Sci.* 7:526–9.

Corzo, A., W. A. Dozier III, R. E. Loar, M. T. Kidd and P. B. Tillman. 2009. Assessing the threonine-to-lysine ratio of female broilers from 14 to 28 days of age. *J. Appl. Poult. Res.* 18:237–43.

Corzo, A., W. A. Dozier III, L. Mejia, C. D. Zumwalt, M. T. Kidd, and P. B. Tillman. 2011. Nutritional feasibility of L-valine inclusion in commercial broiler diets. *J. Appl. Poult. Res.* 20:284–90.

Corzo, A. 2012. Determination of the arginine, tryptophan, and glycine ideal-protein ratios in high-yield broiler chicks. *J. Appl. Poult. Res.* 21:79–87.

Dean, D. W., T. D. Bidner and L. L. Southern. 2006. Glycine supplementation to low protein, amino acid-supplemented diets supports optimum performance of broiler chicks. *Poult. Sci.* 85:288–96.

Dozier, W. A., III, A. Corzo, M. T. Kidd and M. W. Schilling. 2008. Dietary digestible lysine requirements of male and female broilers from forty-nine to sixty-three days of age. *Poult. Sci.* 87:1385–91.

Dozier, W. A., III, A. Corzo, M. T. Kidd, P. B. Tillman and S. L. Branton. 2009. Digestible lysine requirements of male and female broilers from fourteen to twenty-eight days of age. *Poult. Sci.* 88:1676–82.

Dozier, W. A., III, A. Corzo, M. T. Kidd, P. B. Tillman and S. L. Branton. 2010. Digestible lysine requirements of male broilers from 28 to 42 days of age. *Poult. Sci.* 89:2173–83.

Dozier, W. A., III and R. L. Payne. 2012. Digestible Lys requirements of female broilers from 1 to 15 days of age. *J. Appl. Poult. Res.* 21:348–57.

Dozier, W. A., III and Y. Mercier. 2013. Ratio of digestible total sulfur amino acids to lysine of broiler chicks from 1 to 15 days of age. *J. Appl. Poult. Res.* 22:862–71.

Dozier, W. A., III, K. J. Meloche, P. B. Tillman and Z. Jiang. 2015. Growth performance of male broilers fed diets varying in digestible threonine to lysine ratio from 1 to 14 days of age. *J. Appl. Poult. Res.* 24: 457–62.

Fairchild, B. D., M. Czarick, L. A. Harper, J. W. Worley, C. W. Ritz, B. D. Hale and L. P. Naeher. 2009. Ammonia concentrations downstream of broiler operations. *J. Appl. Poult. Res.* 18:630–9.

Ferguson, N. S., R. S. Gates, J. L. Taraba, A. H. Cantor, A. J. Pestcatore, M. L. Straw, M. J. Ford and D. J. Burnham. 1998a. The effect of dietary protein and phosphorus on ammonia concentration and litter concentration in broilers. *Poult. Sci.* 77:1085–93.

Ferguson, N. S., R. S. Gates, J. L. Taraba, A. H. Cantor, A. J. Pestcatore, M. L. Straw, M. J. Ford and D. J. Burnham. 1998b. The effect of dietary crude protein on growth, ammonia concentration and litter composition in broilers. *Poult. Sci.* 77:1481–7.

Fernando, M. A. and B. M. McCraw. 1973. Mucosal morphology and cellular renewal in the intestine of chickens following a single infection of *Eimeria acervulina*. *J. Parasitol.* 59:493–501.

Garcia, A. and A. B. Batal. 2005. Changes in the digestible lysine and sulfur amino acid needs of broiler chicks during the first three weeks posthatching. *Poult. Sci.* 84:1350–5.

Georgen, D. and J. P. Tintgnac. 1982. Process for the preparation of new lysine-containing solid compositions for addition to animal feed. U.S. Patent 4327118 A.

Havenstein, G. B., P. R. Ferket and M. A. Qureshi. 2003a. Growth, livability, and feed conversion of 1957 versus 2001 broilers when fed diets representative 1957 and 2001 broiler diets. *Poult. Sci.* 82:1500–8.

Havenstein, G. B., P. R. Ferket and M. A. Qureshi. 2003b. Carcass composition and yield of 1957 versus 2001 broilers when fed diets representative 1957 and 2001 broiler diets. *Poult. Sci.* 82:1500–8.

Havenstein, G. B., P. R. Ferket, J. L. Grimes, M. A. Qureshi and K. E. Nestor. 2007. Comparsion of the performance of 1966- versus 2003-type turkeys when fed representative 1966 and 2003 turkey diets: growth rate, livability, and feed conversion. *Poult. Sci.* 86:232–40.

Hendrix Genetics. 2010. Nutritional Management Guide Commercials. http://www.isapoultry.com/en/products

Hernandez, F., M. Lopez, S. Martinez, M. D. Megias, P. Catala and J. Madrid. 2012. Effect of low-protein diets and single sex on production performance, plasma metabolites, digestibility, and nitrogen excretion in 1- to 48-day-old broilers. *Poult. Sci.* 91:683–92.

Hubbard. 2015. Management Guide Broiler. http://www.hubbardbreeders.com

Hybrid. 2013. Hybrid Nutrient Guidelines – 2013 Commercial Program. http://www.hybridturkeys.com

Hy-Line International. Management Guide. 2016. http://www.hyline.com

Kamyab, A. and J. D. Firman. 2000. Digestible threonine requirement of female Nicholas poults during the starter period. *J. Appl. Poult. Res.* 9:62–5.

Kerr, B. J., M. T. Kidd, K. M. Halpin, G. W. McWard and C. L. Quarles. 1999. Lysine level increases live performance and breast yield in male broilers. *J. Appl. Poult. Res.* 8:381–90.

Kidd, M. T., P. R. Ferket and J. D. Garlich. 1998. Dietary threonine responses in growing turkey toms. *Poult. Sci.* 77:1550–5.

Kidd, M. T. and P. B. Tillman 2016. Key principles concerning dietary amino acid responses in broilers. *Anim. Feed Sci. Technol.* 221:314–22.

Kristensen, H. H. and C. M. Wathes. 2000. Ammonia and poultry welfare: a review. *World's Poult. Sci.* 56:235–45.

Leeson, S. and J. D. Summers. 2001. Digestion and nutrient availability. In *Scott's Nutrition of the Chicken*, 4th ed. University Books, Guelph, Ontario, Canada, Pages 1–33.

Lemme, A., V. Ravindran and W. L. Bryden. 2004. Ileal digestibility of amino acids in feed ingredients for broilers. *World's Poult. Sci.* 60:423–37.

Lewis, A. J. and H. S. Bayley. 1995. Amino acid bioavailability. In C. B. Ammerman, D. H. Baker and A. J. Lewis (Eds), *Bioavailability of Nutrients for Animals: Amino Acids, Minerals, and Vitamins.* Academic Press Inc., San Diego, CA, pp. 35–65.

Linares, L. B., R. A. Murarolli, E. A. Guaiume, R. B. Shirley, D. Hoehler, D. R. Ledoux and J. D. Firman. 2012. Digestible lysine and threonine requirements of male turkeys from days 8 to 21 and days 29 to 42. *J. Appl. Poult. Res.* 21:384–98.

Mack, S., D. Bercovici, G. De Groote, B. Leclercq, M. Lippens, M. Pack, J. B. Schutte, and S. Van Cauwenberghe. 1999. Ideal amino acid profisoleucine and dietary lysine specification for broiler chickens of 20 to 40 days of age. *Br. Poult. Sci.* 40:257–65.

Matterson, L. D., L. Decker, E. P. Singsen, A. Kozeff, J. Waddell, C. J. Hasbrouck, H. R. Bird, H. Menge and T. D. Runnels. 1953. The value of supplemental methionine in practical chick starter and broiler rations. *Poult. Sci.* 32:817–26.

Mehri, M., A. A. Davarpanah and H. R. Miraei. 2012. Estimation of ideal ratios of methionine and threonine to lysine in starting broiler chicks using response surface methodology. *Poult. Sci.* 91:771–7.

Mejia, L., P. B. Tillman, C. D. Zumwalt and A. Corzo. 2012. Assessment of the threonine-to-lysine ratio of male broilers from 35 to 49 days of age. *J. Appl. Poult. Res.* 21:235–42.

Miles, D. M., S. L. Branton and B. D. Lott. 2004. Atmospheric ammonia is detrimental to the performance of modern commercial broilers. *Poult. Sci.* 83:1650–4.

Moore, D. T., K. Baker and J. D. Firman. 2004. Digestible sulfur amino acid requirement for male turkeys from six to twelve weeks of age. *J. Appl. Poult. Res.* 13:155–62.

Montagne, L., C. Piel and J. P. Lalles. 2004. Effects of diet on mucin kinetics and composition: Nutrition and health implications. *Nutr. Rev.* 62:105–14.

Neto, R. M., M. L. Ceccantini and J. I. M. Fernandes. 2013. Effects of methionine source, arginine:lysine ratio and sodium chloride level in the diets of grower broilers reared under hightemperature conditions. *Brazilian J. Poult. Sci.* 15:151–60.

Novak, C., H. M. Yakout and S. E. Scheideler. 2006. The effect of dietary protein level and total sulfur amino acid:lysine ratio on egg production parameters and egg yield in Hy-Line W-98 hens. *Poult. Sci.* 85:2195–206.

Parsons, C. M. 1985. Influence of cecectomy on digestibility of amino acids by roosters fed distillers dried grains with solubles. *J. Agric. Sci. Camb.* 104:469–72.

Pinchasov, Y., C. X. Mendoca and L. S. Jensen. 1990. Broiler chick response to low protein diets supplemented with synthetic amino acids. *Poult. Sci.* 69:1950–5.

Powell, S., T. D. Bidner and L. L. Southern. 2011. Effects of glycine supplementation at varying levels of methionine and cysteine on the growth performance of broilers fed reduced crude protein diets. *Poult. Sci.* 90:1023–7.

Ravindran, V., L. I. Hew, G. Ravindran and W. L. Bryden. 1999. A comparison of ileal digesta and excreta analysis for the determination of amino acid digestibility in food ingredients for poultry. *Br. Poult. Sci.* 40:266–74.

Roberts, S. A., H. Xin, B. J. Kerr, J. R. Russell and K. Bregendahl. 2007. Effects of dietary fiber and reduced crude protein on nitrogen balance and egg production in laying hens. *Poult. Sci.* 86:1716–25.

Rochell, S. J., C. M. Parsons and R. N. Dilger. 2016. Effects of *Eimeria acervulina* infection severity on growth performance, apparent ileal amino acid digestibility, and plasma concentrations of amino acids, carotenoids, and α1-acid glycoprotein in broilers. *Poult. Sci.* 95:1573–81.

Rostagno, H. S., J. M. R. Pupa and M. Pack. 1995. Diet formulation for broilers based on total versus digestible amino acids. *J. Appl. Poult. Res.* 4:293–9.

Rostagno, H. R., L. F. T. Albino, J. L. Donzele, P. C. Gomes, R. F. de Olveira, D. C. Lopes, A. S. Firiera and S. L. de T. Barreto. 2011. Brazilian Tables for Poultry and Swine. In H. S. Rostagno (Ed.), *Composition of Feedstuffs and Nutritional Requirements*, 3rd ed. Universidade Federal de Vicosa, Dept. Zootecnia, Vicosa, MG, Brazil.

Sakomura, N. K., E. P. Silva, J. C. P. Dorigam, R. M. Gous and N. St-Pierre. 2015. Modeling amino acid requirements of poultry. *J. Appl. Poult. Res.* 24:267–82.

Salter, D. N. and R. J. Fulford. 1974. The influence of gut microflora on the digestion of dietary and endogenous proteins: Studies of the amino acid composition of the excreta of germ-free and conventional chicks. *Br. J. Nutr.* 32:625–37.

Siegert, W., H. Amadi, A. Helmbrecht and M. Rodehutscord. 2015a. A quantitative study of the interactive effects of glycine and serine with threonine and choline on growth performance in broilers. *Poult. Sci.* 94:1557–68.

Siegert, W., H. Amadi and M. Rodehutscord. 2015b. Meta-analysis of the influence of dietary glycine and serine, with consideration of methionine and cysteine, on growth and feed conversion of broilers. *Poult. Sci.* 94:1853–63.

Siegert, W., K. J. Wild, M. Schollenberger, A. Helmbrecht and M. Rodehutscord. 2016. Effect of glycine supplementation in low protein diets with amino acids from soy protein isolate or free amino acids on broiler growth and nitrogen utilisation. *Br. Poult. Sci.* 57:424–34.

Sibbald, I. R. 1979. A bioassay for available amino acids and true metabolizable energy in feedingstuffs. *Poult. Sci.* 58:668–75.

Tavernari, F. C., G. R. Lelis, P. R. O. Carneiro, R. A. Vieira, R. C. Polveiro, J. A. P. Luengas, H. S. Rostagno and L. F. T. Albino. 2012. Effect of different digestible isoleucine/lysine ratios for broiler chickens. *Revista Brasileira de Zootecnia* 41:1699–705.

Tavernari, F. C., G. R. Lelis, R. A. Vieira, H. S. Rostagno, L. F. T. Albino and A. R. Oliveira Neto. 2013. Valine needs in starting and growing Cobb (500) broilers. *Poult. Sci.* 92:151–7.

Waguespack, A. M., S. Powell, T. D. Bidner and L. L. Southern. 2009. The glycine plus serine requirement of broiler chicks fed low-crude protein corn-soybean meal diets. *J. Appl. Poult. Res.* 18:761–5.

Waugh, F. V. 1951. The minimum-cost dairy feed (An application of linear programming). *J. Farm Economics* 33:299–310.

Wecke, C. and F. Liebert. 2013. Improving the reliability of optimal in-feed amino acid ratios based on individual amino acid efficiency data from N balance studies in growing chickens. *Animals* 3:558–73.

Wu, G., M. M. Bryant, P. Gunawardana and D. A. Roland. 2007. Effect of nutrient density on performance, egg components, egg solids, egg quality, and profits in eight commercial leghorn strains during phase one. *Poult. Sci.* 86:691–7.

Chapter 8

Advances in understanding and improving the role of enzymes in poultry nutrition

Bogdan A. Slominski, University of Manitoba, Canada

1 Introduction

The objective of this chapter is to provide the poultry science community with the current status of research on feed enzymes. Emphasis has been given on identifying the key challenges researchers face in terms of current trends in enzyme development, mechanism(s) of action, enzyme efficacy, as well as new directions and development of consensus protocol and/or research approaches. These approaches not only prevent any potential duplication of effort but rather lead to more coordinated research and collaboration between specialists, departments, universities and feed industry research centres around the world. To help the scientific community to meet these challenges, the key drawbacks and opportunities in the application of phytase, carbohydrases, protease and their combinations in poultry nutrition leading to more effective use of feed ingredients and more sustainable production of poultry meat will be discussed.

http://dx.doi.org/10.19103/AS.2016.0011.08

2 Phytase in poultry diets: enzyme efficacy, phytate and non-phytate phosphorus contents and environmental impacts

There is a large body of scientific literature on this subject dating back over 30 years. Thousands of papers have been published on this subject in numerous journals since the mid-1980s, although as indicated by Angel and Sorbara (2014) a high proportion of papers have been lacking information on the content of phytate, phytase levels, their interaction and thus phytase efficacy. In some published studies, however, phytate phosphorus (P) has been analysed and its digestibility, either at the ileal or excreta level, have been determined. Increased P digestibility and utilization and, hence, reduced P excretion into the environment as a result of phytase addition to poultry diets have been demonstrated (Applegate et al., 2003; Penn et al., 2004; Angel et al., 2006; Leytem et al., 2007). The results of several phytate P digestibility studies, however, have shown that the liberation of P from phytic acid by phytase is incomplete, and as illustrated in Fig. 1, averages only 54% of phytate P hydrolysis of what is present in the diet (i.e. 0.15 vs. 0.28%). It must be emphasized, however, that the degree of phytate P release is not entirely due to microbial phytase supplementation but also a consequence of the endogenous intestinal mucosa phytase action (Maenz and Classen, 1998; McCuaig et al., 1972) and, potentially, the intrinsic phytase activity of feed ingredients. Both sources of phytase could contribute to a significant release of phytate P, which, as illustrated in Fig. 1, would average 0.07% of the diet. If this is taken into account, then the liberation of P from phytic acid following microbial phytase supplementation would average only 21% of phytate P (i.e., 0.07 vs. 0.28%). This does not account for the 0.1% reduction in available P content commonly used for phytase-supplemented poultry diets.

It is of interest to note that endogenous phytase activity has been shown to increase with bird age (Maenz and Classen, 1998; Angel et al., 2002; Marounek et al., 2010) and as recently documented, phytase activity recovered from the ileum of 14-day-old broiler chicken was 45 U/kg and was highly correlated with the amount of phytate hydrolysed (i.e. $r = 0.94$; Morgan et al., 2015). It is worthwhile to mention that the degree of phytate P digestion by intestinal phytase could also increase with bird adaptation to P deficiency (Angel and Ashwell, 2012; Li et al., 2015).

The intrinsic phytase activity within plant tissue, on the other hand, is negligible in some feed ingredients, including corn, SBM or canola meal. It is higher in legume seeds (i.e. ~300 U/kg) and substantially higher and of practical importance in hard grains such as wheat, barley, triticale (i.e. ~2000 U/kg) and rye (~7000 U/kg; Godoy et al., 2005; Slominski et al., 2007; Steiner et al., 2007). Although it is believed that plant phytase is more heat labile than its microbial counterpart (Selle and Ravindran, 2007), it has also been documented that it may survive, to some extent, the pelleting process due to the fact that it is located within the grain particles and thus may not be as vulnerable to higher temperatures within the conditioner as some unprotected microbial phytase supplements (Slominski et al., 2007). This could be of importance for poultry producers involved in organic or non-GMO meat production, where the use of conventional phytase supplements is restricted. In this context, a strong positive relationship between plant phytase activity present in the diet and P availability has been documented (Nys et al., 1996).

2.1 Phytase efficacy

From the data presented in Fig. 1, it would appear that the degree of phytate P release is not directly related to the inclusion rate of exogenous phytase but is more likely a consequence of the fact that the phytate molecule is relatively inaccessible for hydrolysis. This would be due to the formation of insoluble phytate-Ca, and other divalent mineral, complexes in the small intestine (Maenz, 2001; Tamim et al., 2004). Therefore, it is believed that phytate hydrolysis takes place mainly in the forestomach (crop, proventriculus, gizzard) where the pH is more conducive to phytase action and the substrate phytate is more water-soluble (Selle and Ravindran, 2007; Li et al., 2015). As a result, the conditions and the digesta transit time in the upper gut are likely to be the important determinants of phytase efficacy.

Various approaches have been undertaken to improve the efficacy of phytase, including (1) reduced dietary Ca levels; (2) the use of single or multi-enzyme preparations; (3) the use of organic acids, mineral chelating agents and vitamin D_3; or (4) high doses of phytase. Key information on these topics along with any potential effects, or no effects, of phytase on amino acids and energy utilization are discussed in several review articles (Angel et al., 2002; Selle and Ravindran, 2007; Singh, 2008; Cowieson et al., 2011; Slominski 2011; Woyengo and Nyachoti, 2011; Selle et al., 2012; Angel and Sorbara, 2014).

Considerable interest in improving phytase efficacy and thus increasing the benefits of phytase application, including those beyond the release of P from plant phytate, has led to many studies focusing on high dosing with phytase. Such an approach has been triggered by the study of Shirley and Edwards (2003), who demonstrated that phytase inclusion above the industry standard (i.e. 12000 U/kg) make almost all of the phytate P of a corn/SBM diet (i.e. 95%) available. As a consequence, extensive work has recently

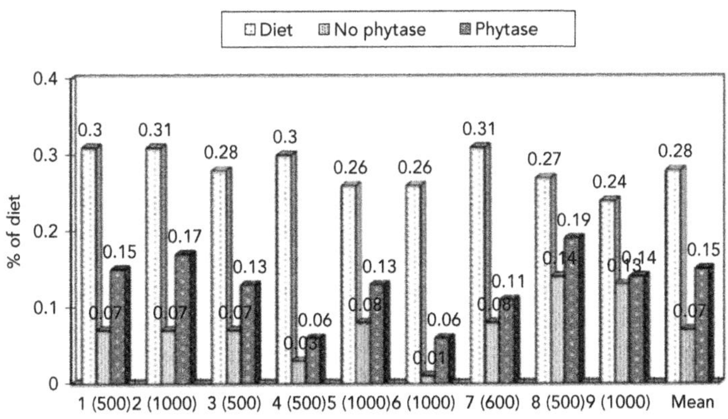

Figure 1 The effect of phytase supplementation on ileal phytate P digestibility (P release) in broiler chickens fed P-deficient control diet and control diet supplemented with phytase.[1]

[1] Calculated based on phytate P digestibility data provided and expressed in actual amounts of phytate P present in the diet. The study numbers refer to P digestibility data reported by (1, 2) Camden et al. (2001), (3) Tamim et al. (2004), (4) Rutherfurd et al. (2004), (5) Olukosi et al. (2007), (6) Leytem et al. (2008), (7) Woyengo et al. (2010), (8) Dilger et al. (2004) and (9) Olukosi et al. (2013). The numbers in parenthesis refer to U/kg of phytase used.

Table 1 Effect of varying levels of phytase supplementation on ileal P digestibility in broiler chickens and weaned pigs

Species/diet type	Total P (% of diet)	Phytase (U/kg)	Ileal P digestibility (%)	Reference
Broiler chickens				
Corn–SBM	0.57	0	32.2[b]	Rutherfurd et al. (2012)
		1000	51.5[a]	
		2000	48.0[a]	
Corn–SBM	0.56	0	48.5	Olukosi and Fru-Nji (2014b)
		1000	37.9	
		2000	46.9	
Weaned pigs				
Corn–barley–SBM	0.36	0	33.5[d]	Kies et al. (2006)
		500	55.3[c]	
		1500	71.5[b]	
		15 000	83.8[a]	
Corn–barley–SBM	0.45	0	32.5[c]	Bento et al. (2012)
		500	62.0[b]	
		1000	66.0[a]	
		2000	67.3[a]	

[a–d] Means within the same study and column with no common superscript differ significantly ($P < 0.05$).

been published on the use of high doses of phytase in broiler chicken and turkey diets. Despite the fact that the analytical methods for phytate P analysis exist, very few studies determined the effect of phytase high dosing on phytate P digestibility. As illustrated in Table 1, a moderate increase, or no effect, in ileal P digestibility was observed following the increase of dietary phytase level from 1000 to 2000 U/kg in broiler chickens and weaned pigs. In one study with weaned pigs (Kies et al., 2006), however, the use of exceptionally high level of phytase (i.e., 15 000 U/kg) significantly increased ileal P digestibility, which is in agreement with the results of the original paper on high dosing by Shirley and Edwards (2003). In many other studies (Table 2), growth performance parameters, including BW gain and FCR, have shown no effect of higher doses of phytase (i.e. 500 vs. 1000 vs. 2000 U/kg). As well, no effect of high doses on toe or tibia ash content was observed. One explanation to such low effects could be the phytate molecule availability for hydrolysis in the upper gut due to its location within the grain or seed and its chemical association with other components or nutrients (Angel and Sorbara, 2014). Another possible explanation could be that due to the use of robust enzyme heat protection technology (i.e. granulation/encapsulation), large granules of the protected phytase are being used, which may result in a heterogeneous and remote enzyme–substrate proximity within the feed matrix and consequently within the GI tract digesta. In this context, the large granule size of some phytase products may prevent the substrate phytate from being hydrolysed effectively in a relatively short period of time within the critical compartments of the upper gut (i.e. crop, proventriculus, gizzard). As an example, it has been demonstrated for two commercial

Table 2 Growth performance of broiler chickens fed low-non-phytate P (NPP) diets supplemented with different levels of phytase

Diet type	Trial length (d)	NPP (% of diet)	Phytase (U/kg)	BW gain (g/bird)	FCR (g feed/ g gain)	Toe or tibia ash (%)	Reference
Corn–SBM	1–21	0.23	0	638[b]	1.42	11.4[b,1]	Rutherfurd et al. (2012)
			1000	695[a]	1.40	12.6[a]	
			2000	680[a]	1.40	12.8[a]	
Corn–SBM	1–25	0.33	0	1151	1.41	–	Gehring et al. (2013)
			500	1165	1.42	–	
			1000	1182	1.41	–	
			2000	1202	1.41	–	
Corn–SBM	1–42	0.30	0	2928	1.68	51.7[b,2]	Walk et al. (2013)
			500	2976	1.68	52.5[a]	
			1000	3004	1.67	52.9[a]	
			1500	3018	1.65	52.6[a]	
Corn–SBM	1–21	0.30	0	791[b]	1.35[a]	48.2[b,2]	Walk et al. (2014)
			500	856[a]	1.32[a]	50.1[a]	
			1000	865[a]	1.30[b]	50.4[a]	
			1500	876[a]	1.29[b]	50.1[a]	
Corn–SBM	7–21	0.40	0	690[b]	1.46[a]	47.8[2]	Olukosi and Fru-Nji (2014a)
		0.38	1000	786[a]	1.37[b]	49.4	
		0.33	2000	809[a]	1.34[b]	51.1	

[a,b] Means within the same study and column with no common superscript differ significantly ($P < 0.05$).
[1] Represent toe ash values.
[2] Represent tibia ash values.

phytase products that due to the large granule size, birds would be required to consume approximately 5 g of feed to receive one particle or piece of enzyme product (Slominski et al., 2007).

The application of multi-carbohydrase preparations or protease in concert with phytase would appear to be a promising solution for increased phytate P availability as it would facilitate the dissociation of the phytate molecule from the fibre components. As well, it may disintegrate the digesta matrix so that the phytate substrate becomes more water-soluble and thus more available for hydrolysis in the upper gut (Jozefiak et al., 2010; Woyengo at al., 2010). It has also been demonstrated that multi-carbohydrate, but not phytase, supplementation increased the phytate P digestibility at the ileal level by 32% from that of the negative control (NC) diet (Boros et al., 2004), most likely due to the improved phytate hydrolysis by endogenous intestinal phytases. Additionally, information on the effect of adding single NSP-degrading carbohydrases to phytase-supplemented diet on phytase efficacy is inconsistent. As reviewed earlier (Slominski, 2011), the addition of xylanase had no effect on phytase efficacy as determined by the degree of apparent P digestibility. This could be attributable to the complexity of the constituent fibre and,

Table 3 Growth performance of broiler chickens fed marginal low-non-phytate P (NPP) diets supplemented with different levels of phytase

Diet type	Trial length (d)	NPP (% of diet)	Phytase (U/kg)	BW gain (g/bird)	FCR (g feed/ g gain)	Mortality (%)	Tibia ash (%)	Reference
Corn–SBM	8–22	0.08	0	303[c]	1.99[a]	–	37.2[c]	Aureli et al. (2011)
			500	596[b]	1.46[b]	–	46.6[b]	
			1000	651[a]	1.45[b]	–	50.0[a]	
			2000	684[a]	1.43[b]	–	50.8[a]	
Corn–SBM	0–18	0.15	0	380	1.53	22.8[a]	32.7[d]	Karimi et al. (2013[a])
			500	478	1.52	3.3[b]	39.1[c]	
			1000	497	1.51	0.0[b]	41.0[b]	
			1500	498	1.51	0.0[b]	42.6[a]	
Corn–SBM	0–18	0.15	0	233[d]	1.84[a]	56.2	24.2[e]	Karimi et al. (2013b)
			500	444[c]	1.50[b]	11.4	31.6[d]	
			1000	510[b]	1.50[b]	2.9	34.6[c]	
			1500	536[a,b]	1.48[b]	1.0	37.0[b]	
			2000	550[a]	1.50[b]	1.9	40.5[a]	

[a–e]Means within the same study and column with no common superscript differ significantly ($P < 0.05$).

more specifically, cell wall structure, which may require a more diversified blend of carbohydrases to be effective in NSP depolymerization. Therefore, increases of microbial phytase efficacy remain a challenge and require further research.

There is, however, a selection of recently published papers, which could serve as examples of poor research on phytase high dosing since they do not meet minimum acceptable scientific and/or animal care standards. As illustrated in Table 3, all three studies were designed and run with exceptionally low levels of available P (i.e. 0.08 and 0.15%). It would not be recommended for any study to reduce the available P level from the required 0.45% to 0.08, or even 0.15%, knowing that the phytase product evaluated can only provide a maximum of 0.1% of available P and that sufficient amount of phytate P may not even be present in the diet for the available P release by phytase to meet the bird requirement. It is quite evident that the birds used in all three studies were in abnormal (i.e. too deficient) nutritional and health states, as documented by high mortality and very poor growth performance, with BW gain values being far lower from those expected for birds of that age, not to mention the health status of the surviving birds within the phytase high-dosing treatments. Data sets such as these create difficulty in making any meaningful conclusions on phytase high dosing.

2.2 Phosphorus, phytase and the environment

Many concerns have recently been raised over the fact that the addition of phytase to broiler diets to reduce P excretion may actually be detrimental to the environment by increasing P solubility in the litter, and subsequently increasing the runoff of P from land applications in situations when phytase is used with excessive levels of inorganic non-phytate P (NPP) in the diet (Waldroup 2002; Angel et al., 2005; McGrath et al., 2005; Powers and Angel, 2008). This would be a consequence of dietary P overformulation as a result of (1) conservative available P requirements, (2) insufficient/variable data on available/NPP content of feed ingredients, and consequently, (3) safety margins for available P content used in diet formulations. As an example, when using tibia ash in broiler chickens (1–21 d) as a response criterion, Waldroup et al. (2000) concluded that the NPP requirement for broiler starters should be 0.39 rather than the generally accepted value of 0.45%. As well, there is a discrepancy in available P requirements for white-egg layers, with values ranging from (on average, 18–70 weeks of age, 100 g feed/bird/day) 0.25% (NRC 1994) to 0.40% (Leeson and Summers, 2005), 0.42% (Shaver White Commercial Management Guide, 2007; Bovans White Commercial Management Guide, 2009) and 0.43% (Lohmann LSL-Lite Commercial Management Guide, 2011). However, promising results have been published on the efficacy of phytase supplementation in laying hen diets. For example, when compared with adequate, as per breeder recommendations, P diets (i.e. ~0.40%), phytase supplementation to low-P diets (0.25% of NPP) resulted in the same production performance (Wu et al., 2006; Hughes et al., 2008; Silversides and Hruby, 2009). Additionally, in a study focused on corn/SBM-based diets containing 0.22% of NPP without supplemental phytase, Ahmadi and Rodehutscord (2012) demonstrated high egg production, egg mass and feed efficiency in layers, with the optimal level of NPP being 0.14% in the presence of phytase at 400 U/kg. It is of interest to note that even at a very low dietary level of NPP in one study (0.15% NPP, Hughes et al., 2008), egg production remained similar and no statistically significant differences were observed when phytase was used. This is very encouraging and indicates that P requirements for layers are not that well established and should be re-evaluated. It should be noted that the use of mash diets that are not subject to heating could be a contributing factor to increased phytase efficacy and the positive results observed in these specific laying hen studies.

To illustrate the importance of the excessive levels of NPP used and the environment, a study with laying hens (34–43 weeks of age) fed diets containing NPP levels as per breeder recommendations has been recently conducted (Rogiewicz and Slominski, 2015). It has been demonstrated that the P retention values averaged 37% for the positive control (PC) diet containing 0.40% of available P and 41% for the phytase-supplemented NC diet containing 0.31% of available P (Table 4). When these data are expressed as actual P content of the diet, the amount of dietary P retained within the hen averaged 0.24 and 0.23% for the PC and the phytase-supplemented NC diet, respectively, and the amount of P excreted averaged 0.41 and 0.32% for the PC and the phytase-supplemented NC

Table 4 Phosphorus balance in laying hens fed an adequate P (control) and a low-P diet supplemented with phytase using the breeder recommended requirement for available P content

Diet	Total P[1] (% of diet)	Non-phytate P[1] (% of diet)	Total P retention (%)	P retained[2] (% of diet)	P excreted	
					% of diet[2]	g/kg manure
Control	0.644	0.432	36.6	0.235	0.408	10.0
Phytase	0.556	0.344	41.2	0.230	0.323	8.3

[1]Represent the determined values; [2]represent actual amount of P present in the diet.

control diet, respectively. Despite the fact that the reduction in P excretion with phytase supplementation was evident and that the difference in P excretion between the PC and the phytase-supplemented NC diets was 0.09% and almost exactly reflecting the 0.1% point reduction in dietary available P content as recommended by the enzyme supplier, the amounts of P excreted were still high and averaged 10.0 and 8.3 g/kg of manure for the positive and the phytase-supplemented NC diets, respectively. From the amount of P excretion when expressed as actual content of the diet (i.e. 0.41 and 0.32%), it would appear evident that the amount of available P in both diets could be reduced by 0.15%, which is in agreement with the study by Hughes et al. (2008).

Another important consideration for more effective P utilization would be the availability of data for total P and NPP contents of feed ingredients. The actual number of analysis is not necessarily small as several sets of data on feed ingredients from different parts of the world exist. As shown in Tables 5 and 6, the contents of NPP within each ingredient may

Table 5 Total, phytate and non-phytate phosphorus (P) contents of cereals grains and their by-products (% DM)

Ingredient and sample origin	*n*	Total P		Phytate P		Non-phytate P[1]	
		Mean	CV	Mean	CV	Mean	Reference
Corn							
USA	10	0.26	7.7	0.17	11.8	**0.09**	Nelson et al. (1968)
Belgium	11	0.28	10.7	0.19	15.8	**0.09**	Eeckhout and De Paepe (1994)
Sri Lanka	4	0.26	3.8	0.22	9.1	**0.04**	Ravindran et al. (1994)
Venezuela	5	0.25	24.0	0.17	11.8	**0.08**	Godoy et al. (2005)
Denmark	3	0.28	0.7	0.20	0.3	**0.08**	Pontoppidan et al. (2007)
USA/Canada	133	0.32	28.7	0.19	13.4	**0.14**	Tahir et al. (2012)
Corn DDGS							
Belgium	3	0.90	–	0.19	–	**0.71**	Eeckhout and De Paepe (1994)
USA	89	0.96	6.5	0.26	27.2	**0.70**	Tahir et al. (2012)
Canada	5	0.83	3.6	0.21	19.0	**0.62**	Rogiewicz and Slominski (2016)
Wheat							
USA	2	0.30	–	0.20	–	**0.10**	Nelson et al. (1968)
Iran	12	0.42	6.8	0.31	6.7	**0.11**	Nahapetian and Bassiri (1976)
Belgium	13	0.33	6.1	0.22	9.1	**0.11**	Eeckhout and De Paepe (1994)
Australia	37	0.34	17.6	0.29	6.9	**0.04**	Selle et al. (2003)
Venezuela	5	0.33	9.1	0.18	1.7	**0.15**	Godoy et al. (2005)
Denmark	3	0.35	0.2	0.26	2.3	**0.10**	Pontoppidan et al. (2007)
Germany	18	0.40	10.0	0.29	12.8	**0.11**	Steiner et al. (2007)
USA/Canada	22	0.42	15.0	0.25	13.9	**0.17**	Tahir et al. (2012)

Table 5 *(Continued)*

Ingredient and sample origin	*n*	Total P Mean	Total P CV	Phytate P Mean	Phytate P CV	Non-phytate P[1] Mean	Reference
Canada	39	0.36	13.9	0.25	16.0	**0.12**	Rogiewicz and Slominski (2016)
Wheat DDGS							
Venezuela	5	1.22	0.8	0.30	10.0	**0.92**	Godoy et al. (2005)
Canada	3	1.05	7.6	0.23	30.4	**0.82**	Rogiewicz and Slominski (2016)
Wheat middlings							
USA	4	1.37	8.0	0.96	11.5	**0.41**	Nelson et al. (1968)
USA/Canada	31	1.31	15.0	0.80	6.4	**0.51**	Tahir et al. (2012)
Bakery by-products	95	0.48	22.1	0.19	27.6	**0.28**	Tahir et al. (2012)
Barley							
USA	5	0.34	5.9	0.19	10.5	**0.15**	Nelson et al. (1968)
Belgium	9	0.37	5.4	0.22	4.5	**0.15**	Eeckhout and De Paepe (1994)
Australia	6	0.30	10.0	0.20	25.0	**0.06**	Selle et al. (2003)
Denmark	3	0.35	0.2	0.19	0.8	**0.16**	Pontoppidan et al. (2007)
Germany	15	0.42	10.0	0.26	11.9	**0.16**	Steiner et al. (2007)
Oats							
USA	9	0.34	5.9	0.19	15.8	**0.15**	Nelson et al. (1968)
Germany	6	0.37	3.8	0.25	6.8	**0.12**	Steiner et al. (2007)
Sorghum							
Belgium	5	0.27	18.5	0.19	21.1	**0.08**	Eeckhout and De Paepe (1994)
Sri Lanka	2	0.41	2.4	0.27	3.7	**0.14**	Ravindran et al. (1994)
Australia	13	0.32	18.8	0.26	19.2	**0.06**	Selle et al. (2003)
Venezuela	5	0.26	3.8	0.17	17.6	**0.09**	Godoy et al. (2005)
Rice							
Sri Lanka	3	0.38	7.9	0.28	7.1	**7.1**	Ravindran et al. (1994)
Venezuela	5	0.12	8.3	0.08	12.5	**12.5**	Godoy et al. (2005)
South Korea	9	0.67	14.9	0.46	13.9	**13.9**	Ahn et al. (2010)
Triticale							
Belgium	6	0.37	5.4	0.25	8.0	**0.12**	Eeckhout and De Paepe (1994)
Germany	12	0.40	8.5	0.28	10.7	**0.12**	Steiner et al. (2007)
Rye	13	0.36	5.3	0.24	9.6	**0.12**	Steiner et al. (2007)

[1]Calculated as the difference between total P and phytate P.

Table 6 Total, phytate and non-phytate phosphorus (P) contents of vegetable protein supplements (% DM)

Ingredient and sample origin	*n*	Total P		Phytate P		Non-phytate P[1]	Reference
		Mean	CV	Mean	CV	Mean	
Soya bean meal							
USA	3	0.66	–	0.38	–	**0.28**	Nelson et al. (1968)
Belgium	15	0.66	4.5	0.35	5.7	**0.31**	Eeckhout and De Paepe (1994)
Sri Lanka	3	0.60	3.3	0.37	2.7	**0.23**	Ravindran et al. (1994)
Australia	22	0.72	6.9	0.49	8.2	**0.23**	Selle et al. (2003)
Venezuela	5	0.57	3.5	0.37	2.7	**0.20**	Godoy et al. (2005)
USA	25	0.76	9.7	0.42	6.1	**0.34**	Manangi and Coon (2006)
Denmark	3	0.66	0.3	0.38	0.5	**0.28**	Pontoppidan et al. (2007)
USA	114	0.84	7.2	0.40	5.6	**0.44**	Tahir et al. (2012)
Canola meal							
Belgium	5	1.12	3.6	0.40	12.5	**0.72**	Eeckhout and De Paepe (1994)
Australia	16	0.95	10.5	0.73	13.7	**0.23**	Selle et al. (2003)
Denmark	3	1.09	0.3	0.76	0.9	**0.34**	Pontoppidan et al. (2007)
USA/Canada	21	1.35	5.1	0.70	4.8	**0.66**	Tahir et al. (2012)
Canada	44	1.12	2.7	0.67	4.4	**0.45**	Adewole et al. (2016)
Lupins							
Belgium	1	0.25	–	0.05	–	**0.20**	Eeckhout and De Paepe (1994)
Australia	4	0.41	–	0.22	–	**0.20**	Selle et al. (2003)
Germany	14	0.57	26.0	0.35	26.3	**0.22**	Steiner et al. (2007)
Poland	4	0.68	0.2	0.48	0.2	**0.20**	Rogiewicz and Slominski (2016)
Sunflower meal							
Australia	2	0.98	–	0.81	–	**0.17**	Selle et al. (2003)
Denmark	3	1.20	0.0	0.59	0.4	**0.61**	Pontoppidan et al. (2007)
Peas							
Australia	3	0.38	–	0.18	–	**0.19**	Selle et al. (2003)
Germany	18	0.41	12.2	0.24	18.8	**0.17**	Steiner et al. (2007)
Poland	3	0.38	10.5	0.25	20.0	**0.13**	Rogiewicz and Slominski (2016)
Faba beans							
Australia	8	0.44	20.5	0.26	26.9	**0.17**	Selle et al. (2003)
Germany	11	0.57	15.8	0.39	20.5	**0.18**	Steiner et al. (2007)
Poland	2	0.52	11.5	0.16	50.0	**0.37**	Rogiewicz and Slominski (2016)
Cottonseed meal	11	1.24	7.3	0.99	8.0	**0.25**	Selle et al. (2003)

[1]Calculated as the difference between total P and phytate P.

range substantially even though the analysed values only represent the mean values, and this could be due to different analytical methods used for both total and NPP analyses. However, the content of NPP of different feed ingredients could still be of value, and after excluding some outliers within each ingredient, the values could be used effectively in diet formulations. As well, the values could be used to update the nutrient specification data, including the upcoming new NRC Nutrient Requirements of Poultry (Applegate and Angel, 2014).

One important issue with using NPP values in ration formulation would be that NPP may not be totally available, while phytate P could be partly available to the bird (Angel et al., 2002). Therefore, it has been suggested that poultry diets be formulated on a digestible or retainable P basis (Adedokun and Adeola, 2013; Mutucumarana et al., 2014). This would require, however, the application of very specific assays and semi-purified diets, with the test ingredient used as a sole source of P.

An interesting protocol for P digestibility measurements has recently been offered by the Working Group of the European Federation of Branches of World's Poultry Science Association (Working Group Report, WPSJ, 2013). It has been recommended that the 'precaecally digestible P' measurements would be used as the method of choice for future determination of available P in broiler chickens. Although the assay is not that different from past measurements of available P at the terminal ileum level, it strictly defines the conditions of the assay, including the age of birds (i.e. 21–28 days old), available and total P levels of the experimental diets, their two- or three-fold increments from any test ingredient evaluated (all being confirmed by chemical analyses), and finally the regression analysis performed on the basis of P and indigestible marker (i.e. TiO_2, Cr_2O_3 or acid-insoluble ash) contents of the diet and terminal ileum digesta. The conditions for testing efficacy of supplemental phytase, including the minimum of phytate P present in the diet (i.e. 0.23%), have also been proposed. This is a very proactive approach and a well-developed protocol. However, the proposed level of available P of 0.15% of the diet, regardless of any further P increments resulting from the test ingredient addition, is too low. As demonstrated in Table 3, and based on the personal experience of the author, such low levels would certainly jeopardise the welfare of fast-growing birds. With the availability of modern analytical techniques, there is no need for such extreme approaches since the levels of available P ranging from 0.25 to 0.35% of the diet would still be sufficient to achieve the meaningful evaluation of any given test ingredient.

Although ingredient evaluation based on digestible P may provide more accurate assessment of P availability, the variability in the NPP content of feed ingredients illustrated in Tables 5 and 6 may suggest that it would be almost impossible to apply such determined digestible P values in practical diet formulations, unless large databases are developed. Examples of such data are summarized in Table 7, and demonstrate that in some studies true digestible or available P contents were almost identical with NPP contents (Lumpkins and Batal, 2005; Martinez Amezcua et al., 2004; Lumpkins and Batal, 2005). In other studies, however, the digestible P contents were considerably higher than those of the NPP contents (Dilger and Adeola, 2006; Mutucumarana et al., 2014). It is of interest to note that in the latter studies the authors concluded that the discrepancy was due to low dietary Ca concentrations in the experimental diets. In addition, NPP contents reported in Table 7 are not always in agreement with the data presented in Tables 5 and 6.

In conclusion, the available P specifications of feed ingredients and P requirements need to be more accurately defined, which, in turn, should allow for the use of more

Table 7 Comparison of calculated non-phytate P and determined available P contents of feed ingredients for broiler chickens (%, as-fed basis)

Ingredient	Total P	Phytate P	Non-phytate P[1]	Available P[2]	Reference
Corn	0.28[3]	0.20[3]	0.08	0.08	Lumpkins and Batal (2005)
Corn	0.25	0.18	0.08	0.17	Mutucumarana et al. (2014)
Corn DDGS	0.74	0.22[4]	0.52	0.53	Martinez Amezcua et al. (2004)
Corn DDGS	0.74	0.27	0.47	0.45	Lumpkins and Batal (2005)
SBM	0.65	0.37	0.28	0.54	Dilger and Adeola (2006)
Canola meal	0.97	0.69	0.28	0.46	Mutucumarana et al. (2014)

[1]Calculated as the difference between total P and phytate P.
[2]Determined in vivo using the P balance or slope ratio techniques.
[3]Values from NRC (1994).
[4]Represents the mean value of phytate P contents presented in Table 5.

accurate available P values and less excessive safety margins in poultry diet formulations, which would minimize P excretion into the environment. Currently, overfeeding of dietary P is a common practice, with excesses of 20–100% over the perceived formulation safety margins needed for P (i.e. 10%) (Powers and Angel, 2008; Applegate and Angel, 2014).

2.3 Future research

1. The effect of fibre-degrading enzymes, proteases and amylases on the efficacy of P liberation in different compartments of the gastrointestinal tract.
2. Validation of the current requirements for digestible P content of poultry diets.
3. Comparison and validation of the methods used for total and phytate P analysis and NPP measurements.
4. Development of simple methods and/or prediction equations for digestible P contents of feed ingredients.

3 Non-starch polysaccharides (NSP) and NSP enzymes: physiological effects, multi-carbohydrase enzymes and prebiotic potential

Non-starch polysaccharides are the major components of dietary fibre in traditional ingredients used in poultry diets, and comprise cellulose and non-cellulosic polysaccharides; the latter often being referred to as hemicelluloses and pectic substances. Such classification is based on their solubility in water and alkaline solutions used in the earlier methods of analysis. In cereal grains, non-cellulosic polysaccharides consist mainly of arabinoxylans and β-glucans, whereas arabinans, arabinogalactans, galactans, galactomannans, mannans and pectic polysaccharides predominate in soya bean meal, canola meal, peas, lupins or flax. The latter fraction has been considered to include diverse types of polysaccharide structures based on linked units of rhamnose, galacturonic acid, galactose and arabinose (i.e. rhamnogalacturonans, galacturonans, arabinans).

In general, NSP are structural polysaccharides and are important components of cell walls. Some structural polysaccharides, however, especially rhamnogalacturonans and highly branched arabonogalactans, can be extended to include several gums and mucilages. It has been suggested that the formation of mucilages and gums involves the apposition of additional sugar residues to the outer chains of polysaccharides already present in the plant. Furthermore, flax mucilage consists of two types of polysaccharides, a neutral arabinoxylan and an acidic pectic-like polysaccharide containing rhamnose, galactose and galacturonic acid residues (Cui et al., 1994).

Non-starch polysaccharides are also part of carbohydrate–protein complexes, including cell wall hydroxyproline-rich glycoproteins, and arabinogalactan proteins. These cell wall proteins are particularly rich in hydroxyproline, while most of the remaining amino acids would include serine, histidine, lysine, tyrosine and valine. In general, carbohydrates account for approximately 40–60% of the weight of all cell wall glycoproteins, with arabinose and galactose being the major carbohydrate constituents and accounting for approximately 90 and 5%, respectively, of the total carbohydrate moieties (Showalter and Varner, 1989). It is of interest to note that the protein component in such glycoproteins is resistant to proteolysis because of its substitution with these 'bulky' carbohydrate groups. In corn grain coats, hydroxyproline is relatively abundant, while in soya bean it could account for 1.6% of the outer layer of the seed coat. In earlier research from our laboratory, structural proteins associated with the cell walls of canola meal accounted for approximately 10% of the total protein content of the meal, and as such should be considered to be highly indigestible.

Maillard reaction products, which are formed during heat treatment of feedstuffs, also represent protein-like components that are resistant to the digestive enzymes of monogastric animals. The Maillard reaction is a heat-catalysed glycation associated with the covalent attachment of reducing sugars to α- or ε-NH_2 groups of amino acids and protein to form glycated proteins. The first glycation product, a glycosylamine or Schiff base, rearranges to a more stable ketoamine or Amadori product, which can then form cross-links with other amino groups. The resulting polymeric aggregates are called advanced glycation end products and are responsible for the brown colour and (possibly) aroma and flavour typical of heated feedstuffs. The end products of the Maillard reaction often resemble lignin polymers and as such become an integral part of dietary fibre and are determined by using the methods of modern fibre analysis (Freedman, 1996).

Dietary fibre represents a range of NSP present in feed ingredients together with lignin, cell wall proteins and Maillard reaction products. Other non-digestible substances, which may be included in a broader definition of the term fibre, are polyphenols (i.e. tannins), resistant starch and oligosaccharides (Trowell et al., 1976; Codex 2009).

Non-starch polysaccharide analysis is based on component sugar (i.e. rhamnose, xylose, arabinose, mannose, galactose and glucose) measurement by gas–liquid chromatography and colorimetry (uronic acids) following their hydrolysis with acids (Englyst and Cummings, 1984, 1988). It is a very laborious and time-consuming methodology and requires skilful laboratory personnel to perform the assay. However, when different sample preparation approaches are employed, the method could be very useful for the determination of a variety of NSP fractions, including cellulose and non-cellulosic polysaccharides,

Table 8 Non-starch polysaccharide (NSP) contents of feed ingredients (%, 8% moisture basis)

Ingredient	*n*	Rhamnose	Arabinose	Xylose	Mannose	Galactose	Glucose	Uronic acids	Total NSP
Corn	7	0.00	1.65	2.30	0.14	0.47	2.69	0.62	7.88
DDGS, corn	20	0.00	4.90	7.56	1.14	1.14	6.79	1.40	22.94
Wheat	37	0.00	2.17	3.67	0.21	0.27	3.16	0.39	9.86
DDGS, wheat	14	0.00	5.12	7.81	1.27	0.87	6.35	1.02	22.45
Wheat shorts	2	0.00	8.28	13.91	0.38	0.81	10.48	1.56	35.42
Rye	3	0.00	3.09	5.21	0.50	0.42	4.43	0.47	14.12
Barley	1	0.00	2.80	5.60	0.40	0.30	8.70	0.60	18.40
Soya bean meal	23	0.13	2.57	1.06	0.68	4.80	3.45	3.28	15.97
Canola meal	114	0.22	4.48	1.92	0.52	1.64	6.51	6.05	21.33
Peas	8	0.04	2.82	1.17	0.09	0.56	5.82	2.59	13.09
Lupin	33	0.05	3.83	3.54	0.59	7.03	12.38	3.97	31.39
Faba bean	25	0.06	2.19	1.48	0.07	0.49	8.40	3.53	16.22
Rice bran	7	0.00	4.19	4.88	0.44	0.98	7.74	1.95	20.18
Palm kernel	6	0.00	1.10	6.58	34.60	2.09	10.64	1.65	56.65
Copra meal	3	0.00	1.13	0.65	26.10	2.87	5.29	0.76	36.80

Source: Rogiewicz and Slominski, 2016.

water-soluble and water-insoluble NSP, or even starch and starch resistant to digestion. In the enzyme development process, the method can be used to effectively determine the degree of NSP depolymerization since the enzyme hydrolysis products would become soluble in 80% ethanol, the solution that is discarded during sample preparation and thus ethanol-soluble components would not be included in the NSP analysis. Another alternative in enzyme evaluation studies could be a colorimetric method of water-soluble carbohydrate analysis following removal of simple sugars, di- and oligosaccharides, and low-molecular weight polysaccharides, including those resulting from enzyme hydrolysis, with 80% ethanol (Slominski et al., 1993). The neutral detergent fibre (NDF) methodology (Goering and Van Soest, 1970) could also be considered for such studies. However, NDF values tend to underestimate important NSP components of feed ingredients due to their partial solubility within solutions used for NDF analysis. Such NSP components would represent the potential targets for enzyme use and, therefore, care must be taken in any data interpretation from the enzyme incubation studies using NDF as an assay methodology.

The NSP content varies not only between different feedstuffs, but also within the same feedstuff due to differences in growing conditions, variety, analytical techniques or, as may be the case for soya bean meal, the degree of hull removal. Typical NSP and their constituent sugar contents of common feedstuffs are summarized in Table 8. The various types of polysaccharides may be rationalized from the component sugar profile. In cereal grains arabinoxylans predominate (Henry, 1987), although significant amounts of β-glucans and cellulose are also present. The relatively high concentration of uronic acid, along with glucose residues, indicates that pectic-type substances and cellulose are the major cell wall constituents of vegetable proteins (Siddiqui and Wood, 1977; Bacic et al., 1988; Bach Knudsen, 1997). It would appear that the bulk of arabinose and galactose not associated with pectic substances derives from arabinan and arabinogalactan, while xylose indicates the presence of xylan and xyloglucan. Although the structural features of polysaccharides containing mannose units are not fully elucidated, it seems probable that these polysaccharides are structurally similar to galactommannans, which have been shown to be widespread in the seeds of leguminous plants (Aspinal, 1982). The main NSP structures, discussed above, are listed in Fig 2.

It should be noted that the structural features of polysaccharides can be somewhat oversimplified by omitting interpolymeric linkages within the cell wall. Specifically, cellulose fibrils make up an important part of the framework of the cell walls and contain about 100 cellulose molecules which are bound together by a matrix of other polymeric polysaccharides. The cell wall then can be compared to cases of reinforced concrete, in which the cellulose fibrils correspond to the steel rods and the matrix material to the concrete.

Although the structures of individual polysaccharide polymers are reasonably well elucidated, little is known about the attachments and linkages between such polymers. The manner in which they are interlinked would determine the properties of the cell walls. Such properties and susceptibility of the cell wall structure to enzymatic hydrolysis in different tissues or species would be affected by the type of covalent, ionic or hydrogen bonding within the cell walls. Therefore, the properties of cell walls cannot be deducted from the properties of the component polymers but rather would depend on how these polymers are interlinked and arranged in space to form the three-dimensional structures of the intact cell wall (Jarvis, 2011).

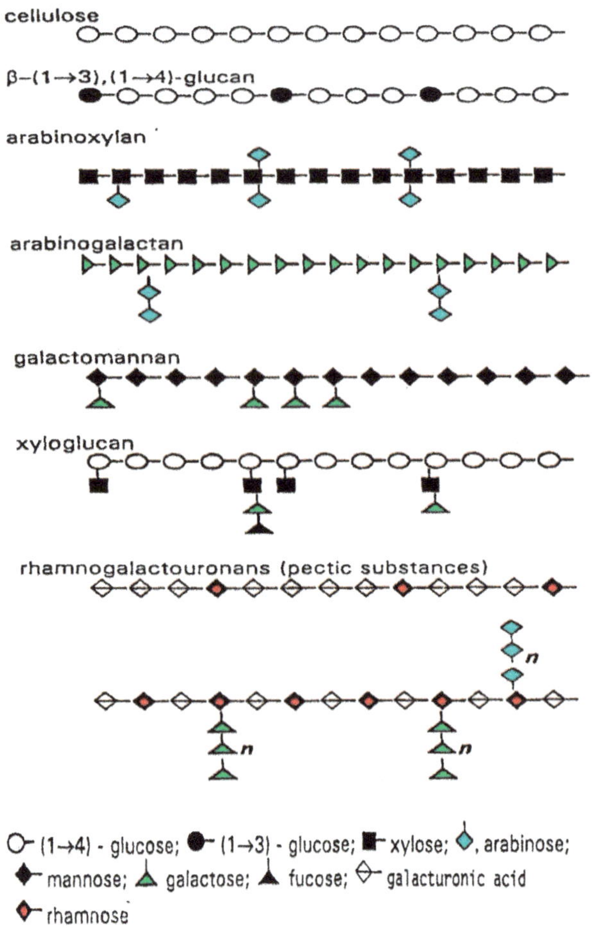

Figure 2 Non-starch polysaccharide structures commonly found in feed ingredients of plant origin (Smits and Annison, 1996). © World's Poultry Science Association, *World's Poultry Science Journal*, Vol. 52, July 1996, C. H. M. Smits and G. Annison.

3.1 Physiological effects of NSP

Non-starch polysaccharides of feed ingredients are poorly utilized by poultry, can pose a viscosity problem, encapsulate nutrients and thus affect nutrient absorption. However, dietary NSP also interact with gut microflora which could be considered beneficial from the point of view of gut health. The water-soluble and viscous β-glucans and arabinoxylans present in barley, rye and wheat interfere with the mixing of digestive enzymes and nutrients and impede digesta movement and transport of hydrolysis products to the intestinal mucosa. As a result, these effects may cause a decrease in animal performance (Graham and Aman, 1991). In addition, management problems related to 'sticky droppings' have been indicated to be directly associated with the high water-holding capacity of β-glucans and arabinoxylans. To counteract such anti-nutritional effects, many commercial preparations of β-glucanase and xylanase have been developed over the past 30 years. In addition to viscosity reduction, the use of effective combinations of NSP-degrading enzymes could reduce the nutrient encapsulating effect of cell walls which, in turn, could contribute to increased protein, starch and fat utilization, as demonstrated by studies on wheat (Tervilä-Wilo et al., 1996; Bedford and Autio, 1996), full-fat oilseeds (Meng et al., 2006; Slominski et al., 2006) or expelled full-fat soya beans (Ayoade et al., 2012).

3.2 Multi-carbohydrase enzymes

For some applications, the conventional approach of eliminating the viscous properties of arabinoxylan, and beta-glucan of wheat, barley or rye with single-enzyme preparations of xylanase or glucanase has proven to be sufficient. This is correct in situations when the grain components of animal diets exert their anti-nutritive effect by increasing gut viscosity and thus interfering with nutrient digestion. This is, however, not always the case in modern animal diets containing corn or wheat not necessarily showing any negative effects due to viscous polysaccharides, as opposed to earlier studies using high-viscosity Australian or European wheats. Only limited studies have been conducted in North America on a certain type of high-viscosity wheat (i.e. Prairie spring wheat) in which the application of single xylanase preparation has proven to be beneficial. However, these were isolated situations more so to facilitate research on the mode of action of xylanase rather than to address the overall need for this single-enzyme supplement. With modern enzyme preparations, barley would not be an exception, since multi-carbohydrase preparations would always contain sufficient amounts of beta-glucanase to eliminate any potential anti-nutritive effects of barley beta-glucans.

Over the past ten years, there has been a debate within the scientific community as to the use of single- versus multi-enzyme preparations. For some, the use of multi-carbohydrase preparations is a logical approach considering the complexity of NSP, and their interlinking within the cell wall structure. In addition, the complexity of NSP of many protein supplements, including soya bean meal, canola meal, peas or lupins is far greater than that of cereal grains and thus would require more diversified enzyme preparations for effective NSP depolymerization. This is because the targeting of a number of linkages within the cell wall structure would require many activities acting in concert as opposed to a single enzyme, the application of which would result in minimal NSP depolymerization, and only on the surface of the cell wall. The best example of this would be a study involving the effect of different enzyme activities on the degradation of NSP of common feed ingredients (Meng et al., 2005). In this study, a very careful in vitro evaluation of single enzymes and enzyme combinations was used to develop a multi-carbohydrase preparation. As a consequence, only the most complex enzyme combinations, containing cellulase, pectinase, galactanase, mannanase, xylanase and glucanase activities, were effective in NSP depolymerization of wheat, corn, SBM, canola meal and peas.

It is generally agreed that the amount of energy derived from NSP hydrolysis, either from small amounts of simple sugars absorbed directly from the small intestine, or the conversion of NSP hydrolysis products, including oligosaccharides and low-molecular-weight polysaccharides, to SCFA in the lower gut may not be that high. Based on some studies, dietary enzyme supplementation may account for ~50–100 kcal/kg of an extra energy generated. This is understandable considering a short digesta transit time and pH conditions in the chicken GI tract unfavourable for most feed enzymes. Consequently, depending on the diet composition and the content of water-soluble NSP, limited NSP depolymerization would occur, with the determined NSP digestibility values ranging from 12 to 30% following enzyme supplementation, compared with that of ~5–8%, often observed for birds fed non-enzyme (i.e. control) supplemented diets. This could be further substantiated by the fact that NSP enzymes would only be effective on the NSP present in the starchy endosperm and aleurone layer of cereal grains or in the embryos of oilseeds, as opposed to the hull or coat fractions of such ingredients, due to their extremely rigid structure and association with other fibre components, including polyphenols and lignin.

In many cases, the application of multi-carbohydrase enzymes would lead to increased fat, protein and phytate P utilization as a result of their release for digestion following enzymatic breakdown of cell wall material. Many studies have recently been published to support this concept, the best example being studies focused on improving the utilization of fat and omega-3 fatty acids from canola seed, flaxseed or *Camelina* seed. In such studies, the multi-carbohydrase preparations were very effective in elimination of the physical barrier of the cell walls and thus facilitated the access of endogenous digestive enzymes (i.e. lipase) to the substrate, which resulted in improved oil (and thus energy) and omega-3 fatty acids deposition (Boros et al., 2004; Meng et al., 2006; Slominski et al., 2006; Jia et al., 2008; Jozefiak et al., 2010; Jankowski et al., 2015; Woyengo et al., 2016). The use of multi-carbohydrase enzyme preparations was also effective at improving growth performance, which almost entirely resulted from the elimination of the cell wall barrier and thus improved dietary energy utilization. On the same note, the use of multi-carbohydrase enzyme preparations in concert with phytase have been shown to facilitate the release and utilization of P, most likely due to dissociation of the phytate molecule from the fibre fraction of the diet (Woyengo et al., 2010).

Several studies have recently demonstrated the beneficial effects of multi-carbohydrase enzymes in many other applications, including extractability of protein from rapeseed meal (Rommi et al., 2014), improving the nutritive value of distillers dried grains with solubles (DDGS) during the fermentation process (Jakobsen et al., 2015), identifying the multi-enzyme preparation with numerous NSP-degrading enzymes for soya bean (Vahjen et al., 2005), improving the nutritive value of rice bran, an ingredient known to be very difficult to target by the conventional xylanase supplements (Liu et al., 2015), or demonstrating the positive effects of the multi-carbohydrase preparations on growth performance of broiler chickens in the practical feeding trials (Avila et al., 2012).

3.3 Prebiotic effect of NSP hydrolysis products

Carbohydrase enzymes have a direct, positive effect on animal performance by improving nutrient digestion and absorption, thereby reducing substrate availability for microbial growth in the ileum (Choct et al., 1999; Bedford and Apajalahti, 2001; Slominski 2011). It is well known that any change to the diet composition will lead to a shift in the profile of the intestinal bacterial community (Apajalahti et al., 2004). In the process of depolymerizing various polysaccharides in the diet, carbohydrase enzymes may produce gluco-, xylo-, manno- and galacto-oligosaccharides (Silva et al., 1983) and low-molecular-weight polysaccharides (Jia et al., 2009a). These in a manner similar to prebiotics, may facilitate the proliferation of beneficial bacteria for gut health such as *Bifidobacterium* and *Lactobacillus*, thereby reducing the abundance of pathogens such as *Clostridium* sp., *Salmonella* sp., *Escherichia coli* sp. and *Campylobacter* sp. (Gibson and Roberfroid, 1995). In this context, these enzyme hydrolysis products may indirectly prohibit the growth of certain pathogenic species by increasing intestinal lumen acidity through an increase in lactic acid production in the lower gut. Concomitantly, non-substrate utilizers, in a highly competitive ecosystem, will be suppressed and can virtually disappear. In this context, the use of lactic acid bacterial cultures such as *Lactobacillus acidophilus* and *Streptococcus faecalis* has shown promising results in suppressing *C. perfringens* proliferation (Fukata et al., 1991) and reducing *C. perfringens*-associated mortality (Hofacre et al., 2003). In addition, certain enzyme hydrolysis products may attract microbes away from the intestinal binding sites by a means of competitive exclusion, thereby reducing colonization and disease and allowing the mucosa to perform its function of secretion, digestion and nutrient absorption.

From the information listed above, it would appear that when NSP are broken down by a blend of carbohydrase enzymes, they acquire the potential to become prebiotics and can, in turn, exert health benefits by improving the intestinal environment. It has been demonstrated that the use of a multi-carbohydrase preparation containing pectinase, cellulase, mannanase, galactanase, galactosidase, xylanase, glucanase and other enzyme activities was effective in promoting growth and feed utilization of broiler chickens fed an antibiotic-free diet and facilitated post-disease compensatory growth of broiler chickens challenged with *C. perfringens*, a causative agent of necrotic enteritis (Jia et al., 2009b). In this study, enzyme supplementation was accompanied by a 1.3 log reduction in *C. perfringens* counts (from 4.3 to 3.0 $\log_{10}$ CFU/g).

Furthermore, when using an advanced 'in situ' experimental model, the infusion of NSP hydrolysis products into living piglet intestinal segments that were experimentally infected with *E. coli* K88 demonstrated that those segments infused with enzyme hydrolysis products had greater fluid absorption than control segments (Kiarie et al., 2008), which under practical conditions could lead to reduced scours and improved recovery post infection. Several studies have recently demonstrated the prebiotic effects of carbohydrase enzyme supplements on gut microbiota, including cecal fermentation of enzymatically degraded products into short-chain fatty acids, particularly butyrate, reduction of *E. coli* numbers in the cecum, or increased glycolytic activities of the intestinal microflora enzymes, α-glucosidase, α-galactosidase and β-galactosidase (Rosin et al., 2007; Zdunczyk et al., 2013; Yacoubi et al., 2015, Khadem et al., 2016).

Thus, it would appear from this research that the benefits to be gained from enzyme supplementation are not only from improved nutrient digestion and feed efficiency, but also from improved gut health and mitigation of enteric infections as a result of prebiotics formed from the hydrolysis of NSP of common feedstuffs.

In conclusion, single xylanase and glucanase enzymes have been successfully applied in poultry programmes in parts of the world where cereal ingredients such as wheat, barley or rye predominate in poultry diets. However, when the same enzyme activities are applied to diets containing corn, SBM, rapeseed/canola meal or other high-fibre ingredients, including expelled meals from oilseeds, or DDGS, with different NSP profiles, no performance responses have been noted. The usefulness of other enzyme preparations to target a variety of indigestible dietary components, however, has often been variable and the results are sometimes disappointing. This is mainly due to the fact that many of the enzymes currently used in poultry feeds have been adopted from the food processing, detergent or textile industries and may not have the desired substrate specificity. Development and evolution of 'tailor-made' enzymes or enzyme mixtures to enable more effective utilization of complex diets by poultry would be valuable. When the constituent NSP of such ingredients are analysed, it becomes clear that an extensive blend of carbohydrases must be supplemented if any significant performance responses are to be achieved. This is likely why minimal performance improvements have been reported for such poultry diets supplemented solely with xylanase, β-glucanase or their combinations. Thus, in order to achieve viable and consistent economic returns in commercial poultry feeding programmes, the correct blends of multiple carbohydrases, including cellulases, pectinases, xylanases, glucanases, mannanases and galactanases must be further improved and/or developed. As per my estimate, over 80% of the publications ever published refer to the effect of the commercial enzyme preparations on growth performance and nutrient digestibility in poultry. Very rarely, the objectives of such studies are to explain the mode of action so as to make recommendations for any future steps in improving the enzyme effectiveness.

3.4 Future research

1. Identification of indigestible components of dietary fibre in plant protein supplements and co-products of their processing as they become more prevalent in poultry diets.
2. Development of more effective enzyme combination to target a variety of indigestible components, including not only NSP but also carbohydrate–protein complexes.
3. New enzyme evaluation studies to determine enzyme efficacy in vivo, mechanism of action and their beneficial effect on nutrient utilization and gut microbiota function.

4 β-Mannanase in poultry nutrition

Beta-mannanase (Mannan *endo*-1,4-β-mannosidase, EC 3.2.1.78) is a very intriguing enzyme and is believed to be in a different category since its proposed mode of action is different than typical dietary carbohydrase enzymes. In addition, the mode of action is determined by the types of β-mannans and/or β-galactomannans present in feed ingredients. As recently reviewed (Shastak et al., 2015), β-mannans are predominant NSP of palm kernel meal, copra meal, sesame meal and guar meal. They are also found, although at much lower concentrations, in SBM, rapeseed/canola meal, cereal grains or DDGS. Palm kernel meal, copra meal and guar meal contain 37, 26 and 9% of mannans, respectively. Soya bean meal contains 1.6%, canola meal 0.5%, while wheat or corn only 0.1%. As a co-product of brewer's yeast (*Saccharomyces cerevisiae*) fermentation, DDGS contain 1.3% of mannans. As mannose, the component sugar of NSP, is present in corn or wheat grain in trace amounts, it has been calculated that DDGS would contain ~6% of yeast biomass (Alizadeh et al., 2016).

It must be emphasized that yeast mannans, and thus those present in DDGS, consist mainly of α-1,6-linked mannose units and are different from plant β-1,4-mannans. Furthermore, locust bean mannan could be different from that of guar gum, and both are different from that of SBM due to the extent of β-mannan backbone substitution by galactose units joined by an α-1,6 glycosidic linkage, and thus the mannose to galactose ratio would ultimately be different. As a result, the mannan contents are often determined on the basis of the amounts of mannose, and could be higher if the galactose, and possibly other component sugar residues (i.e. glucose), were taken into account.

Due to differences in β-mannan and/or β-galactomannan structures, and different specificity of β-mannanase preparations to such structures, different degrees of hydrolysis and the production of different hydrolysis products, including mannobiose, mannotriose,

Table 9 β-Mannanase activity as determined by using different substrates (U/g[1])

β-Mannanase	Locust bean gum	Guar gum	Carob gum
A	54,618	1,782	17,042
B	27,576	2,076	14,143
C	17,922	1,776	33
D	138	21	–
E	59	30	12

[1]One unit of activity is defined as the amount of enzyme required to release one micromole of mannose reducing sugar per minute under the defined assay conditions.

manno-oligosaccharides with different degree of polymerization, and possibly free mannose and galactose, would be expected. Different substrate specificities as determined by using several microbial β-mannanase are illustrated in Table 9. The enzyme preparations evaluated in this study (Rogiewicz et al., 2016) represent the most common β-mannanase preparations currently recommended to the feed industry. It is of interest to note that the highest enzyme activity was achieved with locust bean galactomannan. All enzyme preparations displayed activity, in decreasing order, with carob galactomannan and guar galactomannan.

Several mechanisms of action have been proposed for the positive responses observed to date from dietary β-mannanase supplementation. These include (1) elimination of negative effects associated with viscous properties of guar and copra meals, (2) release of energy from mannan hydrolysis products, (3) elimination of the negative effect of β-mannans on glucose absorption and insulin production, (4) suppression of the pathogenic bacteria proliferation by β-mannan hydrolysis products, or (5) prevention of energy partitioned away from lean tissue growth due to the elimination of a β-mannan-induced innate immune response.

Guar or copra mannans are highly viscus, and like β-glucans and arabinoxylans of cereal grains (see NSP section in this chapter) would interfere with the mixing of digestive enzymes and nutrients, transport of hydrolysis products to the intestinal mucosa with a resultant decrease in nutrient utilization. Numerous studies on the elimination of the adverse effects of poultry diets containing guar meal, copra meal and palm kernel meal by β-mannanase supplementation have been published (Ellis et al., 1995; Kamran et al., 2002; Lee et al., 2003, 2005; Daskiran et al., 2004; Sundu et al., 2006, 2008; Hussain et al., 2012; Utami et al., 2013; Ibuki et al., 2014; Shastak et al., 2015).

4.1 Soya bean galactomannan and β-mannanase

The β-mannan content of SBM is very low and ranges from 1.0 to 1.5% for dehulled and from 1.3 to 2.1% for non-dehulled samples with an estimated average β-galactomannan content of 1.26 and 1.61%, respectively, when using the determined mannose values and the galactose:mannose ratio of 1:1.8 (Hsiao et al., 2006). The difference between the two types of SBM is a clear indication that soya bean galactomannan is a constituent of the cell walls with a higher content in the hull fraction. From this, it could be assumed that its water solubility would be very low, and according to our recent study, water-soluble β-galactomannan content of SBM averages only 0.096% and, as a result, would account for only 10% of the total β-galactomannan content of SBM.

As opposed to the high amounts of β-mannans present in guar or copra meals, this small amount is not likely to contribute to any increased intestinal viscosity in poultry fed corn/SBM-based diets, and one would not expect any impact of β-mannanase supplementation on intestinal viscosity. This has been confirmed in two recent studies with broiler chickens fed with corn/SBM-based diets without and with β-mannanase supplementation. In one study, the viscosity values for the control and the mannanase-supplemented diets were 1.24 versus 1.24 cP, respectively (Latham et al., 2016), while in another study, the corresponding values for the control and the enzyme-supplemented diets averaged only 2.13 and 1.96 cP (Mehri et al., 2010).

From aforementioned studies, it is reasonable to assume that any degradation of such small quantities of substrate by β-mannanase supplementation would not produce as much of an improvement in energy utilization and growth performance as those observed in other studies. This could be further substantiated by the results of an in vitro incubation study conducted in our laboratory when using different β-mannanase preparations. As illustrated in Table 10, a minimal degree of SBM β-galactomannan depolymerization, as evidenced by mannose disappearance, was observed for both SBM and its water-soluble NSP fraction. In this context, even the multi-carbohydrase preparation was not very effective on soya bean β-mannans, although a significant depolymerization of arabinogalactan, the major NSP of SBM, was evident. It is of interest to note that the specific activity of different β-mannanases towards the water-soluble β-mannans of SBM was low, suggesting that even in the soluble form, they are associated with the other polysaccharide polymers.

Numerous studies on the effect of β-mannanase supplementation on growth performance and nutrient utilization in poultry fed corn/SBM-based diets have been conducted, demonstrating some positive effects (Jackson et al., 2004; Cho and Kim, 2013; Williams et al., 2014; Latham et al., 2016), moderate effects, supported by numerical differences between the treatments (Odetallah et al., 2002; Kong et al., 2011; Wu et al., 2005; Mussini et al., 2011), or no effects (Ouhida et al., 2002; Vahjen et al., 2005; Sornlake et al., 2013). Therefore, it is very difficult to pinpoint the exact mechanism that may explain the positive effects associated with dietary β-mannanase supplementation.

It has been proposed that SBM β-galactomannans may trigger the innate immune response by the animal when it detects a pathogen-associated molecular pattern analog, which would induce a metabolically costly stimulation of the innate immunity. As a result, such a complex response may lead to unnecessary losses in dietary energy utilization, which would be minimized by β-mannanase supplementation, allowing the nutrients to be redirected towards optimum performance (Li et al., 2010; Shastak et al., 2015). In fact, the lower serum IgG and IgM concentration by β-mannanase supplementation to the corn/SBM-based diets was observed in one study (Li et al., 2010). However, in the study by Zou et al. (2006), IgA and IgG were not different while the increase in IgM was observed, indicating immune system stimulation following mannanase supplementation.

Another possible mode of action could be that β-mannanase supplementation may release manno-oligosaccharides, which like the yeast bioactives, may reduce the adhesion of pathogenic bacteria to intestine epithelial cells and thus decrease their colonization while, at the same time, promoting the growth of *Bifidobacterium* and *Lactobacillus*. The latter concept has been discussed in an excellent review article by Shastak et al. (2015).

Table 10 Non-starch polysaccharides (NSP) depolymerization following incubation of soya bean meal (SBM) and SBM water-soluble NSP isolate (SBMI) following incubation with different β-mannanase preparations (A to E)

	NSP component sugar (mg/g)							
Substrate/enzyme	Rhamnose	Arabinose	Xylose	Mannose	Galactose	Glucose	Uronic acids	Total NSP
SBM								
Control	2.5	23.5	10.2	6.9	41.7	40.5	33.2	158.5
β-Mannanase								
A	2.8	23.0	10.1	6.1	42.4	39.2	35.1	158.7
B	2.7	21.9	9.9	5.5	40.0	36.8	34.3	151.0
C	3.0	23.7	10.4	5.9	42.0	45.9	32.6	163.5
D	2.7	23.5	11.0	6.6	41.4	40.9	35.1	161.1
E	2.8	23.7	10.7	6.3	42.2	39.5	33.7	158.8
SBMI								
Control	1.6	12.5	1.8	12.3	27.7	6.8	19.0	81.8
β-Mannanase								
A	1.3	10.7	1.6	10.0	24.3	6.5	19.6	74.0
B	1.2	8.1	1.6	8.4	18.9	5.8	19.3	63.3
C	1.4	13.6	1.7	9.6	28.7	6.0	19.2	80.2
D	1.2	10.6	2.2	10.6	24.8	5.7	25.3	80.3
E	1.2	9.9	2.2	10.6	25.9	5.4	25.7	80.8
Multi-carbohydrase	1.2	4.6	1.9	8.3	13.2	4.8	20.0	53.0

In the light of the small amounts of manno-oligosaccharides being released from β-galactomannans of SBM following β-mannanase supplementation (Table 10), it is difficult to believe that such small amounts could contribute to the energy-sparing effect or suppression of the pathogenic bacteria proliferation. Further research in this area is definitely warranted.

4.2 Future research

1. Identification of the main and side activities of β-mannanase preparations.
2. Validation of the β-mannanase specific activity towards SBM galactomannan.
3. Validation of the energy-sparing effect due to elimination of the mannan-induced innate immune response.

5 Starch digestion and supplemental α-amylase

As starch is the most important dietary energy source, any variability in its digestion would significantly affect animal performance. As reviewed by Classen (1996), a strong negative relationship between starch digestibility and AME content exists.

The applications of normal digestive tract enzymes as feed supplements have been driven by the idea that newly hatched chicks may be deficient in key digestive enzymes, including α-amylase. This has been corroborated by Noy and Sklan (1995), who showed that specific activity of α-amylase rapidly increases with age up to 2–3 weeks post-hatch. The secretion of α-amylase observed in this study, however, was parallel to the increase in feed consumption and there was no difference in starch digestion from 4 to 21 days of age. However, the authors found starch digestion to be slightly lower at 85% compared with 95% or higher observed earlier (Riesenfeld et al., 1980; McNab, 1993; Uni et al., 1995), and in some more recent studies (Weurding et al., 2001; Gracia et al., 2003; Boros et al., 2004; Meng and Slominski, 2005; Kaczmarek et al., 2014).

Some differences in starch digestibility exist, but they would most likely be related to the nature of starch from source to source (i.e. peas, common beans, potatoes, soft vs. hard wheat) rather than endogenous α-amylase deficiency. It is of interest to note that no difference between ileal and total tract starch digestibility has been observed (Table 11). Therefore, it could be hypothesized that the small fraction of undigested starch is not fermented to any significant extent in the hindgut of the chicken, which is consistent with the data of Kussaibati et al. (1982), who reported similar amounts of undigested starch in conventional and germ-free birds. Thus, it could be concluded that poultry species have the inherent ability to release sufficient levels of starch-hydrolysing enzymes to facilitate almost complete starch digestion (Moran, 1982). Consequently, supplementation of diets with α-amylase would result in little or no response, which is consistent with the data sets provided in Table 12. As the utilization of the small fraction of undigested starch could be of concern, this would most likely be achieved by using multi-carbohydrase supplements (Boros et al., 2004). Finally, one important consideration worthy of investigation could be if supplemental α-amylase, along with protease, would speed up the softening, disintegration and solubilization of the feed matrix in the crop, which, in turn, could facilitate the improvements in phytase or carbohydrase efficiencies due to improved substrate (i.e., phytate, NSP) accessibility.

Table 11 Ileal and total tract digestibility of starch in broiler chickens

Diet type	Age	Starch digestibility (%)		Reference
		Ileal	Total tract	
Wheat–SBM	29 d	94.4	93.8	Weurding et al. (2001)
Corn–SBM		97.0	97.4	
Barley–SBM		98.2	98.3	
Sorghum–SBM		95.3	95.4	
Corn–Peas–SBM		80.4	81.0	
Corn–SBM	7 d	–	93.5	Gracia et al. (2003)
	28 d	96.4	96.2	
Corn–SBM	14 d	96.8	97.8	Kaczmarek et al. (2014)

Table 12 Starch digestibility in broiler chickens as affected by α-amylase supplementation

Diet type	Age	Starch digestibility (%)		Reference
		Control	α-Amylase	
Corn–SBM	7 d	93.5[b]	96.2[a]	Gracia et al. (2003)
	28 d	96.2[b]	98.0[a]	
Corn–SBM	21 d	96.0	97.2	Meng and Slominski (2005)
Corn–canola meal		96.0	96.3	
Corn–peas		91.6	92.9	
Corn–SBM	14 d	97.8	97.7	Kaczmarek et al. (2014)
Corn–SBM	22 d	89.6	90.2	Stefanello et al. (2015)

[a,b] $P < 0.05$.

6 Microbial protease supplementation

The indigestible portion of protein present in feedstuffs is difficult to characterize and consequently has not been sufficiently investigated. A review of research data, however, indicates that approximately 20% of the total protein content of feedstuffs is not digested in the small intestine of monogastric animal as determined by studies utilizing terminal ileum collection of digesta. As reviewed by Parsons (1996), the digestibility of lysine, methionine and threonine in a variety of feedstuffs for poultry average 80.1, 86.3 and 80.7%, respectively.

By analogy to resistant starch, different types of proteins that are resistant to digestive enzymes can be considered. It would appear that the encapsulation effect of the cell walls could reduce total protein digestion to an even higher extent than that observed for starch. This could be due to the structure and thickness of the cell walls, the starchy endosperm walls being thinner and most likely less lignified than the protein-containing cells of the aleurone layer of the grain. Another protein-like component that is resistant to digestive enzymes could include carbohydrate–protein complexes, including glycoproteins associated with the cell wall structure, or Maillard reaction products.

It should be noted that exogenous proteases are often fed in combination with other enzymes, including amylase and xylanase (Cowieson and Adeola, 2005; Cowieson et al., 2006; Cowieson and Ravindran, 2008; Olukosi et al., 2010); hence, it is difficult to determine the specific effects resulting from protease addition, and therefore the results are of limited value in explaining the mode of action of supplemental proteases.

Table 13 Growth performance and apparent ileal crude protein (CP) digestibility of broiler chickens fed diets without and with monocomponent protease supplementation

Diet type/ treatment	Trail length (d)	Growth performance: BW gain (g/bird/d)	Growth performance: FCR (g feed/g gain)	CP digestibility (%)	Reference
Corn–SBM	6–42				Ouhida et al. (2002)
Control		53.5	1.73	84.3	
Protease[1]		55.2	1.72	83.8	
Corn–SBM	1–42				Odetallah et al. (2005)
Control		62.0[b]	1.86[a]	–	
Protease[2]		64.7[a]	1.83[b]	–	
Corn–SBM	7–22				Angel et al. (2011)
Control		44.1[b]	1.58[a]	77.9[b]	
Protease[3]		45.5[a,b]	1.54[a,b]	82.6[a]	
Protease[4]		47.2[a]	1.50[b]	82.7[a]	
Wheat–SBM	1–34				Kalmendal and Tauson (2012)
Control		68.2	1.47[a]	59.7[6]	
Protease[5]		67.6	1.43[b]	64.3	
Corn–SBM	1–14				Kaczmarek et al. (2014)
Control		26.0	1.40	80.1	
Protease[7]		25.9	1.40	79.6	
Wheat–SBM[8]	7–21				Opoku et al. (2015a)
Control		35.3	1.50	49.9[6]	
Protease[9]		36.3	1.46	52.1	
Corn–SBM–wheat DDGS[8]	1–72				Opoku et al. (2015b)
Control		88.9	2.08	70.0[b]	
Protease[9]		91.5	2.03	81.9[a]	

[a,b] $P < 0.05$; [1]42000 U/kg; [2]600000 U/kg; [3]7500 U/kg; [4]30000 U/kg; [5]15000 U/kg; [6]represents N retention; [7]4000 U/kg; [8]fed to turkey hens; [9]1100 U/kg.

During the first few days of life, the small intestine develops rapidly and this fast growth is accompanied by an increase in digestive enzyme production (Uni et al., 1995). Although newly hatched chicks may be deficient in key digestive enzymes, Noy and Sklan (1995) demonstrated that specific activity of trypsin rapidly increases with age. Although the secretion of trypsin could be directly related to an increase in feed intake, the authors found that protein digestion was lower in the first seven days of life (i.e. ~80%), after which it reached the plateau of 90%.

As indicated earlier, feed ingredients may contain a variety of complex proteins that may not be digested due to reasons other than the deficiency in endogenous proteases. As emphasized by Angel et al. (2011), many publications exist on enzyme blends, and fewer studies have been conducted by using single proteases. The results of such studies are presented in Table 13. Whereas some data indicate inconsistent or no effect results, other publications show improvements in growth performance and nitrogen utilization with protease supplementation. As is the case with any other enzyme supplements, the use of proteases of different origin, specificity or the inclusion level adds to the variability in the results obtained. Therefore, it would have to be agreed as to what future directions are to be undertaken to further improve protein utilization. Will it be just simple microbial protease supplementation with the variable responses or will it be necessary to identify various protein–carbohydrate complexes, along with the enzymes other than proteases, to facilitate the release of amino acids from such complex polymers?

6.1 Future research

1 Identification of indigestible protein polymers of feed ingredients.
2 Development of effective protease and other enzyme combinations, to target a variety of indigestible carbohydrate–protein complexes.

7 Conclusions

In order to define the available P specifications of feed ingredients and P requirements of different poultry species, and thus discourage the use of excessive safety margins in diet formulations and consequently minimize P excretion into the environment, the development of consensus protocol and/or research approaches is needed. As described in the phytase section of this chapter, it has been recommended that the terminal ileum digestible P measurements be used for future determination of available P content of feed ingredients with phytase supplementation. The application of such an assay would provide for a more accurate assessment of P availability and would contribute to the development of databases for each feed ingredient. However, the question remains how different would the in vivo evaluation outcomes be from those of the NPP measurements. Most likely, the in vivo measurements of available P would be as variable as those determined by chemical analyses resulting from variations in available and total P contents of feed ingredients due to cultivar, location and growing conditions, including differences in P fertilization. Therefore, the application of such determined P digestibility values with full confidence may not be possible in practical diet formulations, unless prediction equations are developed on the basis of simple analytical chemistry measurements. However, it must be emphasized that such NPP measurements may not be that different from those of the available P determined in vivo. This could be due to the fact that although the availability of NPP may not be complete, it could be compensated by partial utilization of phytate P. Thus, the available P values determined in vivo could be similar to those determined by chemical analyses. Therefore, attempts should be made to develop the prediction equations for available P contents with and without phytase supplementation based on a simple analysis of NPP.

In the light of the current shift towards the use of high-fibre co-products, targeting the dietary fibre components of poultry diets with enzyme supplementation becomes a challenge. In addition to cost reduction, utilization of co-products of biofuel and food industries would increase the overall fibre content of poultry diets substantially. As well, the potential increase in dietary fibre content of plant origin would result from limiting the use of rendered animal by-products. Therefore, the use of enzyme technology to enhance nutrient utilization from such NSP-rich plant by-products would require the correct and substrate-specific blends of multiple carbohydrases to effectively target the different NSP fractions of feed ingredients. The presence and content of such fractions would most likely increase with age and the potential substrate shifts in different phases of the feeding programmes would occur. Therefore, it would have to be determined if such changes in diet formulation with bird age and the nature of substrates would require different enzyme activities present within any proposed multi-carbohydrase supplement. In addition, current trends towards diets with no animal by-products, as well as some commercial integrators raising larger, older broilers requiring relatively high levels of digestible amino acid to increase breast yield, could serve as arguments in support of any study on the effects that multi-carbohydrase supplementation has on standardized ileal digestibility (SID) of amino acids. It must be emphasized that even small improvements in SID of lysine, threonine or methionine due to enzyme supplementation could represent a significant economic saving.

Multi-carbohydrase supplementation could also have a profound effect on gut microflora, with the NSP hydrolysis products acting as prebiotics and contributing to the overall gut health. This would appear to be of great importance in feeding programmes without antibiotic growth promoters. In some parts of the world, including North America, the poultry industry is currently grasping the reality and urgency of such programmes and is seeking alternative solutions.

The feed industry has in many instances been faced with enzyme products not necessarily developed for feed application but rather for applications in the pulp and paper, bakery, brewery, beverage, confectionary, or pharmaceutical industries, and more recently in the biofuel industries. With the exception of phytase, mannanase and to some extent xylanase, which have been developed specifically for the feed industry, many other enzyme preparations have been proposed for use in animal feeding. However, their mode of action has often been difficult to explain and their application difficult to rationalize. As a consequence, any steps in improving their efficacy have been difficult to achieve. It seems to be the case for enzyme supplements such as amylase, protease, as well as many carbohydrases, including pectinase, cellulase, galactanase and others. Therefore, more fundamental research is needed to identify the indigestible components of feed ingredients as they relate to such enzyme application so as to demonstrate the benefits from their improved utilization. This would require standardization of assay methodologies for specific physiological needs of the bird and should become an integral component of the enzyme validation and application process.

8 Where to look for further information

Although the most recent information on exogenous enzyme research and application has been included in this chapter, the reader is encouraged to revisit the cited publications as well as to explore the topic further in the upcoming research papers published by the peer-reviewed journals, including *Poultry Science, British Poultry Science, Animal Feed Science and Technology, Journal of Agricultural and Food Chemistry*, and others. Many review articles published in the *World's Poultry Science Journal* could serve as good sources of information for graduate students already involved or currently initiating their research programs on exogenous enzymes. Conferences organized by the *Poultry Science* and *World's Poultry Science* Associations around the world could serve as good venues for new information and discussions on the current trends in enzyme development, their mechanism of action and subsequent application.

9 References

Adedokun, S. A., and O. Adeola. 2013. Calcium and phosphorus digestibility: Metabolic limits. *J. Appl. Poult. Res.* 22: 600–8.

Adewole, D. I., A. Rogiewicz, B. Dyck and B. A. Slominski. 2016. Chemical and nutritive characteristics of canola meal from Canadian crushing plants. *Anim. Feed Sci. Technol.* (submitted).

Ahmadi, H., and M. Rodehutscord. 2012. A meta-analysis of responses to dietary nonphytate phosphorus and phytase in laying hens. *Poult. Sci.* 91(8): 2072–8.

Ahn, D. J., J. G. Won, C. M. Rico and S. C. Lee. 2010. Influence of variety, location, growing year, and storage on the total phosphorus, phytate-phosphorus, and phytate-phosphorus to total phosphorus ratio in rice. *J. Agric. Food Chem.* 58: 3008–11.

Alizadeh, M., J. C. Rodriguez-Lecompte, H. Echeverry, G. H. Crow and B. A. Slominski. 2016. Effect of yeast-derived products and distillers dried grains with solubles (DDGS) on antibody-mediated immune response and gene expression of pattern recognition receptors and cytokines in broiler chickens immunized with T-cell dependent antigens. *Poult. Sci.* 95: 507–17.

Angel, R., T. J. Applegate, W. W. Saylor, J. T. Sims, R. Maguire and W. Powers. 2002. Effect of dietary inclusion level of nonphytate phosphorus (nPP) and phytase on total and soluble phosphorus (P) in poult excreta. *Poult. Sci.* 81(Suppl. 1): 92.

Angel, C. R., W. J. Powers, T. J. Applegate, N. M. Tamim and M. C. Christman. 2005. Influence of phytase on water-soluble phosphorus in poultry and swine manure. *J. Environ. Qual.* 34: 563–71.

Angel, R., W. W. Saylor, A. D. Mitchell, W. Powers and T. J. Applegate. 2006. Effect of dietary phosphorus, phytase, and 25-hydroxycholecalciferol on broiler chicken bone mineralization, litter phosphorus, and processing yields. *Poult. Sci.* 85: 1200–11.

Angel, C. R., W. Saylor, S. L. Vieria and N. Ward. 2011. Effects of a monocomponent protease on performance and protein utilization in 7- to 22-day-old broiler chickens. *Poult. Sci.* 90: 2281–6.

Angel R., and C. M. Ashwell. 2012. Epigenetic impacts associated with early phosphorus (P) conditioning on broiler performance. Proceedings of the WPSC, Salvador, Brozil, 5–9 August, pp. 360–4.

Angel, R., and J. O. B. Sorbara. 2014. Why is it important to understand substrates if we are to optimize exogenous enzyme efficacy? *Poult. Sci.* 93: 2375–9.

Apajalahti, J., A. Kettunen and H. Graham. 2004. Characteristics of the gastrointestinal microbial communities, with special reference to the chicken. *World's Poult. Sci. J.* 60: 223–32.

Applegate, T. J., B. C. Joern, D. L. Nussbaum-Wagler and R. Angel. 2003. Water-soluble phosphorus in fresh broiler litter is dependent upon phosphorus concentration fed but not on fungal phytase supplementation. *Poult. Sci.* 82: 1024–9.

Applegate, T. J., and R. Angel. 2014. Nutrient requirements of poultry publication: History and need for an update. *J. Appl. Poult. Res.* 23: 567–75.

Aspinal, O. 1982. Analysis of polysaccharides. In *Food Carbohydrates*, eds D. R. Lineback and G. E. Inglett. AVI, Westport, CT: 357–74.

Aureli, R., M. Umar Faruk, I. Cechova, P. B. Pedersen, S. G. ElvigJoergensen, F. Fru and J. Broz. 2011. The efficacy of a novel microbial 6-phytase expressed in *Aspergillus oryzae* on the performance and phosphorus utilization in broiler chickens. *Int. J. Poult. Sci.* 10: 160–8.

Avila, E. J. Arce, C. Soto, F. Rosas, M. Ceccantini and D. R. McIntyre. 2012. Evaluation of an enzyme complex containing nonstarch polysaccharide enzymes and phytase on the performance of broilers fed a sorghum and soybean meal diet. *J. Appl. Poult. Res.* 21: 279–86.

Ayoade, D. I., E. Kiarie, T. A. Woyengo, B. A. Slominski and C. M. Nyachoti. 2012. Effect of a carbohydrase mixture on ileal amino acid digestibility in extruded full-fat soybeans fed to finishing pigs. *J. Anim. Sci.* 90: 3842–7.

Bach Knudsen, K. E. 1997. Carbohydrate and lignin contents of plant materials used in animal feeding. *Anim. Feed Sci. Technol.* 67: 319–38.

Bacic, A., P. J. Harris and B. A. Stone. 1988. Structure and function of plant cell walls. In *The Biochemistry of Plants*, ed. J. Preiss. Academic Press, New York, NY: 297–371.

Bedford, M. R., and K. Autio. 1996. Microscopic examination of feed and digesta from wheat-fed broiler chickens and its relation to bird performance. *Poult. Sci.* 75 (Suppl. 1): 14.

Bedford, M., and J. Apajalahti. 2001. Microbial interactions in the response to exogenous enzyme utilization. In *Enzymes in farm animal nutrition*, ed. M. R. Bedford and G. G. Partridge. CABI Publishing, Wallingford, UK: 299–314.

Bento, M. H. L., C. Pedersen, P. W. Plumstead, L. Salmon, C. M. Nyachoti and P. Bikker. 2012. Dose response of a new phytase on dry matter, calcium, and phosphorus digestibility in weaned piglets. *J. Anim. Sci.* 90(Suppl 4): 245–7.

Boros, D., B. A. Slominski, W. Guenter, L. D. Campbell and O. Jones. 2004. Wheat by-products in poultry nutrition. Part II. Nutritive value of wheat screenings, bakery by-products and wheat mill run and their improved utilization by enzyme supplementation. *Can. J. Anim. Sci.* 84: 429–35.

Bovans White Commercial Management Guide. 2009. http://www.isapoultry.com.

Camden, B. J., P. C. H. Morel, D. V. Thomas, V. Ravindran and M. R. Bedford. 2001. Effectiveness of exogenous microbial phytase in improving the bioavailabilities of phosphorus and other nutrients in maize-soya-bean meal diets for broilers. *Anim. Sci.* 73: 289–97.

Cho, J. H., and I. H. Kim. 2013. Effects of beta-mannanase supplementation in combination with low and high energy dense diets for growing and finishing broilers. *Livestock Sci.* 154: 137–43.

Choct, M., R. J. Hughes and M. R. Bedford. 1999. Effects of a xylanase on individual bird variation, starch digestion throughout the intestine, and ileal and caecal volatile fatty acid production in chickens fed wheat. *Br. Poult. Sci.* 40: 419–22.

Classen, H. L. 1996. Cereal grain starch and exogenous enzymes in poultry diets. *Anim. Feed Sci. Technol.* 62: 21–7.

Codex 2009. Report on the 30th Session of the Codex Committee on Nutrition and Foods for Special Dietary Uses. ALINORM 09/32/26, Rome: Codex Alimentarius Commission.

Cowieson, A. J., and O. Adeola. 2005. Carbohydrases, protease, and phytase have an additive beneficial effect in nutritionally marginal diets for broiler chicks. *Poult. Sci.* 84: 1860–7.

Cowieson, A. J., D. N. Singh and O. Adeola. 2006. Prediction of ingredient quality and the effect of combination of xylanase, amylase, protease and phytase in the diets of broiler chicks. 1. Growth performance and digestible nutrient intake. *Br. Poult. Sci.* 47: 477–89.

Cowieson, A. J., and V. Ravindran. 2008. Effect of exogenous enzymes in maize-based diets varying in nutrient density for young broilers: Growth performance and digestibility of energy, minerals and amino acids. *Br. Poult. Sci.* 49: 37–44.

Cowieson, A. J., P. Wilcock and M. R. Bedford. 2011. Super-dosing effects of phytase in poultry and other monogastrics. *World's Poult. Sci. J.* 67: 225–36.

Cui, W., G. Mazza and C.G. Biliaderis. 1994. Chemical structure, molecular size distributions, and rheological properties of flaxseed (Linum usitatissimum) gum. *J. Agric. Food Chem.* 42: 1891–5.

Daskiran, M., R. G. Teeter, D. Fodge and H. Y. Hsiao. 2004. An evaluation of endo-beta-mannanase (Hemicell) effects on broiler performance and energy use in diets varying in beta-mannan content. *Poult. Sci.* 83: 662–8.

Dilger, R. N., E. M. Onyango, J. S. Sands and O. Adeola. 2004. Evaluation of microbial phytase in broiler diets. *Poult. Sci.* 83: 962–70.

Dilger, R. N., and O. Adeola. 2006. Estimation of true phosphorus digestibility and endogenous phosphorus loss in growing chicks fed conventional and low-phytate soybean meals. *Poult. Sci.* 85: 661–8.

Eeckhout, W., and M. De Paepe. 1994. Total phosphorus, phytatephosphorus and phytase activity in plant feedstuffs. *Anim. Feed Sci. Technol.* 47: 19–29.

Ellis, P. R., F. G. Roberts, A. G. Low and L. M. Morgan. 1995. The effect of high-molecular-weight guar gum on net apparent glucose absorption and net apparent insulin and gastric inhibitory polypeptide production in the growing pig: relationship to rheological changes in jejunal digesta. *Br. J. Nutr.* 74: 539–56.

Englyst, H. N., and J. H. Cummings. 1984. Simplified method for the measurement of total non-starch polysaccharides by gas-liquid chromatography of constituent sugars as aditol acetates. *Analyst* 109: 937–42.

Englyst, H. N., and J. H. Cummings. 1988. Improved method for the determination of dietary fiber as non-starch polysaccharides in plant foods. *J. Assoc. Off. Anal. Chem.* 71: 808–14.

Freedman M. 1996. Food browning and its prevention: An overview. *J. Agric. Food Chem.* 44(3): 632–53.

Fukata, T., Y. Hadate, E. Baba and A. Arakawa. 1991. Influence of bacteria on *Clostridium perfringens* infections in young chickens. *Avian Dis.* 35: 224–47.

Gehring, C. K., M. R. Bedford and W. A. Dozier. 2013. Extra-phosphoric effects of phytase with and without xylanase in corn-soybean meal-based diets fed to broilers. *Poult. Sci.* 92: 979–91.

Gibson, G. R. and M. B. Roberfroid. 1995. Dietary modulation of the human colonic microbiota: Introducing the concept of prebiotics. *J. Nutr.* 125: 1401–12.

Godoy, S., C. Chicco, F. Meschy and F. Requena. 2005. Phytic phosphorus and phytase activity of animal feed ingredients. *Interciencia* 30: 24–8.

Goering, H. K., and P. J. Van Soest. 1970. Forage fibre analyses (apparatus, reagents, procedures, and some applications). Agric. Handbook No. 379, ARS-USDA, Washington, DC.

Gracia, M. I., M. J. Aranibar, R. Lazaro, P. Medel and G. G. Mateos. 2003. Alpha-amylase supplementation of broiler diets based on corn. *Poult. Sci.* 82: 436–42.

Graham, H., and P. Aman. 1991. Nutritional aspects of dietary fibers. *Anim. Feed Sci. Technol.* 32: 143–58.

Henry, R. J. 1987. Pentosan and (1-3), (1-4)-β-glucan concentrations in endosperm and wholegrain of wheat, barley, oats and rye. *J. Cereal Sci.* 6: 253–8.

Hofacre, C. L., T. Beacorn, S. Collett and G. Mathis. 2003. Using competitive exclusion, mannan-oligosaccharide and other intestinal products to control necrotic enteritis. *J. Appl. Poult. Res.* 12: 60–4.

Hsiao, H.-Y., D. M. Anderson and N. M. Dale. 2006. Levels of β-Mannan in soybean meal. *Poult. Sci.* 85: 1430–2.

Hughes, A. L., J. Dahiya, C. Wyatt and H. Classen. 2008. The efficacy of quantum phytase in a forty-week production trial using White Leghorn laying hens fed corn-soybean meal-based diets. *Poult. Sci.* 87: 1156–61.

Hussain, M., A. U. Rehman and M. F. Khalid. 2012. Feeding value of guar meal and the application of enzymes in improving nutritive value for broilers. *World's Poult. Sci. J.* 68: 253–68.

Ibuki, M., Y. Yoshimoto, M. Inui, K. Fukui, H. Yonemoto, T. Saneyasu, K. Honda and H. Kamisoyama. 2014. Dietary mannanase-hydrolyzed copra meal improves growth and increases muscle weights in growing broiler chickens. *Anim. Sci. J.* 85: 562–68.

Jackson, M. E., K. Geronian, A. Knox, J. McNab and E. McCartney. 2004. A dose-response study with the feed enzyme (3-mannanase in broilers provided with corn- soybean meal based diets in the absence of antibiotic. *Poult. Sci.* 83: 1992–6.

Jakobsen, G. V., B. B. Jensen, K. E. Bach Knudsen and N. Canibe. 2015. Improving the nutritional value of rapeseed cake and wheat dried distillers grain with solubles by addition of enzymes during liquid fermentation. *Anim. Feed Sci. Technol.* 208: 198–213.

Jankowski, J., Z. Zdunczyk, D. Mikulski, J. Naczmanski, J. Juskiewicz, A. Troszynska and B. A. Slominski. 2015. Inclusion of flaxseed in turkey diets decreases the n-6/n-3 PUFA ratio and increases the proportion of biologically active EPA and DHA without affecting meat quality. *Europ. J. Lipid Sci. Technol.* 117: 797–809.

Jarvis, M. C. 2011. Plant cell walls: Supramolecular assemblies. *Food Hydrocolloids.* 25: 257–62.

Jia, W., B. A. Slominski, W. Guenter, A. Humphreys and O. Jones. 2008. The effect of enzyme supplementation on egg production parameters and omega-3 fatty acids deposition in laying hens fed flaxseed and canola seed. *Poult. Sci.* 87: 2005–14.

Jia, W., B. A. Slominski, H. L. Bruce, C. M. Nyachoti and R. O. Jones. 2009a. Enzyme addition facilitates the post-disease compensatory growth of broiler chickens challenged with *Clostridium perfringens*. *Can. J. Anim. Sci.* 89: 369–81.

Jia, W., B. A. Slominski, H. L. Bruce, G. Blank and O. Jones. 2009b. Effects of diet type and enzyme addition on growth performance and gut health of broiler chickens during sub-clinical *Clostridium perfringens* challenge. *Poult. Sci.* 88: 132–40.

Jozefiak, D., A. Ptak, S. Kaczmarek, P. Mackowiak, M. Sassek and B. A. Slominski. 2010. Multi-carbohydrase and phytase supplementation improves growth performance and liver insulin receptor sensitivity in broiler chickens fed diets containing full-fat rapeseed. *Poult. Sci.* 89: 1939–46.

Kaczmarek, S. A., A. Rogiewicz, M. Mogielnicka, A. Rutkowski, R. O. Jones and B. A. Slominski. 2014. The effect of protease, amylase, and nonstarch polysaccharide-degrading enzyme supplementation on nutrient utilization and growth performance of broiler chickens fed corn-soybean meal-based diets. *Poult. Sci.* 93: 1745–53.

Kalmendal, R., and R. Tauson. 2012. Effects of a xylanase and protease, individually or in combination, and an ionophore coccidiostat on performance, nutrient utilization, and intestinal morphology in broiler chickens fed a wheat-soybean meal-based diet. *Poult. Sci.* 91: 1387–93.

Kamran, M., T. N. Pasha, A. Mahmud and Z. Ali. 2002. Effect of commercial enzyme (natugrain) supplementation on the nutritive value and inclusion rate of guar meal in broiler rations. *Int. J. Poult. Sci.* 1: 167–73.

Karimi, A., C. Coto, F. Mussini, S. Goodgame, C. Lu, J. Yuan, M. R. Bedford and P. W. Waldroup. 2013b. Interactions between phytase and xylanase enzymes in male broiler chicks fed phosphorus-deficient diets from 1 to 18 days of age. *Poult. Sci.* 92: 1818–23.

Karimi, A., Y. Min, C. Lu, C. Coto, M. R. Bedford and P. W. Waldroup. 2013a. Assessment of potential enhancing effects of a carbohydrase mixture on phytase efficacy in male broiler chicks fed phosphorus-deficient diets from 1 to 18 days of age. *Poult. Sci.* 92: 192–8.

Khadem, A., M. Lourenço, E. Delezie, L. Maertens, A. Goderis, R. Mombaerts, M. Höfte, V. Eeckhaut, F. Van Immerseel and G. P. J. Janssens. 2016. Does release of encapsulated nutrients have an important role in the efficacy of xylanase in broilers? *Poult. Sci.* doi:10.3382/pew002.

Kiarie, E. G., B. A. Slominski, D. O. Krause and C. M. Nyachoti. 2008. Non-starch polysaccharides hydrolysis products of soybean and canola meal protect against enterotoxigenic *Escherichia coli* in piglets. *J. Nutr.* 138: 502–8.

Kies, A. K., L. H. De Jonge, P. A. Kemme and A. W. Jongbloed. 2006. Interaction between protein, phytate, and microbial phytase. In vitro studies. *J. Agric. Food Chem.* 54: 1753–8.

Kong, C., J. H. Lee and O. Adeola. 2011. Supplementation of β-mannanase to starter and grower diets for broilers. *Can. J. Anim. Sci.* 91: 389–97.

Kussaibati, R., J. Guillaume and B. Leclercq. 1982. The effects of the gut microflora on the digestibility of starch and proteins in young chicks. *Ann. Zootech.* 31: 483–8.

Latham, R. E., M. Williams, K. Smith, K. Stringfellow, S. Clemente, R. Brister and J. T. Lee. 2016. Effect of β-mannanase inclusion on growth performance, ileal digestible energy, and intestinal viscosity of male broilers fed a reduced-energy diet. *J. Appl. Poult. Res.* 25: 40–7.

Leytem, A. B., P. W. Plumstead, R. O. Maguire, P. Kwanyuen and J. Brake. 2007. What aspect of dietary modification in broilers controls litter water-soluble phosphorus: Dietary phosphorus, phytase, or calcium? *J. Environ. Qual.* 36: 453–63.

Leytem, A. B., B. P. Willing and P. A. Thacker. 2008. Phytate utilization and phosphorus excretion by broiler chickens fed diets containing cereal grains varying in phytate and phytase content. *Anim. Feed Sci. Technol.* 146: 160–8.

Lee, J. T., C. A. Bailey and A. L. Cartwright. 2003. β-Mannanase ameliorates viscosity-associated depression of growth in broiler chickens fed guar germ and hull fractions. *Poult. Sci.* 82: 1925–31.

Lee, J. T., S. Connor-Appleton, C. A. Bailey and A. L. Cartwright. 2005. Effects of guar meal by-product with and without P-mannanase hemicell on broiler performance. *Poult. Sci.* 84: 1261–7.

Leeson, S., and J. D. Summers. 2005. *Commercial Poultry Nutrition*. 3rd Ed. Univ. Books, Guelph, Ontario, Canada.

Li, W., R. Angel, S.-W. Kim, E. Jiménez-Moreno, M. Proszkowiec-Weglarz and P. W. Plumstead. 2015. Age and adaptation to Ca and P deficiencies: 2. Impacts on amino acid digestibility and phytase efficacy in broilers. *Poult. Sci.* 94: 2917–31.

Li, Y., X. Chen, Y. Chen, Z. Li and Y. Cao. 2010. Effects of β-mannanase expressed by Pichia pastoris in corn–soybean meal diets on broiler performance, nutrient digestibility, energy utilization and immunoglobulin levels. *Anim. Feed Sci. Technol.* 159: 59–67.

Liu, Q., D. Y. Zhou, L. Chen, R. Q. Dong and S. Zhuang. 2015. Effects of feruloyl esterase, non-starch polysaccharide degrading enzymes, phytase, and their combinations on in vitro degradation of rice bran and nutrient digestibility of rice bran based diets in adult cockerels. *Livestock Sci.* 178: 255–62.

Lohmann LSL-Lite Commercial Management Guide. 2011. http://www.hylinena.com.

Lumpkins, B. S., and A. B. Batal. 2005. The bioavailability of lysine and phosphorus in distillers dried grains with solubles. *Poult. Sci.* 84: 581–6.

Manangi, M. K., and C. N. Coon. 2006. The effect of calcium carbonate particle size and solubility oil the utilization of phosphorus from phytase for broilers. *Poult. Sci.* 85(Suppl 1): 110–110.

Marounek, M., M. Skrivan, O. Rosero and O. Rop. 2010. Intestinal and total tract phytate digestibility and phytase activity in the digestive tract of hens fed a wheat-maize-soyabean diet. *J. Anim. Feed Sci.* 19: 430–9.

McCuaig, L. W., M. I. Davies and I. Motzok. 1972. Intestinal alkaline phosphatase and phytase of chicks: effect of dietary magnesium, calcium, phosphorus and thyroactive casein. *Poult. Sci.* 55: 526–30.

McGrath, J. M., J. T. Sims, R. O. Maguire, W. W. Saylor, C. R. Angel and B. L. Turner. 2005. Broiler diet modification and litter storage: Impacts on phosphorus in litters, soils, and runoff. *J. Environ. Qual.* 34: 1896–1909.

McNab, J. 1993. Digestibility of starch by poultry. *Feed Mix* 1(2): 24–8.

Maenz, D. D., and H. L. Classen. 1998. Phytase activity in the small intestinal brush border membrane of the chicken. *Poult. Sci.* 77: 557–63.

Maenz, D. D. 2001. Enzymatic characteristics of phytases as they relate to their use in animal feeds. In *Enzymes in Farm Animal Nutrition*, ed. M. R. Bedford and G. G. Partridge. CAB International, Wallingford, UK: 61–83.

Martinez Amezcua, C., C. M. Parsons, and S. L. Noll. 2004. Content and relative bioavailability of phosphorus in distillers dried grains with solubles. *Poult. Sci.* 83: 971–6.

Mehri, M., M. Adibmoradi, A. Samie and M. Shivazad. 2010. Effects of β-Mannanase on broiler performance, gut morphology and immune system. *Afr. J. Biotech.* 37: 6221–8.

Meng, X., B. A. Slominski, M. Nyachoti, L. D. Campbell and W. Guenter. 2005. Degradation of cell wall polysaccharides by combinations of carbohydrase enzymes and their effect on nutrient utilization and broiler chicken performance. *Poult. Sci.* 84: 37–47.

Meng, X., and B. A. Slominski. 2005. Nutritive values of corn, soybean meal, canola meal, and peas for broiler chickens as affected by a multicarbohydrase preparation of cell wall degrading enzymes. *Poult. Sci.* 84: 1242–51.

Meng, X., B. A. Slominski, L. D. Campbell, W. Guenter and O. Jones. 2006. The use of enzyme technology for improved energy utilization from full-fat oilseeds. Part I: Canola seed. *Poult. Sci.* 85: 1025–30.

Moran, E. T. 1982. Starch digestion in fowl. *Poult. Sci.* 61: 1257–67.

Morgan, N. K., C. L. Walk, M. R. Bedford and E. J. Burton. 2015. Contribution of intestinal- and cereal-derived phytase activity on phytate degradation in young broilers. *Poult. Sci.* 94: 1577–83.

Mussini, F., C. A. Coto, S. Goodgame, C. Lu, A. J. Karimi, J. Lee and P. W. Waldroup. 2011. Effect of β-mannanase on broiler performance and dry matter output using corn-soybean meal based diets. *Int. J. Poult. Sci.* 10: 778–81.

Mutucumarana, R. K., V. Ravindran, G. Ravindran and A. J. Cowieson. 2014. Measurement of true ileal digestibility and total tract retention of phosphorus in corn and canola meal for broiler chickens. *Poult. Sci.* 93: 412–19.

Nahapetian, A., and A. Bassiri. 1976. Variations in concentrations and interrelationships of phytate, phosphorus, magnesium, calcium, zinc, and iron in wheat varieties during two years. *J. Agric. Food Chem.* 24: 947–50.

National Research Council (NRC). 1994. *Nutrient Requirements of Poultry*. 9th rev. ed. National Academy Press, Washington, DC.

Nelson, T. S., L. W. Ferrara and N. L. Storer. 1968. Phytate phosphorus content of feed ingredients derived from plants. *Poult. Sci.* 47: 1372–4.

Noy, Y., and D. Sklan. 1995. Digestion and absorption in the chick. *Poult. Sci.* 74: 366–73.

Nys, Y., D. Frapin and A. Pointillart. 1996. Occurrence of phytase in plants, animals and microorganisms. In *Phytase in Animal Nutrition and Waste Management*, ed. M. B. Coelho and E. T. Kornegay. BASF, New Jersey: 213–36.

Odetallah, N. H., P. R. Ferket, J. L. Grimes and J. L. McNaughton. 2002. Effect of mannan-endo-1,4-β-mannosidase on the growth performance of turkeys fed diets containing 44% and 48% crude protein soybean meal. *Poult. Sci.* 81: 1322–31.

Odetallah, N. H., J. J. Wang, J. D. Garlich and J. C. H. Shih. 2005. Versazyme supplementation of broiler diets improves market growth performance. *Poult. Sci.* 84: 858–64.

Olukosi, O. A., A. Cowieson and O. Adeola. 2007. Age-related influence of a cocktail of xylanase, amylase, and protease or phytase individually or in combination in broilers. *Poult. Sci.* 86: 77–86.

Olukosi, O. A., A. J. Cowieson and O. Adeola. 2010. Broiler responses to supplementation of phytase and admixture of carbohydrases and protease in maize–soyabean meal diets with or without maize Distillers' Dried Grain with Solubles. *Br. Poult. Sci.* 51: 434–43.

Olukosi, O. A., C. Kong, F. Fru-Nji, K. M. Ajuwon and O. Adeola. 2013. Assessment of a bacterial 6-phytase in the diets of broiler chickens. *Poult. Sci.* 92: 2101–8.

Olukosi, O. A., and F. Fru-Nji. 2014a. The interplay of dietary nutrient level and varying calcium to phosphorus ratios on efficacy of a bacterial phytase: 1. Growth performance and tibia mineralization. *Poult. Sci.* 93: 3037–43.

Olukosi, O. A., and F. Fru-Nji. 2014b. The interplay of dietary nutrient level and varying calcium to phosphorus ratios on efficacy of a bacterial phytase: 2. Ileal and total tract nutrient utilization. *Poult. Sci.* 93: 3044–152.

Opoku, E. Y., H. L. Classen and T. A. Scott. 2015. The effects of extrusion of wheat distillers dried grains with solubles with or without an enzyme cocktail on performance of turkey hen poults. *Poult. Sci.* 94: 185–94.

Opoku, E. Y., H. L. Classen and T. A. Scott. 2015. Effects of wheat distillers dried grains with solubles with or without protease and β-mannanase on the performance of turkey hen poults. *Poult. Sci.* 94: 207–14.

Ouhida, I., J. F. Pérez, M. Anguita and J. Gasa. 2002. Influence of β-Mannanase on broiler performance, digestibility, and intestinal fermentation. *J. Appl. Poult. Res.* 11: 244–9.

Parsons, C. M. 1996. Digestible amino acids for poultry and swine. *Anim. Feed Sci. Technol.* 59: 147–53.

Penn, C. J., G. L. Mullins, L. W. Zelazny, J. G. Warren and J. M. McGrath. 2004. Surface runoff losses of phosphorus from Virginia soils amended with turkey manure using phytase and high available phosphorus corn diets. *J. Environ. Qual.* 33: 1431–9.

Pontoppidan, K., D. Pettersson and A.-S. Sandberg. 2007. The type of thermal feed treatment influences the inositol phosphate composition. *Anim. Feed Sci. Technol.* 132: 137–47.

Powers, W., and R. Angel. 2008. A review of the capacity for nutritional strategies to address environmental challenges in poultry production. *Poult. Sci.* 87: 1929–38.

Ravindran, V., G. Ravindran and S. Sivalogan. 1994. Total and phytate phosphorus contents of various foods and feedstuffs of plant origin. *Food Chem.* 50: 133–6.

Riesenfeld, G., D. Sklan, A. Bar, U. Eisner and S. Hurwitz. 1980. Glucose absorption and starch digestion in the intestine of the chicken. *J. Nutr.* 110: 117–21.

Rogiewicz, A., and B. A. Slominski. 2015. Phosphorus (P) management with phytase supplementation. *Eggspertise*, June 2015, pp. 2–3.

Rogiewicz, A., and B. Slominski. 2016. Phytate and non-phytate phosphorus, and non-starch polysaccharide contents of feed ingredients (unpublished).

Rogiewicz, A., J. Naczmanski, M. Alizadeh and B. A. Slominski. 2016. Beta-mannans of soybean meal: Quantification, in vitro hydrolysis with microbial β-mannanases, and their effect on growth performance and immune status of broiler chickens. *Poult. Sci.* (submitted).

Rommi, K., T. K. Hakala, U. Holopainen, E. Nordlund, K. Poutanen and R. Lantto. 2014. Effect of enzyme-aided cell wall disintegration on protein extractability from intact and dehulled rapeseed (Brassica rapa L. and Brassica napus L.) press cakes. *J. Agric. Food Chem.* 62: 7989–97.

Rosin, E. A., G. Blank, B. A. Slominski and R. A. Holley. 2007. Enzyme supplements in broiler Chicken diets: In vitro and in vivo effects on bacterial growth. *J. Sci. Food Agric.* 87: 1009–20.

Rutherfurd, S. M., T. K. Chung, P. C. Morel and P. J. Moughan. 2004. Effects of microbial phytase on the ileal digestibility of phytate phosphorus, total phosphorus, and amino acids in a low phosphorus diet for broilers. *Poult. Sci.* 83: 61–8.

Rutherfurd, S. M., T. K. Chung, D. V. Thomas, M. L. Zou and P. J. Moughan, 2012. Effect of a novel phytase on growth performance, apparent metabolizable energy, and the availability of minerals and amino acids in a low-phosphorus corn-soybean meal diet for broilers. *Poult. Sci.* 91: 1118–27.

Selle, P. H., A. R. Walker and W. L. Bryden. 2003. Total and phytate-phosphorus contents and phytase activity of Australian sourced feed ingredients for pigs and poultry. *Aust. J. Exp. Agric.* 45: 475–9.

Selle, P., and V. Ravindran. 2007. Microbial phytase in poultry nutrition. *Anim. Feed Sci. Technol.* 135: 1–41.

Selle, P. H., A. J. Cowieson, N. P. Cowieson and V. Ravindran. 2012. Protein–phytate interactions in pig and poultry nutrition: a reappraisal. *Nutr. Res. Rev.* 25: 1–17.

Shastak, Y., P. Ader, D. Feuerstein, R. Ruehle and M. Matuschek. 2015. ß-Mannan and mannanase in poultry nutrition. *World's Poult. Sci. J.* 71: 161–74.

Shaver White Commercial Management Guide. 2009. http://www.isapoultry.com

Shirley, R. B., and H. M. Edwards. 2003. Graded levels of phytase past industry standards improves broiler performance. *Poult. Sci.* 82: 671–80.

Showalter, A. M., and J. E. Varner. 1989. Biology and molecular biology of plant hydroxyproline-rich glycoproteins. In *The Biochemistry of Plants: A Comprehensive Treatise*, Vol. 15: *Molecular Biology*, ed. A. Marcus. Academic Press, New York, 485-520.

Siddiqui, I. R., and P. J. Wood. 1977. Carbohydrates of rapeseed: A review. *J. Sci. Food Agric.* 28: 530–8.

Silva, S. D., K. Hesselman and P. Aman. 1983. Effects of water and beta-glucanase treatment on non-starch polysaccharides in endosperm of low and high viscous barley. *Swed. J. Agric. Res.* 13: 211–19.

Silversides, F. G., and M. Hruby. 2009. Feed formulation using phytase in laying hen diets. *J. Appl. Poult. Res.* 18: 15–22.

Singh, P. K. 2008. Significance of phytic acid and supplemental phytase in chicken nutrition: A review. *World's Poult. Sci. J.* 64: 553–80.

Slominski, B. A. 2011. Recent advances in research on enzymes for poultry diets. *Poult. Sci.* 90: 2013–23.

Slominski, B. A., W. Guenter and L. D. Campbell. 1993. New approach to water-soluble carbohydrate determination as a tool for evaluation of plant cell wall degrading enzymes. *J. Agric. Food Chem.* 41: 2304–8.

Slominski, B. A., X. Meng, L. D. Campbell, W. Guenter and O. Jones. 2006. The use of enzyme technology for improved energy utilization from full-fat oilseeds. Part II: Flaxseed. *Poult. Sci.* 85: 1031–7.

Slominski, B. A., T. Davie, M. C. Nyachoti and O. Jones. 2007. Heat stability of endogenous and microbial phytase during feed pelleting. *Livest. Sci.* 109: 244–6.

Smits, C. H. M., and G. Annison. 1996. Non-starch polysaccharides in broiler nutrition – Towards a physiologically valid approach to their determination. *World's Poult. Sci. J.* 52: 203–21.

Sornlake, W., P. Matetaviparee, N. Rattanaphan, S. Tanapongpipat and L. Eurwilaichitr. 2013. β-Mannanase production by Aspergillus niger BCC4525 and its efficacy on broiler performance. *J. Sci. Food Agric.* 93: 3345–51.

Stefanello, C., S. L. Vieira, G. O. Santiago, L. Kindlein, J. O. B Sorbara and A. J. Cowieson. 2015. Starch digestibility, energy utilization, and growth performance of broilers fed corn-soybean basal diets supplemented with enzymes. *Poult. Sci.* 94: 2472–9.

Steiner, T., R. Mosenthin, B. Zimmermann, R. Greiner and S. Roth. 2007. Distribution of phytase activity, total phosphorus and phytate phosphorus in legume seeds, cereals and cereal by-products as influenced by harvest year and cultivar. *Anim. Feed Sci. Technol.* 133: 320–34.

Sundu, B., A. Kumar and J. Dingle. 2006. Response of broiler chicks fed increasing levels of copra meal and enzymes. *Int. J. Poult. Sci.* 5: 13–18.

Sundu, B., A. Kumar and J. Dingle. 2008. The effect of proportion of crumbled copra meal and enzyme supplementation on broiler growth and gastrointestinal development. *Int. J. Poult. Sci.* 7: 511–15.

Tahir, M., M. Y. Shim, N. E. Ward, C. Smith, E. Foster, A. C. Guney and G. M. Pesti. 2012. The phytate and other nutrient compositions of feed ingredients for poultry. *Poult. Sci.* 91: 928–35.

Tamim, N. M., R. Angel and M. Christman. 2004. Influence of dietary calcium and phytase on phytate phosphorus hydrolysis in broiler chickens. *Poult. Sci.* 83: 1358–67.

Tervilä-Wilo, A., T. Parkkonen, A. Morgan, M. Hopeakoski-Nurminen, K. Poutanen, P. Heikkinen and K. Autio. 1996. In vitro digestion of wheat microstructure with xylanase and cellulase from Trichoderma reesei. *J. Cereal Sci.* 24: 215–25.

Trowell H., D. A. T. Southgate, T. M. S. Wolever, A. R. Leeds, M. A. Gassull and D. J. A. Jenkins. 1976. Dietary Fiber redefined. *Lancet* 1: 967.

Uni, Z., Y. Noy and D. Sklan. 1995. Posthatch changes in morphology and function of the small intestines in heavy- and light-strain chicks. *Poult. Sci.* 74: 1622–9.

Utami, W., A. Meryandini and K. G. Wiryawan. 2013. Characterization of bacterial mannanase for hydrolyzing palm kernel cake to produce manno-oligosaccharides prebiotics. *Media Peternakan* 36: 192–6.

Vahjen, W., T. Busch and O. Simon. 2005. Study on the use of soya bean polysaccharide degrading enzymes in broiler nutrition. *Anim. Feed Sci. Technol.* 120: 259–76.

Waldroup, P. 2002. Phosphorus, phytase and the environment: A retrospective look. In *Proceedings of the Maryland Nutr. Conf. for Feed Manufacturers*. Univ. of Maryland, College Park, MD: 195–200.

Waldroup, P. W., J. H. Kersey, E. A. Saleh, C. A. Fritts, F. Yan, H. L. Stillborn, R. C. Crum Jr. and V. Raboy. 2000. Nonphytate phosphorus requirements and phosphorus excretion of broiler chicks fed diets composed of normal or high available phosphate corn with and without phytase. *Poult. Sci.* 79: 1451–9.

Walk, C. L., M. R. Bedford, T. S. Santos, D. Paiva, J. R. Bradley, H. Wladecki, C. Honaker and A. P. McElroy. 2013. Extra-phosphoric effects of superdoses of a novel microbial phytase. *Poult. Sci.* 92: 719–25.

Walk, C. L., T. T. Santos and M. R. Bedford. 2014. Influence of superdoses of a novel microbial phytase on growth performance, tibia ash, and gizzard phytate and inositol in young broilers. *Poult. Sci.* 93: 1172–7.

Weurding, R. E., A. Veldman, W. A. G. Veen, P. J. van der Aar and M. W. A. Verstegen. 2001. Starch digestion rate in the small intestine of broiler chickens differs among feedstuffs. *J. Nutr.* 131: 2329–35.

Williams, M. P., B. Brown, S. Rao and J. T. Lee. 2014. Evaluation of beta-mannanase and nonstarch polysaccharide-degrading enzyme inclusion separately or intermittently in reduced energy diets fed to male broilers on performance parameters and carcass yield. *J. Appl. Poult. Res.* 23: 715–23.

Woyengo, T. A., B. A. Slominski and R. O. Jones. 2010. Growth performance and nutrient utilization of broiler chickens fed diets supplemented with phytase alone or in combination with citric acid and multicarbohydrase. *Poult. Sci.* 89: 2221–9.

Woyengo, T. A., and C. M. Nyachoti. 2011. Review: Supplementation of phytase and carbohydrases to diets for poultry. *Can. J. Anim. Sci.* 91: 177–92.

Woyengo, T. A., R. Patterson, B. A. Slominski, E. Beltranena and R. T. Zijlstra. Nutritive value of cold-pressed camelina cake with or without supplementation of multi-enzyme in broiler chickens. *Poult. Sci.* (published 18 March 2016, 10.3382/ps/pew098).

Working Group Report, WPSJ. 2013. Determination of phosphorus availability in poultry. *World's Poult. Sci.* 69: 687–98.

Wu, G., Z. Liu, M. Bryant and D. Roland. 2006. Comparison of Natuphos and Phyzyme as phytase sources for commercial layers fed corn-soy diet. *Poult. Sci.* 85: 64–9.

Wu, G., M. M. Bryant, R. A. Voitle and D. A. Roland. 2005. Effects of beta-mannanase in corn-soy diets on commercial leghorns in second-cycle hens. *Poult. Sci.* 84: 894–97.

Yacoubi, N., L. Saulnier, E. Bonnin, L. Rhayat, R. Ducatelle and P. Van Immerseel. 2015. Degradation products of multi-enzyme preparations improve the performance and digestive health of wheat-fed broilers. *Poult. Sci.* 94 (Suppl. 1): 9.

Zdunczyk, Z., J. Jankowski, J. Juskiewicz, D. Mikulski and B. A. Slominski. 2013. Effect of different dietary levels of low-glucosinolate rapeseed (canola) meal and non-starch polysaccharide-degrading enzymes on growth performance and gut physiology of growing turkeys. *Can. J. Anim. Sci.* 93: 353–62.

Zou, X. T., X. J. Qiao and Z. R. Xu. 2006. Effect of β-mannanase (Hemicell) on growth performance and immunity of broilers. *Poult. Sci.* 85: 2176–9.

Chapter 9

Advances in understanding the role of phytate in phosphorus and calcium nutrition of poultry

Markus Rodehutscord, University of Hohenheim, Germany

1 Introduction

All living organisms depend upon a continuous supply of phosphates for the formation of structural body components, such as bones, and several other physiological mechanisms to function. The required amounts have already been estimated, and different scientific committees have published dietary allowances for poultry (NRC, 1994; GfE, 1999; Rostagno, 2011). The phosphate supply along the food chain is maintained by the application of fertilizers and feed phosphates, which are produced largely from rock phosphates. The global rock phosphate stores are limited, and they might face depletion within a century (Cordell et al., 2009). This limitation, together with the accumulation of these stores in only a few countries, is considered one of the greatest challenges facing sustainable food production (Gross, 2010; Neset and Cordell, 2012). Specifically in regard to the farm animal sector, there are two main approaches to address these challenges: (1) exploring possibilities for reduction in feed phosphate application and (2) improving the efficacy of phytase supplements in non-ruminant feeding (Rodehutscord, 2008). These

http://dx.doi.org/10.19103/AS.2016.0011.09

challenges have initiated an increased research regarding the presence and role of inositol phosphates in the gastrointestinal tract.

2 Phytate and phytase

In plant seeds, the primary stored form of phosphorus (P) is phytic acid (*myo*-inositol 1,2,3,4,5,6-hexakis (dihydrogen phosphate); $InsP_6$), and it is usually present as phytate (any salt of $InsP_6$) (Eeckhout and De Paepe, 1994; Ravindran et al., 1994; Rodehutscord et al., 2016). This, together with the significant role of plant seeds and their processing products (e.g. oilseed meals produced from soybeans and rapeseed) in animal feeding, makes $InsP_6$ the most important source of organic P in diets for poultry and other non-ruminants. Poultry diets contain about 0.25–0.30% of $InsP_6$-P, depending on what raw materials are used and how their growing and processing conditions were. *Myo*-inositol pentaphosphates ($InsP_5$) and other inositol phosphates with varying degrees of phosphorylation ($InsP_x$) were found in cereal grains only at marginal levels (Rodehutscord et al., 2016). However, in oilseed meals, especially rapeseed meal, up to one-third of the total InsP-P can be contained as $InsP_5$-P (Pontoppidan et al., 2007), which indicates initiation of phosphorus release from $InsP_6$ during thermal processing of oilseed meals.

Utilization of phosphate from $InsP_6$ and lower forms of InsP by animals requires stepwise cleavage of phosphate groups from the inositol ring. This hydrolysis is catalyzed by phytases (Sandberg and Andlid, 2002) and other phosphatases. Phytases are not specific for $InsP_6$ and can further catalyze hydrolysis of lower $InsP_x$, which can also be catalyzed by other phosphatases that cannot hydrolyze $InsP_6$. The sequence of cleavage yields $InsP_x$ with varying degrees of phosphorylation and *myo*-inositol as intermediate or end products. Nature has evolved specific strategies for different organisms to cleave phosphate from $InsP_6$ under various conditions (Mullaney and Ullah, 2007). Hence, 'phytase' refers to a group of enzymes that differ in catalytic properties, structures, sizes, and origins (Mullaney and Ullah, 2003). Based on stereospecificity of $InsP_6$ hydrolysis, different types of phytases have been recognized by IUPAC-IUB: 3-phytases (E.C.3.1.3.8), 4-phytases (E.C.3.1.3.26), and 5-phytases (E.C.3.1.3.72). This classification is based on the suggestion to use numbering based on D-configuration for *myo*-inositol (IUPAC, 1976). Bacterial phytases, such as *Escherichia coli* phytase, which preferentially generate D-Ins(1,2,3,4,5)P_5, are classified as 6-phytases (Greiner et al., 2000). Phytases that generate D-Ins(1,2,3,5,6)P_5, such as the majority of plant phytases, should be called 4-phytases based on the configuration introduced by IUPAC, but are conventionally also called 6-phytases (based on L-configuration). As the major $InsP_5$ generated by phytases from *Selenomonas cerevisiae* and *Aspergillus niger* has been identified to be D-Ins(1,2,4,5,6)P_5, these phytases are called 3-phytase according to the D-configuration (Konietzny and Greiner, 2002). In principle, all enzymes belonging to this group can be active in the digestive tract of animals and contribute to $InsP_6$ degradation. However, the extent of $InsP_6$ degradation in non-ruminant animals is variable and not complete, which can leave a substantial part of the $InsP_6$-P unutilized. In an attempt to further increase utilization of $InsP_6$-P, recent research tried to understand the $InsP_6$ degradation process better, including modulating factors.

3 $InsP_6$ degradation in the digestive tract

3.1 Extent of $InsP_6$ degradation in the absence of phytase in the diet

When referring to $InsP_6$ degradation in the digestive tract, pigs and different poultry species are often subsumed as non-ruminants or 'monogastrics'. This indicates that the extent and processes of $InsP_6$ breakdown are similar among species. However, this assumption needs a revision.

Studies using pigs fed diets with low intrinsic plant phytase activity found prececal (pc) $InsP_6$ breakdown between 16 and 60% (Sandberg et al., 1993; Seynaeve et al., 2000; Rapp et al., 2001; Kemme et al., 2006; Baumgärtel et al., 2008); this variation is likely due to the presence of differing levels of calcium (Ca) and P in the diets. The only value found to be higher than 40% was measured at very low Ca and P dietary levels. However, regardless of the magnitude of pc $InsP_6$ hydrolysis, post-ileal $InsP_6$ hydrolysis was nearly complete in pigs (Sandberg et al., 1993; Seynaeve et al., 2000; Schlemmer et al., 2001; Baumgärtel et al., 2008). Since P absorption posterior to the ileum does not seem to be relevant, phosphate released from $InsP_6$ in the hindgut remains unavailable to the animal. Consequently, P digestibility in pigs, determined according to the standard protocol (GfE, 1994), hardly exceeded 30% in low-phytase feedstuffs such as corn grain or oilseed meals (Düngelhoef et al., 1994; Rodehutscord et al., 1996).

In contrast, several studies conducted with broiler chickens have reported that pc $InsP_6$ degradation ranged between 62 and 89% (Table 1). All of these studies used diets that were mainly based on corn and soybean meal and thus were very low in plant intrinsic

Table 1 Literature values on $InsP_6$ breakdown[1] in broiler chicken

Main ingredients of the diet	Ca (g/kg)	Total P (g/kg)	Sampling location	$InsP_6$ break down (%)	Reference
Corn–SBM	1.8	4.1	Ileum	69	Tamim and Angel, 2003
Corn–SBM	3.6	3.7	Ileum	75	Applegate et al., 2003
Corn–SBM	1.8	4.0	Ileum	69	Tamim et al., 2004
Corn–SBM	4.3	3.0	Ileum	62	Shastak et al., 2014
Corn	1.3	3.1	Ileum	89	Leytem et al., 2008
Corn–SBM	6.7	4.7	Ileum	74	Zeller et al., 2015a
Corn–SBM	5.4	4.0	Ileum	67	Zeller et al., 2015c
Corn–SBM	6.0	2.9	Excreta	70	Shastak et al., 2014
Corn–SBM	5.0	5.0	Excreta	65	Mohammed et al., 1991
Corn–SBM	6.3	4.3	Excreta	63	Mitchell and Edwards Jr., 1996b

SBM: soybean meal.

[1] This table subsumes studies that have used different analytical assays to detect phytate in general or $InsP_6$ in particular.

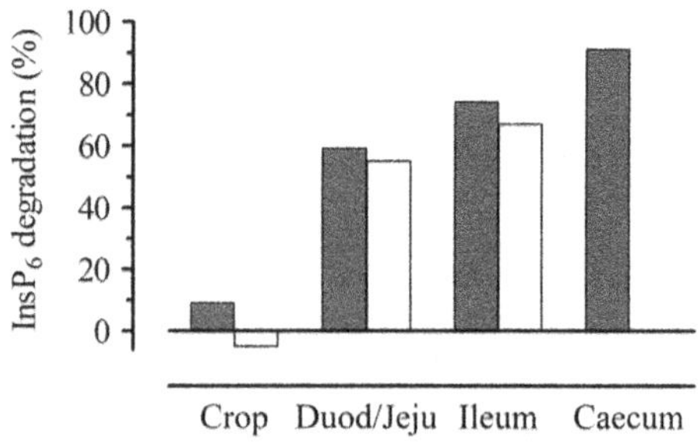

Figure 1 $InsP_6$ degradation measured in different sections of the digestive tract of broiler chickens fed corn–soybean meal-based diets without a phytase supplement. Grey columns (Zeller et al., 2015a); light columns (Zeller et al., 2015c, 2016).

phytase activity. Additionally, none of them used diets that contained any mineral P supplement, and all were low in Ca content. This combination factor is a relatively artificial situation and not representative of practical-type diets. Nevertheless, it demonstrates the high biological potential broiler chickens have to hydrolyze $InsP_6$ in the digestive tract. The extent to which factors such as Ca and P mineral supplementation can suppress the expression of this potential will be addressed later in this chapter.

When using digesta from the intestine of laying hens that had been fed a corn–soybean meal-based diet, substantial breakdown of sodium phytate *in vitro* was observed (Marounek et al., 2008), especially when content from the ceca was used. In the crop of broiler chickens, $InsP_6$ breakdown was very low when diets did not contain phytase (Fig. 1). However, after passage of digesta through the acidic environment of the proventriculus/gizzard, $InsP_6$ breakdown measured in the duodenum/jejunum reached levels of more than 50%. $InsP_6$ breakdown further increased with passage of the digesta to the terminal ileum, and it was about 90% in the ceca. This development in breakdown was achieved without intrinsic plant phytase or supplemented phytase in the diet, thus raising the question: Where do the enzymes catalyzing $InsP_6$ hydrolysis originate?

3.2 Possible relevance of endogenous mucosal phytases

It is often assumed that animals do not possess sufficient endogenous phytase in their intestinal epithelia, whereas there is a growing body of evidence that this assumption is not correct for fowl. Studies that were mainly conducted with broiler chickens but also with laying hens revealed some phytase activity when using purified brush border membrane vesicles from different sections of the small intestine (Maenz and Classen, 1998; Onyango et al., 2006; Huber et al., 2015). Values found in different studies are difficult to compare because of differences in assay details, but variation was found in individual studies. Epithelial phytase was found highest in preparations of the duodenum, and it decreased in the more posterior parts of the small intestine (Maenz and Classen, 1998). There was also some indication in the studies that epithelial phytase secretion is reduced at higher inorganic phosphate (P_i) concentrations in the intestinal lumen (Huber et al., 2015). It was also reduced when diets containing 9 g Ca/kg compared to 4 g Ca/kg were fed (Applegate et al., 2003). However, it was stated that the regulation of endogenous phytase secretion is not well examined, and that the quantitative contribution of endogenous phytase to intestinal $InsP_6$ breakdown cannot be assessed at this time (Huber et al., 2015). Because of their localization, brush border membrane-associated phytase may only have a restricted relevance for hydrolysis occurring in the gut lumen. Furthermore, endogenous enzymes may express different behaviour in the environment of the digestive tract than under the well-standardized *in vitro* assay conditions. The extent of $InsP_6$ breakdown in the duodenum/jejunum of broilers after feeding diets with different supplementary levels of a mineral phosphate and phytase (Zeller et al., 2015c) did not correlate with the endogenous epithelial phytase activity measured in the companion work (Huber et al., 2015). In three experiments using different dietary Ca and vitamin D_3 combinations as well as different bird strains (Applegate et al., 2003), a significant correlation between pc $InsP_6$ breakdown and V_{max} of phytase activity of the brush border membrane vesicles was observed in one, but not in the other two experiments. Hence, while brush border membrane vesicles clearly show that substantial enzyme activity is there, its quantitative relevance for $InsP_6$ degradation in the complex situation present in the lumen of the digestive tract still needs to be elucidated.

3.3 Possible relevance of endogenous microbial phytases

Another potential source of phytase is the microbiota colonizing the digestive tract (Leytem et al., 2008). Gnotobiotic broiler chickens compared with conventional non-gnotobiotic had much higher $InsP_6$ levels in the cecal content (Kerr et al., 2000). In laying hens, the specific phytase activity in ceca content was higher by a factor of more than 10 when compared with content from crop, stomach and small intestine (Marounek et al., 2010). These results indicate the substantial involvement of microbes in $InsP_6$ breakdown in the intestine. *In vitro* studies have shown the $InsP_6$ degrading activity of various bacteria (Konietzny and Greiner, 2002; Vats and Banerjee, 2004). Additionally, lactic acid-producing bacteria isolated from the chicken intestine were identified as possible $InsP_6$ degrading candidates (Raghavendra and Halami, 2009). Among the bacteria in the small intestine of broiler chicken, lactobacilli are the most common (Rehman et al., 2007). Interestingly, broiler chickens fed a diet supplemented with *Lactobacillus* species had increased P retention (Angel et al., 2005). When using deep sequencing techniques, Witzig et al. (2015) recently found lactobacilli to be the dominating bacteria in the crop and small intestine of broiler chickens, but the phylotypes of lactobacilli differed between the crop and the small intestine. In addition to the dominant phylotypes, genes for $InsP_6$ phosphatases were identified in *Bacteroides* spp., *Burkholderia* spp., and three species of the genus *Bifidobacterium*, which are also present in the digestive tract of chickens (Tamayo-Ramos et al., 2012; Stentz et al., 2014). This strongly supports the hypothesis of the gut microbiome contributing to $InsP_6$ breakdown. However, as for epithelial phytase, it is not possible to quantify the relevance of microbial phytase at this time.

Probably the crop microbiota is important for $InsP_6$ breakdown further down the digestive tract. Lactobacilli colonize there and produce enzymes. Hydration and temperature favour bacterial growth and enzyme activity (Svihus et al., 2002). Short retention time of digesta in the crop might cause enzymes produced there to be active mainly in the subsequent sections, especially in the gizzard. Later in the small intestine, additional phytase and other phosphatases from the epithelium get involved, all together contributing to the development of $InsP_6$ breakdown as shown in Figure 1. This remains a hypothesis until experiments involving gnotobiotic birds may offer the possibility to distinguish between $InsP_6$ breakdown by enzymes of mixed endogenous origin or solely epithelial origin.

The microbial population in the ceca has a very high diversity including bacteria known to break down $InsP_6$ (Rehman et al., 2007; Witzig et al., 2015). Correspondingly, *in vitro* phytase activity in ceca content of laying hens was much higher than in anterior sections of the digestive tract after feeding a corn–soybean meal-based diet (Marounek et al., 2008). The $InsP_6$ content in the ceca was higher in gnotobiotic compared to conventional broilers (Kerr et al., 2000). This demonstrated the high impact the microorganism has on $InsP_6$ breakdown in the ceca. In fact, less than 10% of dietary $InsP_6$ were recovered in the ceca of broilers when fed a diet without phytase (Zeller et al., 2015a). However, the relevance of cecal $InsP_6$ hydrolysis for the birds is still unclear. Based on the results of Son et al. (2002), it can be approximated that not more than one-quarter of ileal digesta enters the ceca for fermentation. Post-ileal absorption of P has not been reported to exist. Hence, P released from $InsP_x$ in the ceca is excreted as phosphate or bound in other forms.

3.4 Relevance of intrinsic plant phytase activity

Some plant seeds and feedstuffs obtained from them contain intrinsic phytase activity (Eeckhout and De Paepe, 1994; Rodehutscord et al., 2016). Intrinsic phytase activity is high in cereal grains such as wheat, rye and triticale, while it is lower in legume seeds. Within one type of feedstuff, variation in intrinsic phytase activity can be very high, as recently shown for cereal grains (Rodehutscord et al., 2016). Phytase is hardly detectable in oilseed meals and corn grain. Intrinsic plant phytase – assuming it is not deactivated during processing of the feed – can, in principle, contribute to $InsP_6$ breakdown in the digestive tract. This would primarily take place in the crop prior to initiation of proteolysis in the gizzard. However, it is a matter of debate: How relevant the intrinsic plant phytase activity can be for $InsP_6$ breakdown in the intestine? P retention studies involving wheat-based diets found some positive relationship between the intrinsic phytase activity of the wheat batches that were used and P retention in broiler chickens (Barrier-Guillot et al., 1996; Oloffs et al., 2000). P retention in broiler chickens was also lower when an extruded wheat was fed instead of a non-extruded wheat (Oloffs et al., 1998). Other studies using different grains with different phytase activity did not indicate a relationship between the intrinsic plant phytase activity of the diet and $InsP_6$ hydrolysis in broiler chickens (Juanpere et al., 2004; Leytem et al., 2008). Considering these divergent reports, it was of interest to investigate intrinsic phytase effects along the digestive tract. Following the inclusion of microwave-treated wheat instead of untreated wheat in the diet, both intrinsic phytase activity of the diet and $InsP_6$ breakdown in the crop of broiler chickens were substantially reduced (Zeller et al., 2016) (Fig. 2). However, with further passage through the digestive tract, the differences in $InsP_6$ breakdown between the diets containing microwave-treated or untreated wheat disappeared. These differences indicate that other phytase sources had compensated for the lack in intrinsic plant phytase activity (Zeller et al., 2015b). Other authors also concluded that in comparison to endogenous epithelial and microbial phytase, intrinsic feed phytase seems to contribute very little to $InsP_6$ degradation when measured in the terminal ileum of broiler chickens (Leytem et al., 2008; Shastak et al., 2014). When a corn-based diet with phytase below the detection limit was compared with a wheat-based diet with 660 U/kg phytase, pc $InsP_6$ degradation in broiler chickens was approximately 61%, and there was no difference between the two diets (Shastak et al., 2014). Overall, the discrepancies in the evaluation of intrinsic plant phytase relevance that are obvious from

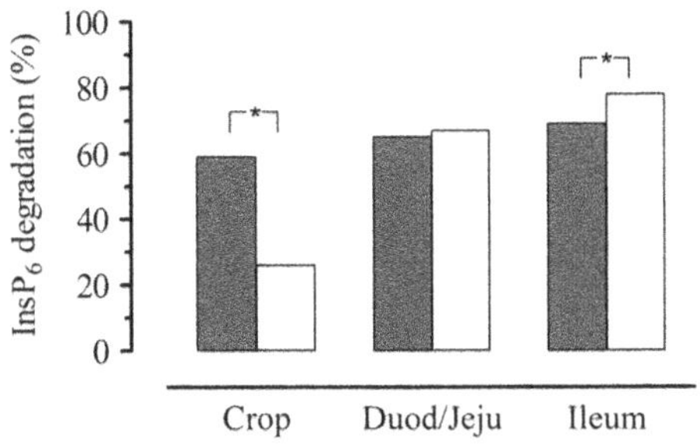

Figure 2 $InsP_6$ degradation measured in different sections of the digestive tract of broiler chickens fed diets that contained untreated wheat (632 U/kg phytase; grey columns) or microwave-treated wheat (121 U/kg phytase, light columns) (Zeller et al., 2015b, 2016). * Significantly different.

this chapter cannot yet be explained. It is possible that they are related to P and Ca mineral inclusion levels of the diets, as addressed later.

4 The relevance of dietary and genetic variation to $InsP_6$ breakdown in broilers

4.1 Dietary variables affecting $InsP_6$ breakdown

$InsP_6$ hydrolysis in the digestive tract is subject to phytate solubility. Dietary factors influencing phytate solubility may therefore have an effect on $InsP_6$ breakdown. It has been found that the level of intestinal $InsP_6$ breakdown in broiler chickens strongly depends upon the Ca concentration of the diet. The level of pc $InsP_6$ disappearance decreased from about 70% to about 20% with increasing Ca from 0.2 to 0.7% (Tamim and Angel, 2003; Tamim et al., 2004). In contrast, pc $InsP_6$ disappearance was only about 25% and not significantly different when 0.7 and 1.0% Ca were used in the diet (Li et al., 2016). This is an indication that Ca does not exhibit negative effects once a certain threshold Ca level is exceeded. But when comparing the different studies, it should be kept in mind that the diets used by Li et al. (2016) included P from meat and bone meal or monocalcium phosphate. The source of Ca may also play a role through differences in Ca solubility. But was it Ca alone that led to low $InsP_6$ disappearance?

It has been described in an *in vitro* assay that P addition to the medium decreased synthesis of phosphatases by *A. ficuum* (Shieh et al., 1969). Such inhibition of enzymes by mineral P may also happen in the digestive tract, with the consequence of reduced $InsP_6$ degradation. When mineral supplements containing both Ca and P were included, $InsP_6$ breakdown was repeatedly found to be significantly reduced (Fig. 3). However, a diminishing effect on $InsP_6$ breakdown was also found to be caused by monosodium phosphate supplementation. If the end product inhibition detected *in vitro* by Shieh et al. (1969) also happens in the digestive tract, then this would partially explain why pc $InsP_6$ breakdown is reduced upon supplementation of a (Ca-free) mineral phosphate. When measured on the excreta level, $InsP_6$ disappearance was about 67% with dietary concentrations of 0.6% Ca and 0.5% P (Delezie et al., 2012). $InsP_6$ disappearance was

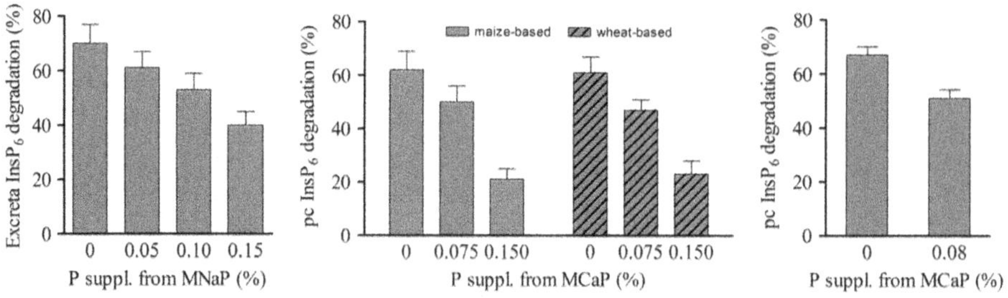

Figure 3 $InsP_6$ degradation of broiler chickens as affected by supplements of anhydrous monosodium phosphate (MNaP) and monocalcium phosphate monohydrate (MCaP) in corn- and wheat-based diets (Shastak et al., 2014; Zeller et al., 2015c).

significantly reduced to 53% upon moderate increase of Ca (0.8%) and P (0.6%) together, but reduction was more pronounced for P than for Ca when they were supplemented separately. This supports the findings reported for the ileum (Fig. 3).

To date, it is unknown what the specific relevance of P and Ca supplements are when supplemented together as calcium phosphates, as is typically done in practical diets. It also is unknown whether enzymes of endogenous epithelial or microbiota origin or exogenous enzymes are affected by such mixed supplements. Taken the high variability of P and Ca levels in practical diets, it is recommended that further research projects aim at quantifying the multiple interactions in order to make implementation in practical feed formulation possible.

Supplementation of 1,25-dihydroxycholecalciferol (1,25-DHCC), alone or in combination with a phytase supplement, increased different criteria of P utilization in broiler chickens than those previously discussed (Mitchell and Edwards Jr., 1996a). The authors suggested that this effect is caused by less phytate precipitation due to an increase in Ca absorption, or by endogenous mucosal phytase being directly increased by the 1,25-DHCC supplement. The authors also mentioned that this effect could be unique to broiler chickens. In accordance with these findings, $InsP_6$ disappearance in another study and measured on the excreta level, was slightly but significantly increased when the cholecalciferol level of the diet was increased from 1250 to 2500 IU/kg (Delezie et al., 2012).

Feed processing can also have an effect on $InsP_6$ degradation by influencing the accessibility of $InsP_6$ for phytase. An *in vitro* study using wheat showed that grinding increased the amount of dialyzable phosphate (Żyła et al., 1999), perhaps because phytase had better access to $InsP_6$, both located differently in the seed. Phytase from ground wheat can also access $InsP_6$ from soybean meal when incubated together (Blaabjerg et al., 2010). Of note, microwave treatment of wheat reduced intrinsic phytase activity and $InsP_6$ degradation in the crop, but increased $InsP_6$ degradation when measured in the terminal ileum (Zeller et al., 2015b). The authors suggested this to be a consequence of disrupted wheat aleurone structures by microwaving that increased accessibility of $InsP_6$ by endogenous epithelial or microbial phytase in the small intestine. Corn ground through 1 mm compared to 3 mm showed higher $InsP_6$ hydrolysis rate upon phytase supplementation *in vitro* (Ton Nu et al., 2014). In line with this observation, phytase supplementation increased pc P digestibility and toe ash in broilers fed a medium but not coarse particle size corn-based diet (Amerah and Ravindran, 2009).

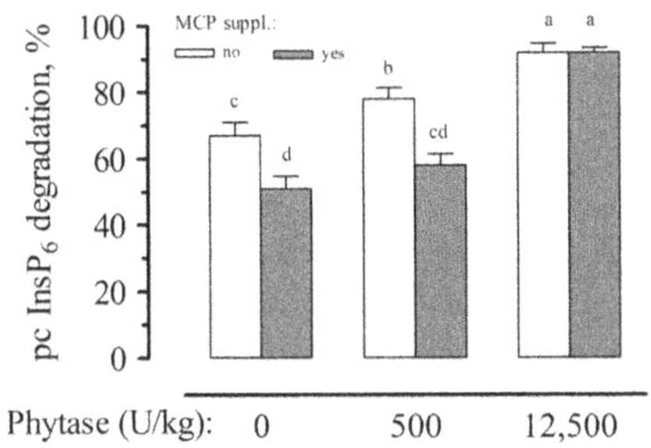

Figure 4 Interactive effects of monocalcium phosphate (MCP) and phytase supplements on $InsP_6$ degradation in the terminal ileum of broiler chickens fed corn–soybean meal-based diets (Zeller et al., 2015c).

4.2 Effects of phytase supplements

Although the broiler chicken's potential of $InsP_6$ breakdown is high, some diets contain mineral supplements that result in the birds not being able to reach this full potential. Therefore, the role of supplemented phytases needs to be evaluated with consideration for the interactions between supplements as they relate to intestinal $InsP_6$ breakdown. Using an *A. niger* phytase, Liebert et al. (1993) found that, when compared to the small intestine, the ratio of phytate-P to total P became much narrower with the supplementation of phytase in the crop and stomach. This suggests that the primary location for phytase activity is the anterior part of the digestive tract. Considering diversification in phytase additives, this hypothesis needs modification. While an *A. niger* phytase was confirmed to substantially increase the $InsP_6$ breakdown in the crop from 9 to 64%, the increase following supplementation of a modified *E. coli* phytase was only up to 31% (Zeller et al., 2015a). However, phytase action in the more distal sections of the digestive tract fully compensated for this difference in the crop, meaning that pc P digestibility was not significantly different or even higher with an *E. coli*-derived phytase than with *A. niger*-derived phytase.

In line with the high level of pc $InsP_6$ breakdown found in broiler chickens fed low-CaP diets, the effects of phytase supplementation at standard industry level are not high when such diets are fed. However, the effects of phytase supplementation at dosages far exceeding the current standards (up to 12 500 FTU/kg of diet) caused a further improvement in pc $InsP_6$ degradation (78–92%) and pc P digestibility (58–71%) (Zeller et al., 2015c). Also of interest, phytase supplementation partly compensated for the inhibitory effects that P mineral sources had on $InsP_6$ breakdown and completely compensated for the inhibitory effects at a very high level of supplementation (Fig. 4).

4.3 Possible effects of animal genetics

As mentioned in Section 3.2, epithelial phytase expression was variable between and within studies, and effects of age, species, strain, and gender of poultry can become relevant (Maenz and Classen, 1998; Applegate et al., 2003; Abudabos, 2012). Differences in P utilization potential between broiler strains are known for a long time (Edwards Jr., 1983). It has been hypothesized that epithelial phytase expression is affected by the bird's genetic background (Beck et al., 2014). Indeed, genomic studies showed significant heritability in the range of 0.10–0.22 for P utilization, P excretion rate and phytate-P bioavailability in broilers and Japanese quail (Zhang et al., 2003; de Verdal et al., 2011; Beck et al., 2016). However, underlying physiological mechanism still needs to be unravelled. Genome effects on epithelial phytase expression are only one potential reason for differences in P utilization. The animal's genes may also affect the abundance of phytase-producing bacteria in the gut microbiota or the expression of phosphate transporters in the intestine, or both.

5 Consequences of changes in $InsP_6$ breakdown for P digestibility

An increase in the intestinal breakdown of $InsP_6$ needs to improve P digestibility in order for there to be a benefit to the bird. Studies at our institute typically measured both $InsP_6$ breakdown and P digestibility in the terminal ileum. A combination of data from different experiments generated some variability in values of both $InsP_6$ degradation and P digestibility in the terminal ileum (Fig. 5). A linear regression analysis applied to the data yielded an estimate for the slope of 0.78; thus, with each increment in $InsP_6$-P disappearance by 1 g, the amount of P digested increased by 0.78 g. It is noteworthy that this is only a rough estimate of what could be referred to as '$InsP_6$-P digestibility', associated with a substantial variation as indicated by the SE of 0.21 estimation. However, the slope of 0.78 is a good indication that a very high proportion of $InsP_6$-P can be absorbed by the animal. The 'non-utilized' proportion (22%) likewise is related to the formation of lower inositol phosphates that are not completely degraded. Isomers of $InsP_5$ and $InsP_4$ were found to exist in the lower ileum regardless of whether phytase was supplemented to the diet or not (Zeller et al., 2015a,b,c).

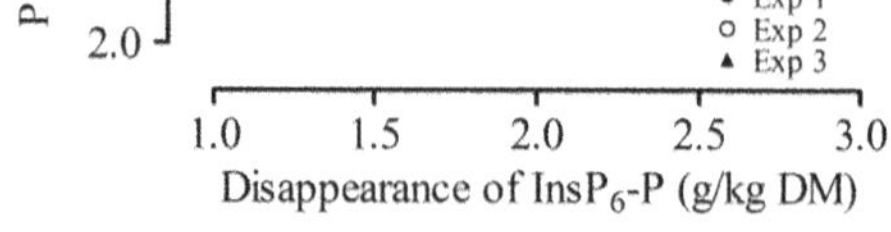

Figure 5 Comparison of $InsP_6$-P that disappeared and P digested at the end of ileum of broiler chickens in three experiments (Zeller et al., 2015a,b,c). Each data point represents a treatment mean. Diets were based on corn or wheat. Some were supplemented with phytase and some were not. A linear regression through all data had a slope of 0.78 (SE 0.21).

6 Brief comparison of broiler chickens and turkeys

When using a low-P basal diet and further diets containing graded levels of MCP, P utilization of the basal diet was higher in broiler chickens than turkeys and ducks. However, utilization of the supplemented mineral P was highest in ducks, followed by turkeys and broiler chickens (Rodehutscord and Dieckmann, 2005). Adebiyi and Olukosi (2015) studied pc P digestibility and P retention of wheat-DDGS using the regression approach as recently recommended by WPSA (2013). Estimated digestibility and retention values were very high in broiler chickens (94% and 92%) and lower in turkeys (76% and 71%). These results were based on total P. However, they indicate that differences in $InsP_6$ degradation in the digestive tract and release of phosphate may have contributed to the differing results among the bird species. When low-P diets based on different corn genotypes were fed to 4-week-old turkeys, pc $InsP_6$ degradation was in the range of 6–15% and pc P digestibility in the range of 22–28% (Ingelmann et al., 2015). With phytase supplementation, the respective values in this study were found to be higher (32–38% pc $InsP_6$ degradation and 37–42% pc P digestibility). Irrespective of phytase inclusion, the levels found by Ingelmann et al. (2015) were remarkably lower than those reported in broiler chicken studies that were conducted with similar type of diets. Although such comparisons of studies need care because they were not conducted under *ceteris paribus* conditions, the differences substantiate the hypothesis that there are species-specific differences in $InsP_6$ breakdown. As mentioned before, endogenous mucosal phytase activity was detected in the small intestine of broiler chickens, and at this time, it is not known if this also occurs in turkeys or ducks, thus leading to differences in phosphate release from $InsP_6$. However, contributing enzymes could also have originated from gut microbes, and these could be different between the species.

7 Conclusions

Broiler chickens have the potential to utilize a substantial part of $InsP_6$-P. To what extent this potential can be reached largely depends upon the inclusion level and source of minerals Ca and P. Diminishing effects of Ca and P mineral supplements can be compensated for by phytase supplements. Increasing interest in minimizing the application of feed phosphates in practical feeding makes it imperative to better understand the interacting factors related to $InsP_6$ breakdown in the digestive tract. There is some evidence that gastrointestinal $InsP_6$ breakdown is lower in turkeys than in broiler chickens; however, the reasons for this difference, as well as consequences for practical application, need to be further investigated. For achieving rapid progress, it will be necessary for more laboratories to establish the analytical protocols for determination of inositol phosphates and phytases in different matrices (feed, digesta content, excreta) and to run them routinely with high accuracy and repeatability. Faster progress might also be achieved by coordinating projects and animal trials in this field of research on an international level and strengthening the network of collaborating institutions.

8 Where to look for further information

Neset, T.-S. S. and Cordell, D. (2012), Global phosphorus scarcity: identifying synergies for a sustainable future, *Journal of the Science of Food and Agriculture*, 92, 2–6.

Rodehutscord, M. (2009), Approaches and challenges for evaluating phosphorus sources for poultry. Invited paper. In WPSA UK-branch (ed.), *Proceedings of the 17th European Symposium on Poultry Nutrition*, 23–27 August 2009, Edinburgh, 2–6.

Shastak, Y. and Rodehutscord, M. (2013), Determination and estimation of phosphorus availability in growing poultry and their historical development, *World's Poultry Science Journal*, 69, 569–85.

Shastak, Y., Zeller, E., Witzig, M., Schollenberger, M. and Rodehutscord, M. (2014), Effects of the composition of the basal diet on the evaluation of mineral phosphorus sources and interactions with phytate hydrolysis in broilers, *Poultry Science*, 93, 2548–59.

Zeller, E., Schollenberger, M., Kühn, I. and Rodehutscord, M. (2015a), Hydrolysis of phytate and formation of inositol phosphate isomers without or with supplemented phytases in different segments of the digestive tract of broilers, *Journal of Nutritional Science*, 4, e1.

9 References

Abudabos, A. M. (2012), Phytate phosphorus utilization and intestinal phytase activity in laying hens, *Italian Journal of Animal Science*, 11, 41–6.

Adebiyi, A. and Olukosi, O. (2015), Determination in broilers and turkeys of true phosphorus digestibility and retention in wheat distillers dried grains with solubles without or with phytase supplementation, *Animal Feed Science and Technology*, 207, 112–19.

Amerah, A. M. and Ravindran, V. (2009), Influence of particle size and microbial phytase supplementation on the performance, nutrient utilisation and digestive tract parameters of broiler starters, *Animal Production Science*, 49, 704–10.

Angel, R., Dalloul, R. A. and Doerr, J. (2005), Performance of broiler chickens fed diets supplemented with a direct-fed microbial, *Poultry Science*, 84, 1222–31.

Applegate, T. J., Angel, R. and Classen, H. L. (2003), Effect of dietary calcium, 25-hydroxycholecalciferol, or bird strain on small intestinal phytase activity in broiler chickens, *Poultry Science*, 82, 1140–8.

Barrier-Guillot, B., Casado, P., Maupetit, P., Jondreville, C., Gatel, F. and Larbier, M. (1996), Wheat phosphorus availability: 2 – *In vivo* study in broilers and pigs; relationship with endogenous phytasic activity and phytic phosphorus content in wheat, *Journal of the Science of Food and Agriculture*, 70, 69–74.

Baumgärtel, T., Metzler, B. U., Mosenthin, R., Greiner, R. and Rodehutscord, M. (2008), Precaecal and postileal metabolism of P, Ca and N in pigs as affected by different carbohydrate sources fed at low level of P intake, *Archives of Animal Nutrition*, 62, 169–81.

Beck, P., Rodehutscord, M., Bennewitz, J. and Bessei, W. (2014), A pilot study of the genetic variation of phosphorus utilization in young Japanese quail (*Coturnix japonica*), *Poultry Science*, 93, 1916–21.

Beck, P., Piepho, H.-P., Rodehutscord, M. and Bennewitz, J. (2016), Inferring relationships between phosphorus utilization, feed per gain, and bodyweight gain in an F_2 cross of Japanese quail using recursive models, *Poultry Science*, 95, 764–73.

Blaabjerg, K., Carlsson, N. G., Hansen-Møller, J. and Poulsen, H. D. (2010), Effect of heat-treatment, phytase, xylanase and soaking time on inositol phosphate degradation *in vitro* in wheat, soybean meal and rapeseed cake, *Animal Feed Science and Technology*, 162, 123–34.

Cordell, D., Drangert, J.-O. and White, S. (2009), The story of phosphorus: Global food security and food for thought, *Global Environmental Change*, 19, 292–305.

de Verdal, H., Narcy, A., Bastianelli, D., Chapuis, H., Meme, N., Urvoix, S., Le Bihan-Duval, E. and Mignon-Grasteau, S. (2011), Improving the efficiency of feed utilization in poultry by selection. 2. Genetic parameters of excretion traits and correlations with anatomy of the gastro-intestinal tract and digestive efficiency, *BMC Genetics*, 12, 71.

Delezie, E., Maertens, L. and Huyghebaert, G. (2012), Consequences of phosphorus interactions with calcium, phytase, and cholecalciferol on zootechnical performance and mineral retention in broiler chickens, *Poultry Science*, 91, 2523–31.

Düngelhoef, M., Rodehutscord, M., Spiekers, H. and Pfeffer, E. (1994), Effects of supplemental microbial phytase on availability of phosphorus contained in maize, wheat and triticale to pigs, *Animal Feed Science and Technology*, 49, 1–10.

Edwards Jr., H. M. (1983), Phosphorus. 1. Effect of breed and strain on utilization of suboptimal levels of phosphorus in the ration, *Poultry Science*, 62, 77–84.

Eeckhout, W. and De Paepe, M. (1994), Total phosphorus, phytate-phosphorus and phytase activity in plant feedstuffs, *Animal Feed Science and Technology*, 47, 19–29.

GfE (1994), Mitteilungen des Ausschusses für Bedarfsnormen der Gesellschaft für Ernährungsphysiologie: Die Bestimmung des verdaulichen Phosphors beim Schwein, *Proceedings of the Society of Nutrition Physiology*, 2, 113–19.

GfE (1999), *Energie- und Nährstoffbedarf landwirtschaftlicher Nutztiere. 7. Empfehlungen zur Energie- und Nährstoffversorgung der Legehennen und Masthühner (Broiler)*. DLG-Verlag, Frankfurt am Main.

Greiner, R., Carlsson, N.-G. and Alminger, M. L. (2000), Stereospecificity of myo-inositol hexakisphosphate dephosphorylation by a phytate-degrading enzyme of *Escherichia coli*, *Journal of Biotechnology*, 84, 53–62.

Gross, M. (2010), Fears over phosphorus supplies, *Current Biology*, 20, R386.

Huber, K., Zeller, E. and Rodehutscord, M. (2015), Modulation of small intestinal phosphate transporter by dietary supplements of mineral phosphorus and phytase in broilers, *Poultry Science*, 94, 1009–17.

Ingelmann, C.-J., Witzig, M., Schollenberger, M., Möhring, J., Kühn, I. and Rodehutscord, M. (2015), Influence of different maize genotypes in diets without and with added phytase on precaecal $InsP_6$ hydrolysis and P digestibility in turkeys, *Proceedings of the Society of Nutrition Physiology*, 25, 34.

IUPAC (1976), Nomenclature of cyclitols. Recommendations, 1973, *Biochemical Journal*, 153, 23–31.

Juanpere, J., Pérez-Vendrell, A. M. and Brufau, J. (2004), Effect of microbial phytase on broilers fed barley-based diets in the presence or not of endogenous phytase, *Animal Feed Science and Technology*, 115, 265–79.

Kemme, P. A., Schlemmer, U., Mroz, Z. and Jongbloed, A. W. (2006), Monitoring the stepwise phytate degradation in the upper gastrointestinal tract of pigs, *Journal of the Science of Food and Agriculture*, 86, 612–22.

Kerr, M. J., Classen, H. L. and Newkirk, R. W. (2000), The effects of gastrointestinal tract micro-flora and dietary phytase on inositol hexaphosphate hydrolysis in the chicken, *Poultry Science*, 79 (Suppl. 1), 11 (Abstract).

Konietzny, U. and Greiner, R. (2002), Molecular and catalytic properties of phytate-degrading enzymes (phytases), *International Journal of Food Science & Technology*, 37, 791–812.

Leytem, A. B., Willing, B. P. and Thacker, P. A. (2008), Phytate utilization and phosphorus excretion by broiler chickens fed diets containing cereal grains varying in phytate and phytase content, *Animal Feed Science and Technology*, 146, 160–8.

Li, W., Angel, R., Kim, S.-W., Brady, K., Yu, S. and Plumstead, P. W. (2016), Impacts of dietary calcium, phytate, and nonphytate phosphorus concentrations in the presence or absence of phytase on inositol hexakisphosphate (IP6) degradation in different segments of broilers digestive tract, *Poultry Science*, 95, 581–9.

Liebert, F., Wecke, C., Schöner, F. J. (1993), Phytase activities in different gut contents of chickens as dependent on level of phosphorus and phytase supplementations. In C. Wenk and M. Boessinger (eds), *Proceedings of the 1st Symposium on Enzymes in Animal Nutrition, Kartause Ittingen, Switzerland*, 13–16 October 1993, 202–5.

Maenz, D. D. and Classen, H. L. (1998), Phytase activity in the small intestinal brush border membrane of the chicken, *Poultry Science*, 77, 557–63.

Marounek, M., Skřivan, M., Dlouhá, G. and Břeňová, N. (2008), Availability of phytate phosphorus and endogenous phytase activity in the digestive tract of laying hens 20 and 47 weeks old, *Animal Feed Science and Technology*, 146, 353–9.

Marounek, M., Skřivan, M., Rosero, O. and Rop, O. (2010), Intestinal and total tract phytate digestibility and phytase activity in the digestive tract of hens fed a wheat-maize-soyabean diet, *Journal of Animal and Feed Sciences*, 19, 430–9.

Mitchell, R. D. and Edwards Jr., H. M. (1996a), Additive effects of 1,25-dihydroxycholecalciferol and phytase on phytate phosphorus utilization and related parameters in broiler chickens, *Poultry Science*, 75, 111–19.

Mitchell, R. D. and Edwards Jr., H. M. (1996b), Effect of phytase and 1,25-dihydroxycholecalciferol on phytate utilization and the quantitative requirement for calcium and phosphorus in young broiler chickens, *Poultry Science*, 75, 95–110.

Mohammed, A., Gibney, M. J. and Taylor, T. G. (1991), The effects of dietary levels of inorganic phosphorus, calcium and cholecalciferol on the digestibility of phytate-P by the chick, *British Journal of Nutrition*, 66, 251–9.

Mullaney, E. J. and Ullah, A. H. J. (2003), The term phytase comprises several different classes of enzymes, *Biochemical and Biophysical Research Communications*, 312, 179–84.

Mullaney, E. J. and Ullah, A. H. J. (2007), Phytases: Attributes, catalytic mechanisms and applications. In B. L. Turner, A. E. Richardson and E. J. Mullaney (eds), *Inositol Phosphates Linking Agriculture and the Environment* (eds), 97–110. CABI Publishing, Wallingford, UK.

Neset, T.-S. S. and Cordell, D. (2012), Global phosphorus scarcity: identifying synergies for a sustainable future, *Journal of the Science of Food and Agriculture*, 92, 2–6.

NRC (1994), *Nutrient Requirements of Poultry*. National Academy Press, Washington, DC.

Oloffs, K., Dolbusin, A. and Jeroch, H. (1998), Einfluß von mikrobieller und nativer Weizenphytase auf die Phosphor-Verwertung bei Broilern, *Archiv für Geflügelkunde*, 62, 260–3.

Oloffs, K., Cossa, J. and Jeroch, H. (2000), Die Bedeutung der korneigenen (nativen) Phytaseaktivität im Weizen für die Phosphor-Verwertung bei Broilern und Legehennen, *Archiv für Geflügelkunde*, 64, 157–61.

Onyango, E. M., Asem, E. K. and Adeola, O. (2006), Dietary cholecalciferol and phosphorus influence intestinal mucosa phytase activity in broiler chicks, *British Poultry Science*, 47, 632–9.

Pontoppidan, K., Pettersson, D. and Sandberg, A.-S. (2007), The type of thermal feed treatment influences the inositol phosphate composition, *Animal Feed Science and Technology*, 132, 137–47.

Raghavendra, P. and Halami, P. M. (2009), Screening, selection and characterization of phytic acid degrading lactic acid bacteria from chicken intestine, *International Journal of Food Microbiology*, 133, 129–34.

Rapp, C., Lantzsch, H.-J. and Drochner, W. (2001), Hydrolysis of phytic acid by intrinsic plant or supplemented microbial phytase (*Aspergillus niger*) in the stomach and small intestine of minipigs fitted with re-entrant cannulas. 3. Hydrolysis of phytic acid (IP_6) and occurence of hydrolysis products (IP_5, IP_4, IP_3 and IP_2), *Journal of Animal Physiology and Animal Nutrition*, 85, 420–30.

Ravindran, V., Ravindran, G. and Sivalogan, S. (1994), Total and phytate phosphorus contents of various foods and feedstuffs of plant origin, *Food Chemistry*, 50, 133–6.

Rehman, H. U., Vahjen, W., Awad, W. A. and Zentek, J. (2007), Indigenous bacteria and bacterial metabolic products in the gastrointestinal tract of broiler chickens, *Archives of Animal Nutrition*, 61, 319–35.

Rodehutscord, M. (2008), Ansatzpunkte zur Schonung der begrenzten Phosphorressourcen, *Archiv für Tierzucht*, 51, 39–48.

Rodehutscord, M. and Dieckmann, A. (2005), Comparative studies with three-week-old chickens, turkeys, ducks, and quails on the response in phosphorus utilization to a supplementation of monobasic calcium phosphate, *Poultry Science*, 84, 1252–60.

Rodehutscord, M., Faust, M. and Lorenz, H. (1996), Digestibility of phosphorus contained in soybean meal, barley, and different varieties of wheat, without and with supplemental phytase fed to pigs and additivity of digestibility in a wheat-soybean-meal diet, *Journal of Animal Physiology and Animal Nutrition*, 75, 40–8.

Rodehutscord, M., Rückert, C., Maurer, H. P., Schenkel, H., Schipprack, W., Bach Knudsen, B. E., Schollenberger, M., Laux, M., Eklund, M., Siegert, W. and Mosenthin, R. (2016), Variation in chemical composition and physical characteristics of cereal grains from different genotypes, *Archives of Animal Nutrition*, 70, 87–107.

Rostagno, H. S. (2011), *Brazilian Tables for Poultry and Swine: Composition of feedstuffs and Nutritional Requirements*, 3rd edition. Universidade Federal de Vicosa, Departamento de Zootecnia.

Sandberg, A.-S. and Andlid, T. (2002), Phytogenic and microbial phytases in human nutrition, *International Journal of Food Science & Technology*, 37, 823–33.

Sandberg, A.-S., Larsen, T. and Sandström, B. (1993), High dietary calcium level decreases colonic phytate degradation in pigs fed a rapeseed diet, *Journal of Nutrition*, 123, 559–66.

Schlemmer, U., Jany, K.-D., Berk, A., Schulz, E. and Rechkemmer, G. (2001), Degradation of phytate in the gut of pigs - pathway of gastro-intestinal inositol phosphate hydrolysis and enzymes involved, *Archives of Animal Nutrition*, 55, 255–80.

Seynaeve, M., Janssens, G., Hesta, M., Nevel, C. V. and Wilde, R. O. D. (2000), Effects of dietary Ca/P ratio, P level and microbial phytase supplementation on nutrient digestibilities in growing pigs: breakdown of phytic acid, partition of P and phytase activity along the intestinal tract, *Journal of Animal Physiology and Animal Nutrition*, 83, 193–204.

Shastak, Y., Zeller, E., Witzig, M., Schollenberger, M. and Rodehutscord, M. (2014), Effects of the composition of the basal diet on the evaluation of mineral phosphorus sources and interactions with phytate hydrolysis in broilers, *Poultry Science*, 93, 2548–59.

Shieh, T. R., Wodzinski, R. J. and Ware, J. H. (1969), Regulation of the formation of acid phosphatases by inorganic phosphate in Aspergillus ficuum, *Journal of Bacteriology*, 100, 1161–5.

Son, J. H., Ragland, D. and Adeola, O. (2002), Quantification of digesta flow into the caeca, *British Poultry Science*, 43, 322–4.

Stentz, R., Osborne, S., Horn, N., Li, A. W. H., Hautefort, I., Bongaerts, R., Rouyer, M., Bailey, P., Shears, S. B., Hemmings, A. M., Brearley, C. A. and Carding, S. R. (2014), A bacterial homolog of a eukaryotic inositol phosphate signaling enzyme mediates cross-kingdom dialog in the mammalian gut, *Cell Reports*, 6, 646–56.

Svihus, B., Hetland, H., Choct, M. and Sundby, F. (2002), Passage rate through the anterior digestive tract of broiler chickens fed on diets with ground and whole wheat, *British Poultry Science*, 43, 662–8.

Tamayo-Ramos, J. A., Sanz-Penella, J. M., Yebra, M. J., Monedero, V. and Haros, M. (2012), Novel phytases from *Bifidobacterium pseudocatenulatum* ATCC 27919 and *Bifidobacterium longum* subsp. *infantis* ATCC 15697, *Applied and Environmental Microbiology*, 78, 5013–15.

Tamim, N. M. and Angel, R. (2003), Phytate phosphorus hydrolysis as influenced by dietary calcium and micro-mineral source in broiler diets, *Journal of Agricultural and Food Chemistry*, 51, 4687–93.

Tamim, N. M., Angel, R. and Christman, M. (2004), Influence of dietary calcium and phytase on phytate phosphorus hydrolysis in broiler chickens, *Poultry Science*, 83, 1358–67.

Ton Nu, M. A., Blaabjerg, K. and Poulsen, H. D. (2014), Effect of particle size and microbial phytase on phytate degradation in incubated maize and soybean meal, *Animal*, 8, 534–41.

Vats, P. and Banerjee, U. C. (2004), Production studies and catalytic properties of phytases (*myo*-inositolhexakisphosphate phosphohydrolases): an overview, *Enzyme and Microbial Technology*, 35, 3–14.

Witzig, M., Camarinha-Silva, A., Green-Engert, R., Hoelzle, K., Zeller, E., Seifert, J., Hoelzle, L. E. and Rodehutscord, M. (2015), Spatial variation of the gut microbiota in broiler chickens as affected by dietary available phosphorus and assessed by T-RFLP analysis and 454 pyrosequencing, *PLoS ONE*, 10, e0143442.

WPSA, Working Group 2 Nutrition of the European Federation of Branches of WPSA (2013), Determination of phosphorus availability in poultry, *World's Poultry Science Journal*, 69, 687–98.

Zeller, E., Schollenberger, M., Kühn, I. and Rodehutscord, M. (2015a), Hydrolysis of phytate and formation of inositol phosphate isomers without or with supplemented phytases in different segments of the digestive tract of broilers, *Journal of Nutritional Science*, 4, e1.

Zeller, E., Schollenberger, M., Kühn, I. and Rodehutscord, M. (2015b), Effect of diets containing enzyme supplements and microwave-treated or untreated wheat on inositol phosphates in the small intestine of broilers, *Animal Feed Science and Technology*, 204, 42–51.

Zeller, E., Schollenberger, M., Kühn, I. and Rodehutscord, M. (2016), Dietary effects on inositol phosphate breakdown in the crop of broilers, *Archives of Animal Nutrition*, 70, 57–71.

Zeller, E., Schollenberger, M., Witzig, M., Shastak, Y., Kühn, I., Hoelzle, L. E. and Rodehutscord, M. (2015c), Interactions between supplemented mineral phosphorus and phytase on phytate hydrolysis and inositol phosphates in the small intestine of broilers, *Poultry Science*, 94, 1018–29.

Zhang, W., Aggrey, S. E., Pesti, G. M., Edwards Jr., H. M. and Bakalli, R. I. (2003), Genetics of phytate phosphorus bioavailability: heritability and genetic correlations with growth and feed utilization traits in a randombred chicken population, *Poultry Science*, 82, 1075–9.

Żyła, K., Gogol, D., Koreleski, J., Świątkiewicz, S. and Ledoux, D. R. (1999), Simultaneous application of phytase and xylanase to broiler feeds based on wheat: in vitro measurements of phosphorus and pentose release from wheats and wheat-based feeds, *Journal of the Science of Food and Agriculture*, 79, 1832–40.

Chapter 10

Probiotics, prebiotics and other feed additives to improve gut function and immunity in poultry

Robert Moore, RMIT University, Australia

1 Introduction

1.1 The use of probiotics, prebiotics and other feed additives: an overview

Modern poultry production is highly efficient and the most ecologically sustainable way to produce high-quality meat protein for human consumption. Of all the terrestrially produced meats, broiler chickens have the lowest requirements for input resources and land (Herrero and Thornton, 2013). These are very important considerations for global food security. In most countries the market for poultry products is highly competitive, and so, optimal performance is a key consideration for growers; a small change in the efficiency of production can make the difference between a profitable and an unprofitable flock. Genetic improvement has been the major driver of the increase in productivity (Zuidhof et al., 2014), but improved nutrition has also been very effective in improving performance, with optimally balanced diets using a variety of agricultural products supplemented with vitamin, mineral and amino acid mixes (Havenstein et al., 2003). The addition of exogenous enzymes has also been widely adopted as a means to maximise the nutrition and energy extracted from feed (Kiarie et al., 2013).

http://dx.doi.org/10.19103/AS.2016.0011.10

Good gut health is also required to ensure the best and most even performance across a flock. 'Gut health' is a somewhat nebulous term that has been described by Bischoff (2011) as 'covers aspects of the gastrointestinal (GI) tract, such as the effective digestion and adsorption of food, the absence of GI illness, normal and stable intestinal microbiota, effective immune status and a state of well-being' (Bischoff, 2011). In the past, antimicrobial growth promoters (AGPs) have been widely used to assist gut health and maximise performance. However, because of concerns that AGP use in animals may cause the spread of antibiotic-resistant bacteria and antibiotic resistance genes into the human population and human bacterial pathogens, their use in animals is now being moderated. With the reduction in AGP use, resulting from legislated restrictions on their use in Europe and from industry-led efforts to reduce their use in other parts of the globe (Dibner and Richards, 2005), alternative methods to maximise gut health and productivity are being sought. Probiotics and prebiotics are important products that may be used to address this need. As well as improving general gut health, some of these products are also being used to reduce the carriage of serious foodborne pathogens such as *Salmonella, Campylobacter, Clostridia* and *Listeria* – pathogens which may not adversely affect chicken health, but which are a threat to human consumers of chicken products and which are hence a major industry concern, with much effort going into developing methods to protect consumers from infection (De Knegt et al., 2015; Gould et al., 2013; Sasaki et al., 2014; Wagenaar et al., 2015).

There are many recent reviews of pre- and probiotics use in chickens (e.g. Ajuwon, 2015; Applegate et al., 2010; Dhama et al., 2014; Hajati and Rezaei, 2010; Kabir, 2009; Schneitz, 2005). Rather than restate material well covered in these reviews, this chapter aims to briefly highlight some of the ways that pre- and probiotics have been identified and their activities characterised, and then goes on to identify gaps in our knowledge and new opportunities to better understand modes of action and develop a new generation of more effective and proven products.

1.2 Probiotic, prebiotic and symbiotic: definitions

The definition of what a probiotic is has subtly changed over the years, but the most common currently accepted definition, as stated in an FAO/WHO report (FAO/WHO, 2006), is, 'Live microorganisms which when administered in adequate amounts confer a health benefit on the host.' This definition has also been ratified by the expert panel of the International Association for Probiotics and Prebiotics (Hill et al., 2014). Prebiotics have been defined as 'a selectively fermented ingredient that allows specific changes, both in the composition and/or activity in the gastrointestinal microflora that confers benefits upon host well-being and health' (Roberfroid, 2007), but the definition is often expanded to include products that, without themselves being fermented, may change the composition and/or activity of the microbiota. Synbiotics are products that combine both prebiotic and probiotic into a single formulation.

There are many commercially available probiotics (Table 1) and prebiotics, plus a large and a rapidly growing list of experimental products reported in the research literature. The probiotic products fall into four main categories: (1) competitive exclusion products, both undefined and defined; (2) single strain products; (3) multi-strain products, which may overlap with defined competitive exclusion products; and (4) synbiotics that combine a probiotic with a prebiotic. Probiotics have also been called direct-feed microbials.

Table 1 Commercially available probiotic products for poultry

Product	Supplier	Organism(s)	References[a]	Notes
Alterion	Adisseo	*Bacillus subtilis*		
Biacton	Biacton	*Lactobacillus farciminis*	(European Food Safety Authority (EFSA), 2006)	
Clostat™	Kemin	*Bacillus subtilis*	(Teo and Tan, 2007)	
CYLACTIN®	DSM	*Enterococcus faecium*		
Enviva®PRO	Danisco	*3 Bacillus subtilis*	(Lee et al., 2010)	
FloraMax®-B11	Pacific Vet Group	3 *Lactobacillus delbrueckii* subsp. *bulgaricus* 3 *Lactobacillus fermentum* 2 *Lactobacillus casei* 2 *Lactobacillus cellobiosus*[b] 1 *Lactobacillus helveticus*	(Higgins et al., 2005; Wolfenden et al., 2007)	
FloraStart®	Pacific Vet Group	*Lactobacillus plantarum* TY036 *Enterococcus faecium* MFF109		
Gallipro®	Chr Hansen	*Bacillus subtilis* *Bacillus licheniformis*	(Sadeghi et al., 2014)	
Lactina	Lactina Ltd	*Lactobacillus acidophilus* *Lactobacillus delbrueckii* subsp. *bulgaricus* *Lactobacillus helveticus* *Lactobacillus lactis*[c] *Streptococcus thermophilus* *Enterococcus faecium*		
LACTOMALT D2 AVI	Centro Sperimentale del Latte	*Lactobacillus acidophilus* D2/CSL		
Lacto-Sacc	Bird & Pet Nutrition	*Saccharomyces cerevisia* *Lactobacillus acidophilus* *Enterococcus faecium*		
MIX AVI 10	Centro Sperimentale del Latte	*Lactobacillus acidophilus* D2/CSL *Lactobacillus plantarum* 14D/CSL		
PoultriMax®	Chr Hansen	*Lactobacillus acidophilus* *Propionibacterium freudenreichii*		
PoultryStar®	Biomin	*Enterococcus faecium* *Pediococcus acidilactici* *Lactobacillus reuteri* *Lactobacillus salivarius* *Bifidobacterium animalis*	(Mountzouris et al., 2009)	Also includes fructo-oligosaccharides

(Continued)

Table 1 *(Continued)*

Product	Supplier	Organism(s)	References[a]	Notes
PrimaLac®	Star-Labs	*Lactobacillus acidophilus* *Lactobacillus casei* *Bifidobacterium bifidum* *Enterococcus faecium*	(Talebi et al., 2008)	
Protexin® and Protexin®Boost	International Animal Health Products (VetaFarm)	*Lactobacillus acidophilus* *Lactobacillus delbrueckii* subsp. *bulgaricus* *Lactobacillus plantarum* *Lactobacillus rhamnosus* *Bifidobacterium bifidum* *Enterococcus faecium* *Streptococcus salivarius* subspecies *thermophilus*	(Kabir et al., 2004)	
Provita Avian	Provita Eurotech	*Bacillus subtilis* *Bacillus licheniformis*		Includes inulin and vitamins
Sporulin®	Pacific Vet Group	3 *Bacillus subtilis*	(Wolfenden et al., 2010)	
Syncra®AVI	Danisco	*Bacillus subtilis*	(Murugesan et al., 2014)	Includes xylanase, amylase and protease

[a]The references provided are only examples; it is not a comprehensive list. Many commercial products are not specifically referenced in the scientific literature. This may be because no peer reviewed work has been published or because the work is not published with the commercial product name.
[b]*Lactobacillus cellobiosus* is no longer a valid name; all *L. cellobiosus* should now be called *Lactobacillus fermentum* (Dellaglio et al., 2004).
[c]*Lactobacillus lactis* is not a valid name; it is unclear what genus of bacteria is being referred to.

2 Prebiotics

The basis of the prebiotic concept is that the products modify the gut microbiota in beneficial ways such as reducing pathogen load and/or increasing the numbers of short-chain-fatty-acid (SCFA)-producing bacteria. Such changes have generally been found to produce benefits for bird health and productivity, although it has also been noted that some of the prebiotics can alter the microbiota but this does not necessarily translate to better bird performance (Geier et al., 2009).

The most widely studied and used class of prebiotics are the non-digestible carbohydrates – non-digestible by the host, but degraded by some bacteria. The rich energy source represented by the carbohydrates survives passage through the upper GIT and promotes the growth of various classes of bacteria that can utilise the complex carbohydrates. Many of the bacteria that utilise the complex carbohydrates produce large quantities of SCFAs, which mediate many of the positive effects of the prebiotics.

Evidence is mounting that organic acids and the fibre that induces the growth of bacteria that produce them are important in the maintenance of gut health in both humans and production animals (Koh et al., 2016). Onrust *et al.* have argued that the use of prebiotics to stimulate the growth of butyrate-producing bacteria, or indeed the direct use of such butyrate-producing bacteria as probiotics, would be advantageous within the poultry industry (Onrust et al., 2015). Some oligosaccharides, such as raffinose and arabinoxylans, may have anti-nutritive activities, but their partially hydrolysed derivatives can show prebiotic activity (Choct and Annison, 1992; Eeckhaut et al., 2008; Iji and Tivey, 1998). Plants rich in various prebiotic oligosaccharides – such as chicory root, onions, garlic, and Jerusalem artichokes – have been widely investigated as sources for these beneficial feed additives. The evidence for the effectiveness of one of the most widely used probiotics – inulin – has recently been reviewed, and it was found that there are quite mixed experiences of its effectiveness, depending on source, dose and length of use (Bucław, 2016).

Much of the work to demonstrate the positive effects that the complex carbohydrate prebiotics have on poultry health and productivity was done in the decades before the rapid advances in microbiota analysis, which are enabled by next-generation DNA sequencing (NGS), had been made. Hence, although the interaction between carbohydrates, GIT bacteria and the host is critical to the success of prebiotic use, there are relatively few detailed studies of the GIT microbiota dynamics following prebiotic use. Some studies have undertaken very limited analysis of specific bacterial populations within the GIT of prebiotic-treated chickens (Xu et al., 2003), but very few have used the power of the NGS 16S ribosomal RNA gene analysis methods to generate a more complete understanding of the microbiota changes that follow prebiotic application. This is an area that needs considerable research attention to characterise the changes occurring and to understand the general characteristics of a highly functional, beneficial GIT microbiota.

A meta-analysis of 78 studies that had investigated the effects of prebiotics on *Salmonella* colonisation of chickens found an overall effect of approximately 65% reduction (Totton et al., 2012). Lactose addition in feed was the most effective prebiotic for reduction of caecal colonisation by *Salmonella*, and the organic acid studies were found to indicate considerable heterogeneity in outcomes. Hanning et al. found some production improvement in slow-growing birds, but not in fast growing birds, when they tested the effects of GOS and FOS supplementation (Hanning et al., 2012). Park et al. found very little change in the gut microbiota when they tested the effects of yeast cell wall extract (Park et al., 2016).

3 The efficacy of probiotics

The efficacy of probiotic products can be difficult to assess. Conventional pharmaceutical products used in the poultry industry, such as antimicrobials and anticoccidials, have clear modes of action that can be observed both *in vitro* an *in vivo*. When used at appropriate dosages they have predictable and reproducible effects that can be clearly measured in terms of bacterial and *Eimeria* loads and disease outcomes. However, this is not necessarily true of probiotics. The exact mode of action of probiotics is generally not defined and the expected effects of probiotic use are not always clearly delineated or easily measurable. For example, 'improvement in gut health' is multifaceted and the particular aspect of gut health that the probiotic is expected to improve is often not stated in the product information. On top of these considerations is the fact that, as live biologicals, the products themselves are more susceptible to significant changes in potency and efficacy due to variations in manufacturing, formulation, storage and use. Perhaps the biggest challenge in maintaining the efficacy of probiotics is in reproducibly controlling the interaction of the live probiotic with the highly complex and variable gut microbiota within which the probiotic must function.

Despite these difficulties there are a number of probiotic products, in particular competitive exclusion products, which have a history of use within the poultry industry. For a few of these products there is a substantial body of scientific literature that describes their successful use to improve the health and/or productivity of birds. As with most issues explored in the scientific literature, care must be taken when interpreting published information; it is generally positive results that are published and it is relatively rare to see negative results published. Some negative or neutral findings do get published; for example, Sarangi and colleagues reported no significant effect of the prebiotic, probiotic, and synbiotic on growth and carcass characteristics (Sarangi et al., 2016). Similarly, Houshmand et al. found no significant effect of a prebiotic or probiotic on broiler performance, intestinal villus height, crypt depth, gut pH or dietary apparent metabolisable energy (Houshmand et al., 2011), but such negative findings are unusual in the scientific literature. Thus it is often difficult to judge the overall reliability and reproducibility of outcomes following probiotic use. The potential user, at times, has to rely on anecdotal accounts and the industry experience of others. A comprehensive meta-analysis of studies that have investigated the efficacy of competitive exclusion products in reducing the carriage of *Salmonella* spp. found that complex undefined experimental competitive exclusion products performed better than most defined commercial products, expect for Preempt™ and Broilact® (Kerr et al., 2013) which also gave good results. Through meta-analysis it was found that spray application on chicks at the hatchery was as effective as direct oral gavage.

Probiotics should not be regarded as 'magic bullets' that can increase the health and/or productivity of any flock. Birds with high levels of gut health are unlikely to show any improvement in performance following probiotic treatment. Probiotics are more likely to be effective in flocks that have suboptimal performance in which there is scope to reduce the level of gut dysbiosis, reduce the carriage of pathogens, improve energy capture from feed or improve immune function.

4 Effects of probiotics and prebiotics

Probiotics and prebiotics have been reported to have significant effects on a wide range of poultry growth, productivity, quality, immune and health outcomes. The range of biological effects that have been noted following the use of prebiotics is largely overlapping with those observed following the use of probiotics. Although there is good statistical evidence for the effectiveness of some products, there is relatively little data addressing biophysical mechanisms underpinning some of the observed benefits. It is likely that in many cases the mechanisms of action are overlapping and interacting rather than being restricted to one clear mode of action.

4.1 Competitive exclusion

The earliest microbial products proposed for use in chickens, the competitive exclusion (CE) products, were originally designed to provide newly hatched chicks with a healthy mature intestinal microbiota (Nurmi and Rantala, 1973). Competitive exclusion products, as the name implies, are mainly directed towards reducing (excluding) the carriage of specific pathogens such as *Salmonella* species, *Campylobacter* species, *Escherichia coli, Listeria monocytogenes* and *Clostridium perfringens* (Corrier et al., 1995; Hofacre et al., 1998; Hume et al., 1998; Soerjadi et al., 1981). It is postulated that the bacteria in CE products directly compete with the target pathogens for occupancy of ecological niches within the gut. The concept is that if the CE bacteria establish in those niches, the niches are no longer available for the undesirable pathogens to colonise and hence the pathogens are 'competitively excluded'. Therefore it would be expected that particular CE products would be most effective against particular, rather than a broad range of, pathogens, although with complex multi-strain CEs, broad protection may be possible. One question that has not been adequately addressed is whether the CE products may affect colonisation by non-target commensal organisms residing in the gut. Such natural residents of the gut may be beneficial and so it would be undesirable to reduce those populations. The efficacy of some CE products may also be enhanced by the influence they have on the local gut environment, for instance lowering the pH and by competing with pathogens for nutrients. Some CE products may also include strains that have direct antimicrobial activity against the target pathogens.

There are a number of different types of CE products. The first to be used were undefined mixtures of bacteria derived from the gut digesta, caeca, mucosal scrapings, or faeces of healthy donor birds. They have been used as directly prepared products or have been amplified using some sort of culturing procedure. Although a number of such products have been shown to be effective in reducing pathogen colonisation, they have a number of problems. Because the microbial contents of such preparations are undefined, it is difficult to ensure consistency of manufacture and to exclude the possibility that undesirable agents will be present. Thus it is difficult, if not impossible, to have such products registered for general use in some jurisdictions, for example, in Europe (European Food Safety Authority (EFSA), 2005). A number of semi-defined CE products have been used. For these products the major classes of bacteria have been defined and the major constituents can be assessed and monitored. The most acceptable products are those which are fully defined with all the components cultured and characterised (Klose et al., 2006). Such products can be fully quality controlled in terms of content and viability. Some single strain probiotic products have also been claimed to be efficacious CE agents.

4.2 Improvements in growth performance

A major motivation in using pre- and probiotic products is to improve the growth performance of birds, whether in terms of FCR and body weight gains in broilers or egg output in layers. Animal trials demonstrating performance gains following pre- and probiotic use have been reported by many authors and for a wide range of both commercial and experimental treatments. Blajman *et al.* have recently published a meta-analysis of randomised controlled trials that have showed the effects of probiotics on broiler growth performance (Blajman et al., 2014) and have found that, among the reports analysed, there was some evidence that the probiotics used did improve body weight gain. However, probiotics are not a homogenous group of products and so it is difficult to draw general conclusions from the wide range of different probiotic strains that have been used; there may be some probiotics that are quite effective, whereas others are not.

To understand the circumstances under which pre- and probiotics may be most effectively used, and to provide a rational basis on which new products could be developed, it would be very valuable to understand the underlying molecular mechanism of their action. There are many possible mechanisms by which pre- and probiotics may improve growth performance. For example, the altered gut microbiota induced by the products may supply enhanced levels of appropriate digestive enzymes or induce higher enzyme production in the host, which in turn can help to better breakdown feed and make nutrients available to the bird (Wang and Gu, 2010). The altered microbiota may also produce more short-chain fatty acids such as acetate, butyrate, and propionate (Onrust et al., 2015). Butyrate is a major energy source for intestinal cells and may also induce the expression of host defence proteins, making birds more resistant to pathogens (Sunkara et al., 2011). There is also evidence that butyrate increases antioxidant capacity (Wu et al., 2016), supresses inappropriate inflammatory activity and effects insulin sensitivity, which could in turn alter energy partitioning to different organs (Mátis et al., 2015). There are some indications that improved nutrient retention may be facilitated by an increase in the size of the small intestine and the villus height to crypt depth ratio, which is induced by some probiotics (Awad et al., 2009; Sen et al., 2012).

4.3 Stimulation of the immune system

A number of studies have found that probiotics can enhance immune function in treated birds as demonstrated by alterations in the abundance of certain types of immune cells and changes in cytokine profiles (Chen et al., 2012; Madej and Bednarczyk, 2016; Yitbarek et al., 2015) and changes in immune responses to model antigens (Brisbin et al., 2011). Stringfellow et al. (2011) noted some increases in monocyte and heterophil activity (oxidative burst) following treatment with a probiotic containing four different strains. The effects were noted when the birds were 1 and 2 weeks of age, but were not sustained at 3 weeks of age (Stringfellow et al., 2011). The state of activation of the immune system can have significant impacts on the health and productivity of birds. An activated immune system is required to fight pathogens that can depress performance, but a state of chronic inflammation can be a drain on energy and productivity. Therefore, optimal productivity relies on the maintenance of immune balance such that birds are protected from pathogenic assaults, but do not divert excessive energy into unnecessary inflammatory or autoimmune responses. Recent findings in mouse models of human autoimmune diseases have shown that ingestion of fibre, in particular the indigestible complex carbohydrates, modulates the composition of gut microbiota and increases the production of organic acids, resulting in the suppression and control of inappropriate autoimmune activity (Macia et al., 2015; Thorburn et al., 2015). It is likely that these findings are very relevant to the immune effects of pre- and probiotics in poultry and may indicate some of the underlying molecular mechanisms that can lead to immune modulation.

Probiotics have also been reported to influence the production of antibodies. Haghighi et al. (2006) reported an increase in serum and secreted antibodies in broilers at 14 days of age following a single oral dose of a probiotic containing strains of *Lactobacillus acidophilus*, *Bifidobacterium bifidum* and *Streptococcus faecalis* (Haghighi et al., 2006). The same *L. acidophilus* strain has been shown to increase the specific antibody response to a model antigen (keyhole limpet hemocyanin) (Brisbin et al., 2011). Talebi et al. (2015)

reported an increase in specific humoral antibodies to Newcastle Disease Virus, Infectious Bursal Disease Virus and Infectious Bronchitis vaccines following continuous treatment of broiler birds with a symbiotic product that contained a strain of *Enterococcus faecium* and several prebiotic components (Talebi et al., 2015).

4.4 Short chain fatty acid production

As emphasised in the productivity and immune sections above, it is becoming increasingly apparent that one of the main drivers of productivity and immune enhancement induced by pre- and probiotic products is the production of short-chain fatty acids (SCFAs). Fibre-rich prebiotics can encourage the growth of SCFA-producing bacteria; some specific probiotic organisms are high-level SCFA producers, and some prebiotics directly supply SCFAs. The positive influence of dietary fibre on host physiology has become apparent from a variety of studies in model systems (Koh et al., 2016). The gut microbiota produces a number of SCFA including acetate, butyrate and propionate. Butyrate is an important energy source for gut enterocytes and is beneficial in maintaining gut barrier function and the structure of intestinal villi. Butyrate has also been shown to induce antimicrobial host defence proteins (Sunkara et al., 2011). Because of these beneficial effects probiotics that stimulate the production of butyrate and prebiotic products containing different forms of butyrate are being increasingly used in the poultry industry. There are many butyrate-containing commercial prebiotic products marketed to chicken growers, but not all are equally effective; small variations in composition can affect the bioavailability of butyrate and the level that can be achieved in the gut, both factors which influence the effects that the additive can exert. Butyrate is a small molecule that is readily absorbed and metabolised in the upper GIT, but for its full effectiveness, it is desirable to deliver butyrate to the lower GIT. Therefore butyrate is most effectively used in a protected form, for example using different coating technologies (Eshak et al., 2016; van den Borne et al., 2015), or in derivatives such as tributyrin (Li et al., 2015). However, research investigating where products are released and how that influences functionality is sorely lacking in the literature.

4.5 Increased effectiveness of mucosal barrier

The intestinal mucosal barrier is the first line of defence against pathogens and also restricts the passage of food antigens. Therefore a strong and properly functioning gut barrier can have important positive effects on bird productivity by reducing pathogen load and by minimising inappropriate inflammatory responses. Some probiotics have been shown in experimental systems to enhance gut barrier function. A number of mechanisms have been suggested as exerting an influence, including direct expression of antimicrobial proteins, induction of host expression of antimicrobial proteins, immune modulation, and induction of mucin production (Smirnov et al., 2005; Tsirtsikos et al., 2012).

4.6 Changes in organ morphology

Pre- and probiotics can change gut morphology, most notably inducing increases in villus height and crypt depth, and altering both the cellular composition of the intestine (Biloni et al., 2013; Calik and Ergün, 2015; Hutsko et al., 2016; Lee et al., 2010) and the gut weight (Wang et al., 2016). To date, researchers have only looked at static measures of morphology, and more dynamic measurement of intestinal turnover or functional cellular programming changes within the GIT have yet to be investigated. Changes in other organs, such as spleen, thymus and cecal tonsils, have also been noted (Awad et al., 2009; Yurong et al., 2005).

4.7 Other effects of pre- and probiotics

A wide range of other effects of pre- and probiotics have been demonstrated and reported in the scientific literature. *pH effects:* Some probiotics may increase pathogen resistance and increase productivity by decreasing the pH of the gut digesta. Proliferation of lactic acid bacteria and other SCFA-producing bacteria can lead to acid production, a lowering of pH, and thus an increase in the activity of some host and microbiota digestive enzymes. A number of potential pathogens are less able to proliferate under low pH conditions. *Protection from stress:* Protection from cold and heat stress has been noted (Huff et al., 2015; Sohail et al., 2013) as has protection from oxidative stress (Anwar et al., 2012; Liao et al., 2015). *Product quality:* Improvement in meat quality including sensory characteristics, as well as in egg and shell quality (Abdelqader et al., 2012; Cesari et al., 2014; Mikulski et al., 2012) and egg production (Gallazzi et al., 2008), has been observed (Maiorano et al., 2012; Pelicano et al., 2003; Zheng et al., 2014). *Environmental improvement:* probiotic use can reduce ammonia and urea excretion and other noxious substances and thus improve manure quality (Ahmed et al., 2014; Park et al., 2016; Terada et al., 1994). *Effects on non-bacterial pathogens:* Some probiotic products may also partially alleviate some of the negative effects of *Eimeria* infection (Abdelrahman et al., 2014; Ritzi et al., 2014). *Haematological changes:* Changes in aspects of blood chemistry have been noted following probiotic use, including increases in serum calcium, potassium and iron, and reduction in triglycerides (Capcarova et al., 2010, 2011; Islam et al., 2009).

5 Selection, delivery and action of probiotic bacteria

5.1 Modes of delivery

Direct oral gavage of probiotics has been used in experimental settings but does not present a realistic option for large-scale treatment of birds in commercial operations. Most currently available probiotics require regular dosing, sometimes on a daily basis; hence, the only practical large-scale administration methods are delivery via the feed or drinking water. Spray application in hatchers to inoculate the newly emerging chicks has also been successful, but does not address the need for repeated dosing. *In ovo* application has also been investigated with some positive results (de Oliveira et al., 2014; Edens et al., 1997; Madej and Bednarczyk, 2016), but it is likely that great caution will be needed as hatchability rates can be severely affected (Cox et al., 1992). The use of probiotic strains *in ovo* may present a risk of inappropriate proliferation and organ invasion of probiotic strains in the rich growth environment of the egg.

5.2 Microbial strains used

A wide variety of bacterial species have been used in probiotic products, both experimental and commercial. *Lactobacillus* and *Bifidobacterium* species of bacteria are popular targets for development as probiotics because of their generally regarded as safe (GRAS) status

Table 2 Microbial species that have been used as probiotics in chickens

Species	Example reference
Lactobacillus	
acidophilus	Gallazzi et al. (2008)
agilis	Lan et al. (2003)
casei	Fajardo et al. (2012)
delbrueckii	Capcarova et al. (2011)
farciminis	European Food Safety Authority (2006)
fermentum	Capcarova et al. (2010)
johnsonii	La Ragione et al. (2004)
helveticus	Capcarova et al. (2011)
paracasei	Cean et al. (2015)
plantarum	Peng et al. (2016)
reuteri	Edens et al. (1997)
rhamnosus	Cean et al. (2015)
salivarius	Zhang et al. (2007)
Bacillus	
amyloliquefaciens	Ahmed et al. (2014)
cereus	Gil de los santos et al. (2012)
coagulans	Wang and Gu (2010)
laterosporus	Wolfenden et al. (2011)
licheniformis	Liu et al. (2012)
subtilis	Harrington et al. (2016)
Bifidobacterium	
animalis	Klose et al. (2006)
bifidum	Bitterncourt et al. (2011)
longum	Santini et al. (2010)
Enterococcus	
faecium	Ghareeb et al. (2012)
faecalis	Han et al. (2013)
Saccharomyces	
boulardii	Gil de los santos et al. (2005)
cerevisiae	Salim et al. (2013)
Streptococcus	
cristus	Zhang et al. (2007)
thermophilus	Capcarova et al. (2011)

(*Continued*)

Table 2 *(Continued)*

Species	Example reference
Aspergillus oryzae	Murugesan et al. (2014)
Candida pintolopessi	Kabir et al. (2004)
Clostridium butyricum	Liao et al. (2015)
Escherichia coli	Huff et al. (2006)
Lactococcus lactis	Brzóska et al. (2012)
Pediococcus acidilactici	Ghareeb et al. (2012)
Pichia pastoris	Gil de los santos et al. (2012)

(Burdock et al., 2006; Feord, 2002). A number of *Bacillus* sp., as well as a number of other bacterial and fungal species, have also been used. Table 2 details some of the microbial species that have been investigated for utility as probiotics. Sometimes particular species are discussed as if all isolates of that species may have probiotic properties, but it should be understood that it is specific isolates that have probiotic activity, rather than a whole class of organism; it is the strain level rather than the species level that is relevant. Thus, although two probiotics may use the same bacterial species they may have quite different efficacy, depending on the specific strains that are incorporated into the product.

5.3 Problems with nomenclature

Some bacterial species claimed to have probiotic activity are misnamed or misidentified. For example *Lactobacillus sporogenes* has been claimed to change some egg qualities (Panda et al., 2008), yet this is not a valid species name. It is unclear what bacterium this study has used. *Streptococcus faecium* has been used when presumably the strain used was actually an isolate of *Enterococcus faecium* (Bitterncourt et al., 2011). Cean et al. (2015) named two potential probiotic strains *Lactobacillus lactis* when presumably they are *Lactococcus lactis*. Similarly, a recent review of poultry probiotics has listed a number of bacteria as commonly used in probiotics, yet they list a number of invalid names, for example *Lactobacillus sporogenes*, *Lactobacillus bulgaricus* and *Lactobacillus cellobiosus* (Dhama et al., 2014). The importance of correct probiotic nomenclature has been highlighted by Hill et al. (2016).

5.4 Selection criteria for probiotics

Most probiotic strains have been initially identified and characterised using various *in vitro* tests such as acid and bile resistance, antimicrobial compound production, and adherence to cultured eukaryotic cells (Babot et al., 2014; Blajman et al., 2015; Bujnakova et al., 2014; Dec et al., 2014; Ehrmann et al., 2002; Garriga et al., 1998; Grimoud et al., 2010; Gusils et al., 1999; Kizerwetter-Świda and Binek, 2016; Noohi et al., 2016; Santini et al., 2010; Taheri et al., 2009). Strains must of course be non-pathogenic and ideally have GRAS status. In considering strains with probiotic potential it is also important to choose those with appropriate biotechnological characteristics that make them suitable for large-scale production, packaging, distribution and use. While all the effective CE products are derived from chickens, some of the probiotics sold into the poultry market are derived from other animals and are claimed to have effects across multiple species, including chickens, turkeys, pigs and cattle. Given the vast differences in GIT environments and resident microbiota between different species, particularly between monogastrics and ruminants (Furet et al., 2009; Lee et al., 2011), it is difficult to understand how a general probiotic can be effective. Perhaps the observed effects are based on the probiotics' efficacy more as a dead feed additive than as a live probiotic. Generally the strains investigated for probiotic activity in poultry have been derived from the target host, and in any new development, it is wise to continue this trend (Rantala and Nurmi, 1973).

5.5 Probiotics: working within the gut microbiota

Probiotics function within the complex milieu of the gut microbiota. The gut microbiota consists predominantly of a great diversity of bacterial strains, as well archea, eukaryotes (e.g. yeasts) and viruses. The intimate interaction of the gut microbiota with the host immune system has profound effects on the regulation of host metabolism, nutrient usage and growth performance (Kogut, 2013). The phylogenetic composition of the microbiota varies along the length of the gastrointestinal tract; relatively sparse and simple, dominated by *Lactobacillus* species in the upper small intestine and much more dense and complex in the caecum and lower GIT (Stanley et al., 2014). The communities that make up the microbiota are strongly influenced by feed composition and initial patterns of colonisation that occur in the earliest hours of a chicken's post-hatch life (Pan and Yu, 2014; Stanley et al., 2013). Although a healthy gut microbiota has some general characteristics in terms of phylogenetic composition and complexity, the fine detail of what constitutes a healthy microbiota can be quite variable. Different flocks of chickens grown under identical conditions can have distinctly different microbiotas, but still perform equally well (Stanley et al., 2013). Much work still remains to be done to understand the parameters that define an optimal gut microbiota that maximises gut health and bird productivity. Because there can be many different microbiota structures that result in healthy chickens, it is, currently, difficult to determine if microbiota perturbations induced by the use of particular probiotics and prebiotics are advantageous, neutral or deleterious to the gut microbiota. The ability to comprehensively characterise the composition of the gut microbiota has been greatly enhanced in recent years by rapid advances in DNA sequencing technology, which have supplied high-throughput culture-free methods for microbiota characterisation. Such methods are well suited to identifying the changes in the gut microbiota brought about by the application of probiotics and prebiotics.

How probiotics interact with the other bacteria within the gut microbiota to produce the end effects on the bird are largely unknown but are likely to be important aspects of the functionality of each probiotic. It can therefore be extrapolated that the intrinsic variabilities in gut microbiota are likely to cause variations in how effective a particular probiotic or prebiotic can be. For example, a probiotic that has beneficial effects within the microbiota established in corn feed chickens may have quite different effects within the very different microbiota established in wheat feed birds.

Of course the main function of CE products is to remodel the gut microbiota so that certain specific pathogens, such as *Salmonella* and *Campylobacter* are reduced in abundance. It is difficult to see how the effects of CE preparations can be specific only for the target pathogenic strains of bacteria. There are usually no in-built intrinsic mechanisms (e.g. very specific antimicrobial proteins) that are likely to limit their actions to only the targets, so it is highly likely that they will also have effects on other members of the gut microbiota.

Most characterisations of the impact of CE agents have only investigated changes in the populations of particular targets with few studies investigating the overall impact on the structure of the whole microbiota. With the rapid advances in microbiota characterisation methods there is now the opportunity to more closely investigate the relationship between effective prebiotic and probiotic administration and the structure and composition of the GIT microbiota; this is an area of research that is ripe for development and there are real opportunities for the suppliers of prebiotic products to provide definitive proof that the products do remodel the GIT microbiota in desirable ways, for example by reducing the carriage of potential pathogens or increasing the number of SCFA-producing bacteria. As our knowledge of what makes a good and bad microbiota develops, it should become increasingly important to demonstrate that newly introduced probiotics, in particular CE products, favour the development of a good microbiota structure.

Microbiota development in chickens raised in modern high biosecurity facilities is somewhat problematic. In the wild, under natural conditions, newly hatched chicks acquire the first bacteria that commence the establishment of the GIT microbiota directly from parent birds and the nest environment. This gives chicks the opportunity to acquire native GIT microbiota that has evolved with birds over millennia. In the modern hatchery, chicks are not exposed to parents, nests, or dirty natural environments. They therefore have no opportunity to acquire normal native microbiota. Instead they become populated with a random collection of foreign bacteria derived from contaminated surfaces within the hatchery, transport boxes, and placement facilities; from the workers in the facilities; and from the first batches of feed and water that are offered to the birds. It has been hypothesised that this random colonisation is at least part of the cause of the large flock to flock variation in GIT microbiota that has been noted (Stanley et al., 2013). The early competitive exclusion products that took whole microbiota from healthy adult birds to inoculate chicks represent one way to tackle this issue. However, young animals are not normally populated with a full adult microbiota but rather undergo a series of microbiota changes over the first several weeks of life (Lu et al., 2003; Oakley et al., 2014). This development of microbiota over time may be important for its functional interaction with the host. The timing of prebiotic and probiotic intervention in this programmed development of the microbiota may be critical to the optimal application of the products.

6 Questions and opportunities regarding the use of probiotics

There are three major types of concern that have been commonly expressed regarding the use of probiotics: (i) concerns about effectiveness and reproducibility of action, (ii) concerns about lack of knowledge regarding mechanisms of action, and (iii) concerns about how users can make an informed choice about which product to use from among the many available ones.

6.1 Effectiveness and reproducibility

Anecdotally, users of probiotics have found that the results achieved can be somewhat variable. This is hardly surprising when the microbial context within which probiotics function is considered. The gut microbiota of animals is highly variable, both between different groups of animals and between individuals in a commonly raised group. A probiotic must work within the context of the gut microbiota to deliver the claimed benefit. It is evident that the currently available probiotics do not efficiently colonise and persist in treated birds as they must be repeatedly dosed, often on a daily basis or as a permanent inclusion in the feed. There are opportunities to develop a new generation of probiotics that are better able to colonise and persist and therefore able to deliver more long-term benefit with less intense delivery.

6.2 Lack of knowledge of the mechanisms of probiotic action

Some of the effects of pre- and probiotics have been well documented but the molecular mechanisms underpinning their action are largely unknown. Are effects directly caused by probiotics/prebiotics or are they secondary actions – does a probiotic need to colonise to be truly effective or can its effect be seen in modification of the gut microbiota?

Although a number of mechanisms of action of probiotics have been suggested (see above), these are largely speculative as there is little definitive information about the specific mechanism involved. Presumably a greater understanding of the molecular basis of probiotic action would open up opportunities to select and develop more appropriate strains tailored for particular action.

6.3 How to make informed choices about what probiotics to use and when

With much still remaining to be learnt about the modes of action of pre- and probiotics, it is a great challenge for poultry producers to determine whether use of such products will help their profitability. There are several pertinent questions for which we do not currently have clear rational answers, as we do not have much data to inform us:

- Can prebiotic or probiotic inclusion improve the performance of flocks that are already healthy and performing well, or are they really only of use in sub-optimal flocks and/or certain environmental conditions and production stages/situations?
- How can poultry producers make a rational comparison and choice between available products?
- Can the 'right' choice be made by taking into account the line of birds being used, the feed formulation, water quality, environmental factors (e.g. temperature, humidity), the health, welfare and microbial history of previous batches of birds within the facility?
- Are probiotics suitable for general broad use across the industry, or could they be more efficaciously used in a more 'personalised medicine' approach to flock health?

The microbiotas in different batches of birds are likely to be very different and provide variable metabolic backgrounds against which probiotics and prebiotics are expected to function. For example if a prebiotic works by selectively enhancing a specific group of bacteria (e.g. SCFA-producing classes), but if that group of bacteria is not present in the resident microbiota, then no amount of supplementation will help. Perhaps the best solution to this potential issue is the use of appropriate synbiotics in which a probiotic is added that can take advantage of the prebiotic used in the formulation.

7 New frontiers and future research directions in probiotic development

7.1 New methods of strain screening and selection

There may be opportunities to develop more effective probiotics with longer-lasting action. Currently most of the commercially available probiotics need to be continually dosed, but it may be possible to develop probiotics that persist within the GIT microbiota and hence only need to be applied once. The standard methods that have been employed to identify potential probiotic strains of bacteria have relied on *in vitro* methods for the initial screening of bacterial isolates. While the *in vitro* screening processes have been used for many years and have evidently been successful in identifying useful probiotic strains it is surprising to find that there is very little compelling proof of the effectiveness of the in vitro screening procedures that have been used (Morelli, 2000). In most studies the *in vitro* tests have been applied to a collection of strains and then a few with properties judged to be favourable have been selected for *in vivo* testing in birds. Very few studies have included any negative control strains in the *in vivo* assessment.

Therefore, it is impossible to determine if the *in vitro* screen is effective or whether most strains within a collection would show some degree of benefit if dosed at a suitable level. *In vitro* and *in vivo* grown bacteria can have very different characteristics, as evidenced by differences in gene expression patterns, and hence there is some reason to be sceptical that *in vitro* tests are a good indicator of *in vivo* characteristics (Hautefort and Hinton, 2000; Tuntufye et al., 2012). It would be valuable to have definitive proof that the *in vitro* screening methods are more likely than random sampling of strains to identify strains with desirable probiotic properties. Therefore there may be opportunities to develop new screening methods to identify specific properties of potential probiotic strains. Perhaps there may be opportunities to develop high-throughput screening methods that can be used *in vivo* or at least define which *in vitro* tests do truly indicate real *in vivo* performance.

7.2 New types of bacteria for use as probiotics

The bacterial strains used in probiotics are most commonly the lactic acid bacteria and the other microbial species indicated in Table 2. At the moment the strict anaerobes are generally not used in products, because of difficulties with growth, formulation, distribution and application, even though there are some species which would otherwise be very interesting to investigate. For instance, *Faecalibacterium prausnitzii* is normally present in healthy chicken GIT microbiota (Stanley et al., 2016) and has been shown in mammalian systems to interact favourably with the host to influence immune system development (Miquel et al., 2013). The segmented filamentous bacteria have been shown to have a very strong influence on some aspects of immune development, but it is only recently that this group of bacteria has been cultured; it would still be challenging to grow such strains as probiotics for commercial use (Ericsson et al., 2014). The colonisation levels of some of these bacteria may be influenced by conventional probiotics (Liao et al., 2012). Until technical solutions become available, perhaps in the form of affordable encapsulation technology or special media that can protect strict anaerobes, such strains, unfortunately, cannot be developed as viable probiotics, although the hope is that they will be useable in the future.

Within the GIT microbiota, many strains that cannot currently be cultured as pure isolates have been identified by the culture-free methods, which are now routinely applied to microbiota analysis. With advances in culturing methods some of these strains may become available for biotechnological development and some may represent good candidates for use as probiotics. There are fundamental developments in bacteriology which we can anticipate will contribute new types of probiotics in the years to come.

7.3 Future research directions

- Continuing efforts are needed to define the mechanisms by which specific prebiotics and probiotics positively influence poultry health and productivity. Such research will tell us under what circumstances the products are likely to be effective and may point to better ways to identify and test new products.
- Research needs to be carried out to definitively demonstrate the value, or the lack of it, of the range of *in vitro* tests that have been widely employed to screen for potential probiotic strains of microorganisms. Despite the use of these tests there are few, if any, well-controlled experiments that have unequivocally demonstrated their value.
- Given the uncertain value of the battery of *in vitro* tests that are currently conducted, it would be interesting to develop *in vivo* testing methods to directly screen potentially probiotic strains by performance in animals.
- There is scope to develop more long-lasting probiotics that do not require the daily dosing that is a feature of many current products.
- Probiotic strains of microorganisms could be engineered to have additional beneficial features. However, the regulatory environment may make acceptance of such products difficult to achieve.
- New production and formulation technologies are required if a new generation of strictly anaerobic bacteria are to be successfully adopted as probiotics. Such bacteria have a number of desirable effects on the host, but are difficult to produce and apply on a commercial scale.
- Ongoing research is needed to further characterise the microbiota within the GIT, the dynamics of its establishment and development, effects on health and productivity, and interactions with pre- and probiotics. Understanding the interaction between native and introduced microbes is important for enhancing the gut health of poultry.

8 Where to look for further information

Further information on pre- and probiotics can be found in the scientific literature and from the websites of manufacturers and suppliers of commercial products. These sources need to be read with a critical eye to assess the quality of the information supplied and the conclusions that have been drawn. There is also a growing volume of popular press articles on gut health and microbiota. Many of these are relevant as general background to inform and assist with the understanding of the significant role that gut bacteria play in health.

9 References

Abdelqader, A., Irshaid, R. and Al-Fataftah, A.-R. (2012). Effects of dietary probiotic inclusion on performance, eggshell quality, cecal microflora composition, and tibia traits of laying hens in the late phase of production. *Trop. Anim. Health Prod.* 45, 1017–24. doi:10.1007/s11250-012-0326-7.

Abdelrahman, W., Mohnl, M., Teichmann, K., Doupovec, B., Schatzmayr, G., Lumpkins, B. and Mathis, G. (2014). Comparative evaluation of probiotic and salinomycin effects on performance and coccidiosis control in broiler chickens. *Poult. Sci.* 93, 3002–8. doi:10.3382/ps.2014-04212.

Ahmed, S. T., Islam, M. M., Mun, H.-S., Sim, H.-J., Kim, Y.-J. and Yang, C.-J. (2014). Effects of *Bacillus amyloliquefaciens* as a probiotic strain on growth performance, cecal microflora, and fecal noxious gas emissions of broiler chickens. *Poult. Sci.* 93, 1963–71. doi:10.3382/ps.2013-03718.

Ajuwon, K. M. (2015). Toward a better understanding of mechanisms of probiotics and prebiotics action in poultry species. *J. Appl. Poult. Res.* 25, 277–83. doi:10.3382/japr/pfv074.

Anwar, H., Rahman, Z. U., Javed, I. and Muhammad, F. (2012). Effect of protein, probiotic, and symbiotic supplementation on serum biological health markers of molted layers. *Poult. Sci.* 91, 2606–13. doi:10.3382/ps.2012-02172.

Applegate, T. J., Klose, V., Steiner, T., Ganner, A. and Schatzmayr, G. (2010). Probiotics and phytogenics for poultry: Myth or reality? *J. Appl. Poult. Res.* 19, 194–210. doi:10.3382/japr.2010-00168.

Awad, W. A., Ghareeb, K., Abdel-Raheem, S. and Bohm, J. (2009). Effects of dietary inclusion of probiotic and synbiotic on growth performance, organ weights, and intestinal histomorphology of broiler chickens. *Poult. Sci.* 88, 49–56. doi:10.3382/ps.2008-00244.

Babot, J. D., Argañaraz-Martínez, E., Saavedra, L., Apella, M. C. and Perez Chaia, A. (2014). Selection of indigenous lactic acid bacteria to reinforce the intestinal microbiota of newly hatched chicken: relevance of *in vitro* and *ex vivo* methods for strains characterization. *Res. Vet. Sci.* 97, 8–17. doi:10.1016/j.rvsc.2014.06.001.

Biloni, A., Quintana, C. F., Menconi, A., Kallapura, G., Latorre, J., Pixley, C., Layton, S., Dalmagro, M., Hernandez-Velasco, X., Wolfenden, A., Hargis, B. M. and Tellez, G. (2013). Evaluation of effects of EarlyBird associated with FloraMax-B11 on *Salmonella* Enteritidis, intestinal morphology, and performance of broiler chickens. *Poult. Sci.* 92, 2337–46. doi:10.3382/ps.2013-03279.

Bischoff, S. C. (2011). 'Gut health': a new objective in medicine? *BMC Med.* 9, 24. doi:10.1186/1741-7015-9-24.

Bitterncourt, L. C., da Silva, C. C., Garcia, P. D. S. R., Donato, D. C. Z., de Albuquerque, R. and Araújo, L. F. (2011). Influence of a probiotic on broiler performance. *Rev. Bras. Zootec.* 40, 2739–43. doi:10.1590/S1516-35982011001200018.

Blajman, J., Gaziano, C., Zbrun, M. V., Soto, L., Astesana, D., Berisvil, A., Scharpen, A. R., Signorini, M. and Frizzo, L. (2015). *In vitro* and *in vivo* screening of native lactic acid bacteria toward their selection as a probiotic in broiler chickens. *Res. Vet. Sci.* 101, 50–6. doi:10.1016/j.rvsc.2015.05.017.

Blajman, J. E., Frizzo, L. S., Zbrun, M. V., Astesana, D. M., Fusari, M. L., Soto, L. P., Rosmini, M. R. and Signorini, M. L. (2014). Probiotics and broiler growth performance: a meta-analysis of randomised controlled trials. *Br. Poult. Sci.* 55, 483–94. doi:10.1080/00071668.2014.931930.

Brisbin, J. T., Gong, J., Orouji, S., Esufali, J., Mallick, A. I., Parvizi, P., Shewen, P. E. and Sharif, S. (2011). Oral treatment of chickens with Lactobacilli influences elicitation of immune responses. *Clin. Vaccine Immunol.* 18, 1447–55. doi:10.1128/CVI.05100-11.

Brzóska, F., Śliwiński, B. and Stecka, K. (2012). Effect of *Lactococcus lactis* vs. *Lactobacillus* spp. bacteria on chicken body weight, mortality, feed conversion and carcass quality. *Ann. Anim. Sci.* 12, 549–59. doi:10.2478/v10220-012-0046-y.

Bucław, M. (2016). The use of inulin in poultry feeding: a review. *J. Anim. Physiol. Anim. Nutr.* doi:10.1111/jpn.12484.

Bujnakova, D., Strakova, E. and Kmet, V. (2014). *In vitro* evaluation of the safety and probiotic properties of Lactobacilli isolated from chicken and calves. *Anaerobe* 29, 118–27. doi:10.1016/j.anaerobe.2013.10.009.

Burdock, G. A., Carabin, I. G. and Griffiths, J. C. (2006). The importance of GRAS to the functional food and nutraceutical industries. *Toxicology* 221, 17–27. doi:10.1016/j.tox.2006.01.012.

Calik, A. and Ergün, A. (2015). Effect of lactulose supplementation on growth performance, intestinal histomorphology, cecal microbial population, and short-chain fatty acid composition of broiler chickens. *Poult. Sci.* 94, 2173–82. doi:10.3382/ps/pev182.

Capcarova, M., Hascik, P., Kolesarova, A., Kacaniova, M., Mihok, M. and Pal, G. (2011). The effect of selected microbial strains on internal milieu of broiler chickens after peroral administration. *Res. Vet. Sci.* 91, 132–7. doi:10.1016/j.rvsc.2010.07.022.

Capcarova, M., Weiss, J., Hrncar, C., Kolesarova, A. and Pal, G. (2010). Effect of *Lactobacillus fermentum* and *Enterococcus faecium* strains on internal milieu, antioxidant status and body weight of broiler chickens: probiotics and internal milieu of chickens. *J. Anim. Physiol. Anim. Nutr.* 94, e215–24. doi:10.1111/j.1439-0396.2010.01010.x.

Cean, A., Stef, L., Simiz, E., Julean, C., Dumitrescu, G., Vasile, A., Pet, E., Drinceanu, D. and Corcionivoschi, N. (2015). Effect of human isolated probiotic bacteria on preventing *Campylobacter jejuni* colonization of poultry. *Foodborne Pathog. Dis.* 12, 122–30. doi:10.1089/fpd.2014.1849.

Cesari, V., Mangiagalli, M. G., Giardini, A., Galimberti, P., Carteri, S., Gallazzi, D. and Toschi, I. (2014). Egg quality and productive performance of laying hens fed different levels of skimmed milk powder added to a diet containing *Lactobacillus acidophilus*. *Poult. Sci.* 93, 1197–201. doi:10.3382/ps.2013-03518.

Chen, C.-Y., Tsen, H.-Y., Lin, C.-L., Yu, B. and Chen, C.-S. (2012). Oral administration of a combination of select lactic acid bacteria strains to reduce the *Salmonella* invasion and inflammation of broiler chicks. *Poult. Sci.* 91, 2139–47. doi:10.3382/ps.2012-02237.

Choct, M. and Annison, G. (1992). Anti-nutritive effect of wheat pentosans in broiler chickens: Roles of viscosity and gut microflora. *Br. Poult. Sci.* 33, 821–34. doi:10.1080/00071669208417524.

Corrier, D. E., Nisbet, D. J., Scanlan, C. M., Hollister, A. G., Caldwell, D. J., Thomas, L. A., Hargis, B. M., Tomkins, T. and Deloach, J. R. (1995). Treatment of commercial broiler chickens with a characterized culture of cecal bacteria to reduce Salmonellae colonization. *Poult. Sci.* 74, 1093–101. doi:10.3382/ps.0741093.

Cox, N. A., Bailey, J. S., Blankenship, L. C. and Gildersleeve, R. P. (1992). *In ovo* administration of a competitive exclusion culture treatment to broiler embryos. *Poult. Sci.* 71, 1781–4. doi:10.3382/ps.0711781.

De Knegt, L. V., Pires, S. M. and Hald, T. (2015). Attributing foodborne salmonellosis in humans to animal reservoirs in the European Union using a multi-country stochastic model. *Epidemiol. Infect.* 143, 1175–86. doi:10.1017/S0950268814001903.

de Oliveira, J. E., van der Hoeven-Hangoor, E., van de Linde, I. B., Montijn, R. C. and van der Vossen, J. M. B. M. (2014). *In ovo* inoculation of chicken embryos with probiotic bacteria and its effect on posthatch *Salmonella* susceptibility. *Poult. Sci.* 93, 818–29. doi:10.3382/ps.2013-03409.

Dec, M., Puchalski, A., Urban-Chmiel, R. and Wernicki, A. (2014). Screening of *Lactobacillus* strains of domestic goose origin against bacterial poultry pathogens for use as probiotics. *Poult. Sci.* 93, 2464–72. doi:10.3382/ps.2014-04025.

Dhama, K., Tiwari, R., Khan, R. U., Chakraborty, S., Gopi, M., Karthik, K., Saminathan, M., Desingu, P. A. and Sunkara, L. T. (2014). Growth promoters and novel feed additives improving poultry production and health, bioactive principles and beneficial applications: The trends and advances-a review. *Int. J. Pharmacol.* 10, 129–59. doi:10.3923/ijp.20l4.129.159.

Dibner, J. J. and Richards, J. D. (2005). Antibiotic growth promoters in agriculture: history and mode of action. *Poult. Sci.* 84, 634–43. doi:10.1093/ps/84.4.634.

Edens, F. W., Parkhurst, C. R., Casas, I. A. and Dobrogosz, W. J. (1997). Principles of *ex ovo* competitive exclusion and *in ovo* administration of *Lactobacillus reuteri*. *Poult. Sci.* 76, 179–96. doi:10.1093/ps/76.1.179.

Eeckhaut, V., Immerseel, F. V., Dewulf, J., Pasmans, F., Haesebrouck, F., Ducatelle, R., Courtin, C. M., Delcour, J. A. and Broekaert, W. F. (2008). Arabinoxylooligosaccharides from wheat bran inhibit *Salmonella* colonization in broiler chickens. *Poult. Sci.* 87, 2329–34. doi:10.3382/ps.2008-00193.

Ehrmann, M. A., Kurzak, P., Bauer, J. and Vogel, R. F. (2002). Characterization of lactobacilli towards their use as probiotic adjuncts in poultry. *J. Appl. Microbiol.* 92, 966–75. doi:10.1046/j.1365-2672.2002.01608.x.

Ericsson, A. C., Hagan, C. E., Davis, D. J. and Franklin, C. L. (2014). Segmented filamentous bacteria: commensal microbes with potential effects on research. *Comp. Med.* 64, 90–9.

Eshak, M. G., Elmenawey, M. A., Atta, A., Gharib, H. B., Shalaby, B. and Awaad, M. H. H. (2016). The efficacy of Na-butyrate encapsulated in palm fat on performance of broilers infected with necrotic enteritis with gene expression analysis. *Vet. World* 9, 450–7. doi:10.14202/vetworld.2016.450-457.

European Food Safety Authority (EFSA) (2006). Opinion of the Panel on additives and products or substances used in animal feed (FEEDAP) on the safety of the micro-organism product 'Biacton' (*Lactobacillus farciminis*) for chickens for fattening, turkeys and laying hens for use as a feed additive in accordance with Council Directive 70/524/EEC. *EFSA J.* 4, 377. doi:10.2903/j.efsa.2006.377.

European Food Safety Authority (EFSA) (2005). Opinion of the Scientific Committee on a request from EFSA related to a generic approach to the safety assessment by EFSA of microorganisms used in food/feed and the production of food/feed additives: Opinion of the Scientific Committee on a request from EFSA related to a generic approach to the safety assessment by EFSA of microorganisms used in fo. *EFSA J.* 3, 226. doi:10.2903/j.efsa.2005.226.

Fajardo, P., Pastrana, L., Méndez, J., Rodríguez, I., Fuciños, C. and Guerra, N. P. (2012). Effects of feeding of two potentially probiotic preparations from lactic acid bacteria on the performance and faecal microflora of broiler chickens. *Sci. World J.* 2012, 562635. doi:10.1100/2012/562635.

FAO/WHO (2006). Probiotics in food: Health and nutritional properties and guidelines for evaluation. FAO food and nutrition paper 85. FAO, Rome.

Feord, J. (2002). Lactic acid bacteria in a changing legislative environment. *Anton. Leeuw.* 82, 353–60. doi:10.1023/A:1020648018998.

Furet, J.-P., Firmesse, O., Gourmelon, M., Bridonneau, C., Tap, J., Mondot, S., Doré, J. and Corthier, G. (2009). Comparative assessment of human and farm animal faecal microbiota using real-time quantitative PCR. *FEMS Microbiol. Ecol.* 68, 351–62. doi:10.1111/j.1574-6941.2009.00671.x.

Gallazzi, D., Giardini, A., Mangiagalli, G. M., Marelli, S., Ferrazzi, V., Orsi, C. and Cavalchini, G. L. (2008). Effects of *Lactobacillus acidophilus* D2/CSL on laying hen performance. *Ital. J. Anim. Sci.* 7, 27–37. doi:10.4081/ijas.2008.27.

Garriga, M., Pascual, M., Monfort, J. M. and Hugas, M. (1998). Selection of lactobacilli for chicken probiotic adjuncts. *J. Appl. Microbiol.* 84, 125–32. doi:10.1046/j.1365-2672.1997.00329.x.

Geier, M. S., Torok, V. A., Allison, G. E., Ophel-Keller, K. and Hughes, R. J. (2009). Indigestible carbohydrates alter the intestinal microbiota but do not influence the performance of broiler chickens. *J. Appl. Microbiol.* 106, 1540–8. doi:10.1111/j.1365-2672.2008.04116.x.

Ghareeb, K., Awad, W. A., Mohnl, M., Porta, R., Biarnés, M., Böhm, J. and Schatzmayr, G. (2012). Evaluating the efficacy of an avian-specific probiotic to reduce the colonization of *Campylobacter jejuni* in broiler chickens. *Poult. Sci.* 91, 1825–32. .doi:10.3382/ps.2012-02168

Gil de los santos, J. R., Storch, O. B., Fernandes, C. G. and Gil-turnes, C. (2012). Evaluation in broilers of the probiotic properties of *Pichia pastoris* and a recombinant *P. pastoris* containing the *Clostridium perfringens* alpha toxin gene. *Vet. Microbiol.* 156, 448–51. doi:10.1016/j.vetmic.2011.11.019.

Gil de los santos, J. R., Storch, O. B. and Gil-turnes, C. (2005). *Bacillus cereus* var. toyoii and *Saccharomyces boulardii* increased feed efficiency in broilers infected with *Salmonella enteritidis. Br. Poult. Sci.* 46, 494–7. doi:10.1080/00071660500181461.

Gould, L. H., Walsh, K. A., Vieira, A. R., Herman, K., Williams, I. T., Hall, A. J. and Cole, D. (2013). Surveillance for foodborne disease outbreaks – United States, 1998-2008. *Morb. Mortal. Wkly. Rep. Surveill. Summ.* 62, 1–34.

Grimoud, J., Durand, H., Courtin, C., Monsan, P., Ouarné, F., Theodorou, V., and Roques, C. (2010). In vitro screening of probiotic lactic acid bacteria and prebiotic glucooligosaccharides to select effective synbiotics. *Anaerobe* 16, 493–500. doi:10.1016/j.anaerobe.2010.07.005.

Gusils, C., Pérez Chaia, A., González, S. and Oliver, G. (1999). Lactobacilli isolated from chicken intestines: potential use as probiotics. *J. Food Prot.* 62, 252–6.
Haghighi, H. R., Gong, J., Gyles, C. L., Hayes, M. A., Zhou, H., Sanei, B., Chambers, J. R. and Sharif, S. (2006). Probiotics stimulate production of natural antibodies in chickens. *Clin. Vaccine Immunol.* 13, 975–80. doi:10.1128/CVI.00161-06.
Hajati, H. and Rezaei, M. (2010). The application of prebiotics in poultry production. *Int. J. Poult. Sci.* 9, 298–304.
Han, W., Zhang, X. L., Wang, D. W., Li, L. Y., Liu, G. L., Li, A. K. and Zhao, Y. X. (2013). Effects of microencapsulated *Enterococcus fecalis* CG1.0007 on growth performance, antioxidation activity, and intestinal microbiota in broiler chickens. *J. Anim. Sci.* 91, 4374–82. doi:10.2527/jas.2012-5956.
Hanning, I., Clement, A., Owens, C., Park, S. H., Pendleton, S., Scott, E. E., Almeida, G., Gil, F. G. and Ricke, S. C. (2012). Assessment of production performance in 2 breeds of broilers fed prebiotics as feed additives. *Poult. Sci.* 91, 3295–9. doi:10.3382/ps.2012-02557.
Harrington, D., Sims, M. and Kehlet, A. B. (2016). Effect of *Bacillus subtilis* supplementation in low energy diets on broiler performance. *J. Appl. Poult. Res.* 25, 29–39. doi:10.3382/japr/pfv057.
Hautefort, I. and Hinton, J. C. (2000). Measurement of bacterial gene expression *in vivo*. *Philos. Trans. R. Soc. B Biol. Sci.* 355, 601–11.
Havenstein, G. B., Ferket, P. R. and Qureshi, M. A. (2003). Growth, livability, and feed conversion of 1957 versus 2001 broilers when fed representative 1957 and 2001 broiler diets. *Poult. Sci.* 82, 1500–8. doi:10.1093/ps/82.10.1500.
Herrero, M. and Thornton, P. K. (2013). Livestock and global change: Emerging issues for sustainable food systems. *Proc. Natl. Acad. Sci. U. S. A.* 110, 20878–81. doi:10.1073/pnas.1321844111.
Higgins, S. E., Torres-Rodriguez, A., Vicente, J. L., Sartor, C. D., Pixley, C. M., Nava, G. M., Tellez, G., Barton, J. T. and Hargis, B. M. (2005). Evaluation of intervention strategies for idiopathic diarrhea in commercial turkey brooding houses. *J. Appl. Poult. Res.* 14, 345–8. doi:10.1093/japr/14.2.345
Hill, C., Guarner, F., Reid, G., Gibson, G. R., Merenstein, D. J., Pot, B., Morelli, L., Canani, R. B., Flint, H. J., Salminen, S., Calder, P. C. and Sanders, M. E. (2014). Expert consensus document: The International Scientific Association for Probiotics and Prebiotics consensus statement on the scope and appropriate use of the term probiotic. *Nat. Rev. Gastroenterol. Hepatol.* 11, 506–14. doi:10.1038/nrgastro.2014.66.
Hill, C., Scott, K., Klaenhammer, T. R., Quigley, E., and Sanders, M. E. (2016). Probiotic nomenclature matters. *Gut Microbes* 7, 1–2. doi:10.1080/19490976.2015.1127484.
Hofacre, C. L., Froyman, R., Gautrias, B., George, B., Goodwin, M. A. and Brown, J. (1998). Use of Aviguard and other intestinal bioproducts in experimental *Clostridium perfringens*-associated necrotizing enteritis in broiler chickens. *Avian Dis.* 42, 579–84.
Houshmand, M., Azhar, K., Zulkifli, I., Bejo, M. H. and Kamyab, A. (2011). Effects of nonantibiotic feed additives on performance, nutrient retention, gut pH, and intestinal morphology of broilers fed different levels of energy. *J. Appl. Poult. Res.* 20, 121–8. doi:10.3382/japr.2010-00171.
Huff, G. R., Huff, W. E., Enders, C., Sonnenborn, U., Rath, N. C., Farnell, M. B., Solis de los santos, F. and Donoghue, A. M. (2006). Oral treatment with the probiotic *Escherichia coli* Nissle 1917 improves body weight and modulates the stress response of poultry in respiratory challenges with avian pathogenic *E. coli*. EPC 2006-12th Eur. Poult. Conf. Verona Italy 10–14 September 2006.
Huff, G. R., Huff, W. E., Rath, N. C., El-Gohary, F. A., Zhou, Z. Y. and Shini, S. (2015). Efficacy of a novel prebiotic and a commercial probiotic in reducing mortality and production losses due to cold stress and *Escherichia coli* challenge of broiler chicks. *Poult. Sci.* 94, 918–26. doi:10.3382/ps/pev068.
Hume, M. E., Byrd, J. A., Stanker, L. H. and Ziprin, R. L. (1998). Reduction of caecal *Listeria monocytogenes* in Leghorn chicks following treatment with a competitive exclusion culture (PREEMPT™). *Lett. Appl. Microbiol.* 26, 432–6. doi:10.1046/j.1472-765X.1998.00371.x.
Hutsko, S. L., Meizlisch, K., Wick, M. and Lilburn, M. S. (2016). Early intestinal development and mucin transcription in the young poult with probiotic and mannan oligosaccharide prebiotic supplementation. *Poult. Sci.* 95, 1173–8. doi:10.3382/ps/pew019.

Iji, P. A. and Tivey, D. R. (1998). Natural and synthetic oligosaccharides in broiler chicken diets. *World Poult. Sci. J.* 54, 129–43. doi:10.1079/WPS19980010.

Islam, M., Rahman, M., Kabir, S., Kamruzzaman, S. and Islam, M. (2009). Effects of probiotics supplementation on growth performance and certain haemato-biochemical parameters in broiler chickens. *Bangladesh J. Vet. Med.* 2, 39-43. doi:10.3329/bjvm.v2i1.1933.

Kabir, S. M. L. (2009). The role of probiotics in the poultry industry. *Int. J. Mol. Sci.* 10, 3531–46. doi:10.3390/ijms10083531.

Kabir, S. M. L., Rahman, M. M., Rahman, M. B., Rahman, M. M. and Ahmed, S. U. (2004). The dynamics of probiotics on growth performance and immune response in broilers. *Int. J. Poult. Sci.* 3, 361–4. doi:10.3923/ijps.2004.361.364.

Kerr, A. K., Farrar, A. M., Waddell, L. A., Wilkins, W., Wilhelm, B. J., Bucher, O., Wills, R. W., Bailey, R. H., Varga, C., McEwen, S. A. and Rajić, A. (2013). A systematic review-meta-analysis and meta-regression on the effect of selected competitive exclusion products on *Salmonella* spp. prevalence and concentration in broiler chickens. *Prev. Vet. Med.* 111, 112–25. doi:10.1016/j.prevetmed.2013.04.005.

Kiarie, E., Romero, L. F. and Nyachoti, C. M. (2013). The role of added feed enzymes in promoting gut health in swine and poultry. *Nutr. Res. Rev.* 26, 71–88. doi:10.1017/S0954422413000048.

Kizerwetter-Świda, M. and Binek, M. (2016). Assessment of potentially probiotic properties of *Lactobacillus* strains isolated from chickens. *Pol. J. Vet. Sci.* 19, 15–20. doi:10.1515/pjvs-2016-0003.

Klose, V., Mohnl, M., Plail, R., Schatzmayr, G. and Loibner, A.-P. (2006). Development of a competitive exclusion product for poultry meeting the regulatory requirements for registration in the European Union. *Mol. Nutr. Food Res.* 50, 563–71. doi:10.1002/mnfr.200500166.

Kogut, M. H. (2013). The gut microbiota and host innate immunity: Regulators of host metabolism and metabolic diseases in poultry? *J. Appl. Poult. Res.* 22, 637–46. doi:10.3382/japr.2013-00741.

Koh, A., De Vadder, F., Kovatcheva-Datchary, P. and Bäckhed, F. (2016). From dietary fiber to host physiology: Short-chain fatty acids as key bacterial metabolites. *Cell* 165, 1332–45. doi:10.1016/j.cell.2016.05.041.

La Ragione, R. M., Narbad, A., Gasson, M. J. and Woodward, M. J. (2004). *In vivo* characterization of *Lactobacillus johnsonii* FI9785 for use as a defined competitive exclusion agent against bacterial pathogens in poultry. *Lett. Appl. Microbiol.* 38, 197–205. doi:10.1111/j.1472-765X.2004.01474.x.

Lan, P. T. N., Binh, L. T. and Benno, Y. (2003). Impact of two probiotic *Lactobacillus* strains feeding on fecal lactobacilli and weight gains in chicken. *J. Gen. Appl. Microbiol.* 49, 29–36. doi:10.2323/jgam.49.29.

Lee, J. E., Lee, S., Sung, J. and Ko, G. (2011). Analysis of human and animal fecal microbiota for microbial source tracking. *ISME J.* 5, 362–5. doi:10.1038/ismej.2010.120.

Lee, K. W., Lee, S. H., Lillehoj, H. S., Li, G. X., Jang, S. I., Babu, U. S., Park, M. S., Kim, D. K., Lillehoj, E. P., Neumann, A. P., Rehberger, T. G. and Siragusa, G. R. (2010). Effects of direct-fed microbials on growth performance, gut morphometry, and immune characteristics in broiler chickens. *Poult. Sci.* 89, 203–16. doi:10.3382/ps.2009-00418.

Liao, N., Yin, Y., Sun, G., Xiang, C., Liu, D., Yu, H. D. and Wang, X. (2012). Colonization and distribution of segmented filamentous bacteria (SFB) in chicken gastrointestinal tract and their relationship with host immunity. *FEMS Microbiol. Ecol.* 81, 395–406. doi:10.1111/j.1574-6941.2012.01362.x.

Liao, X. D., Ma, G., Cai, J., Fu, Y., Yan, X. Y., Wei, X. B. and Zhang, R. J. (2015). Effects of *Clostridium butyricum* on growth performance, antioxidation, and immune function of broilers. *Poult. Sci.* 94, 662–7. doi:10.3382/ps/pev038.

Liu, X., Yan, H., Lv, L., Xu, Q., Yin, C., Zhang, K., Wang, P. and Hu, J. (2012). Growth performance and meat quality of broiler chickens supplemented with *Bacillus licheniformis* in drinking water. *Asian-Australas. J. Anim. Sci.* 25, 682–9. doi:10.5713/ajas.2011.11334.

Lu, J., Idris, U., Harmon, B., Hofacre, C., Maurer, J. J. and Lee, M. D. (2003). Diversity and succession of the intestinal bacterial community of the maturing broiler chicken. *Appl. Environ. Microbiol.* 69, 6816–24. doi:10.1128/AEM.69.11.6816-6824.2003.

Macia, L., Tan, J., Vieira, A. T., Leach, K., Stanley, D., Luong, S., Maruya, M., Ian McKenzie, C., Hijikata, A., Wong, C., Binge, L., Thorburn, A. N., Chevalier, N., Ang, C., Marino, E., Robert, R., Offermanns, S., Teixeira, M. M., Moore, R. J., Flavell, R. A., Fagarasan, S. and Mackay, C. R. (2015). Metabolite-sensing receptors GPR43 and GPR109A facilitate dietary fibre-induced gut homeostasis through regulation of the inflammasome. *Nat. Commun.* 6, 6734. doi:10.1038/ncomms7734.

Madej, J. P. and Bednarczyk, M. (2016). Effect of *in ovo*-delivered prebiotics and synbiotics on the morphology and specific immune cell composition in the gut-associated lymphoid tissue. *Poult. Sci.* 95, 19–29. doi:10.3382/ps/pev291.

Maiorano, G., Sobolewska, A., Cianciullo, D., Walasik, K., Elminowska-Wenda, G., Sławińska, A., Tavaniello, S., Żylińska, J., Bardowski, J. and Bednarczyk, M. (2012). Influence of *in ovo* prebiotic and synbiotic administration on meat quality of broiler chickens. *Poult. Sci.* 91, 2963–9. doi:10.3382/ps.2012-02208.

Mátis, G., Kulcsár, A., Turowski, V., Fébel, H., Neogrády, Z. and Huber, K. (2015). Effects of oral butyrate application on insulin signaling in various tissues of chickens. *Domest. Anim. Endocrinol.* 50, 26–31. doi:10.1016/j.domaniend.2014.07.004.

Mikulski, D., Jankowski, J., Naczmanski, J., Mikulska, M. and Demey, V. (2012). Effects of dietary probiotic (*Pediococcus acidilactici*) supplementation on performance, nutrient digestibility, egg traits, egg yolk cholesterol, and fatty acid profile in laying hens. *Poult. Sci.* 91, 2691–700. doi:10.3382/ps.2012-02370.

Miquel, S., Martín, R., Rossi, O., Bermúdez-Humarán, L., Chatel, J., Sokol, H., Thomas, M., Wells, J. and Langella, P. (2013). *Faecalibacterium prausnitzii* and human intestinal health. *Curr. Opin. Microbiol.* 16, 255–61. doi:10.1016/j.mib.2013.06.003.

Morelli, L. (2000). *In vitro* selection of probiotic lactobacilli: a critical appraisal. *Curr. Issues Intest. Microbiol.* 1, 59–67.

Mountzouris, D. K. C., Balaskas, C., Xanthakos, I., Tzivinikou, A. and Fegeros, K. (2009). Effects of a multi-species probiotic on biomarkers of competitive exclusion efficacy in broilers challenged with *Salmonella* Enteritidis. *Br. Poult. Sci.* 50, 467–78. doi:10.1080/00071660903110935.

Murugesan, G. R., Gabler, N. K. and Persia, M. E. (2014). Effects of direct-fed microbial supplementation on broiler performance, intestinal nutrient transport and integrity under experimental conditions with increased microbial challenge. *Br. Poult. Sci.* 55, 89–97. doi:10.1080/00071668.2013.865834.

Noohi, N., Ebrahimipour, G., Rohani, M., Talebi, M. and Pourshafie, M. R. (2016). Evaluation of potential probiotic characteristics and antibacterial effects of strains of Pediococcus species isolated from broiler chickens. *Br. Poult. Sci.* 1–7. doi:10.1080/00071668.2016.1169247.

Nurmi, E. and Rantala, M. (1973). New aspects of *Salmonella* infection in broiler production. *Nature* 241, 210–11. doi:10.1038/241210a0.

Oakley, B. B., Buhr, R. J., Ritz, C. W., Kiepper, B. H., Berrang, M. E., Seal, B. S. and Cox, N. A. (2014). Successional changes in the chicken cecal microbiome during 42 days of growth are independent of organic acid feed additives. *BMC Vet. Res.* 10, 282. doi:10.1186/s12917-014-0282-8.

Onrust, L., Ducatelle, R., Van Driessche, K., De Maesschalck, C., Vermeulen, K., Haesebrouck, F., Eeckhaut, V. and Van Immerseel, F. (2015). Steering endogenous butyrate production in the intestinal tract of broilers as a tool to improve gut health. *Front. Vet. Sci.* 2. doi:10.3389/fvets.2015.00075.

Pan, D. and Yu, Z. (2014). Intestinal microbiome of poultry and its interaction with host and diet. *Gut Microbes* 5, 108–19. doi:10.4161/gmic.26945.

Panda, A. K., Rama Rao, S. S., Raju, M. V. and Sharma, S. S. (2008). Effect of probiotic (*Lactobacillus sporogenes*) feeding on egg production and quality, yolk cholesterol and humoral immune response of White Leghorn layer breeders. *J. Sci. Food Agric.* 88, 43–47. doi:10.1002/jsfa.2921.

Park, J. W., Jeong, J. S., Lee, S. I. and Kim, I. H. (2016). Effect of dietary supplementation with a probiotic (*Enterococcus faecium*) on production performance, excreta microflora, ammonia emission, and nutrient utilization in ISA brown laying hens. *Poult. Sci.* 241. doi:10.3382/ps/pew241.

Park, S. H., Lee, S. I., and Ricke, S. C. (2016). Microbial populations in naked neck chicken ceca raised on pasture flock fed with commercial yeast cell wall prebiotics via an Illumina MiSeq platform. *PLoS ONE* 11, e0151944. doi:10.1371/journal.pone.0151944.

Pelicano, E. R. L., Souza, P. de, Souza, H. de, Oba, A., Norkus, E. A., Kodawara, L. M. and Lima, T. de (2003). Effect of different probiotics on broiler carcass and meat quality. *Rev. Bras. Ciênc. Avícola* 5, 207–14. doi:10.1590/S1516-635X2003000300009.

Peng, Q., Zeng, X. F., Zhu, J. L., Wang, S., Liu, X. T., Hou, C. L., Thacker, P. A. and Qiao, S. Y. (2016). Effects of dietary *Lactobacillus plantarum* B1 on growth performance, intestinal microbiota, and short chain fatty acid profiles in broiler chickens. *Poult. Sci.* 95, 893–900. doi:10.3382/ps/pev435.

Rantala, M. and Nurmi, E. (1973). Prevention of the growth of *Salmonella* infantis in chicks by the flora of the alimentary tract of chickens. *Br. Poult. Sci.* 14, 627–30. doi:10.1080/00071667308416073.

Ritzi, M. M., Abdelrahman, W., Mohnl, M. and Dalloul, R. A. (2014). Effects of probiotics and application methods on performance and response of broiler chickens to an *Eimeria* challenge. *Poult. Sci.* 93, 2772–8. doi:10.3382/ps.2014-04207.

Roberfroid, M. (2007). Prebiotics: The concept revisited. *J. Nutr.* 137, 830S–837S.

Sadeghi, A. A., Shawrang, P. and Shakorzadeh, S. (2014). Immune response of *Salmonella* challenged broiler chickens fed diets containing Gallipro, a *Bacillus subtilis* probiotic. *Probiotics Antimicrob. Prot.* 7, 24–30. doi:10.1007/s12602-014-9175-1.

Salim, H. M., Kang, H. K., Akter, N., Kim, D. W., Kim, J. H., Kim, M. J., Na, J. C., Jong, H. B., Choi, H. C., Suh, O. S. and Kim, W. K. (2013). Supplementation of direct-fed microbials as an alternative to antibiotic on growth performance, immune response, cecal microbial population, and ileal morphology of broiler chickens. *Poult. Sci.* 92, 2084–90. doi:10.3382/ps.2012-02947.

Santini, C., Baffoni, L., Gaggia, F., Granata, M., Gasbarri, R., Di Gioia, D. and Biavati, B. (2010). Characterization of probiotic strains: An application as feed additives in poultry against *Campylobacter jejuni*. *Int. J. Food Microbiol.* 141, S98–S108. doi:10.1016/j.ijfoodmicro.2010.03.039.

Sarangi, N. R., Babu, L. K., Kumar, A., Pradhan, C. R., Pati, P. K. and Mishra, J. P. (2016). Effect of dietary supplementation of prebiotic, probiotic, and synbiotic on growth performance and carcass characteristics of broiler chickens. *Vet. World* 9, 313–19. doi:10.14202/vetworld.2016.313-319.

Sasaki, Y., Haruna, M., Murakami, M., Hayashida, M., Takahashi, N., Urushiyama, T., Ito, K. and Yamada, Y. (2014). Contamination of poultry products with *Listeria monocytogenes* at poultry processing plants. *J. Vet. Med. Sci.* 76, 129–32. doi:10.1292/jvms.13-0267.

Schneitz, C. (2005). Competitive exclusion in poultry – 30 years of research. *Food Control* 16, 657–67. doi:10.1016/j.foodcont.2004.06.002.

Sen, S., Ingale, S. L., Kim, Y. W., Kim, J. S., Kim, K. H., Lohakare, J. D., Kim, E. K., Kim, H. S., Ryu, M. H., Kwon, I. K. and Chae, B. J. (2012). Effect of supplementation of *Bacillus subtilis* LS 1-2 to broiler diets on growth performance, nutrient retention, caecal microbiology and small intestinal morphology. *Res. Vet. Sci.* 93, 264–8. doi:10.1016/j.rvsc.2011.05.021.

Smirnov, A., Perez, R., Amit-Romach, E., Sklan, D. and Uni, Z. (2005). Mucin dynamics and microbial populations in chicken small intestine are changed by dietary probiotic and antibiotic growth promoter supplementation. *J. Nutr.* 135, 187–92.

Soerjadi, A. S., Stehman, S. M., Snoeyenbos, G. H., Weinack, O. M., and Smyser, C. F. (1981) Some measurements of protection against paratyphoid *Salmonella* and *Escherichia coli* by competitive exclusion in chickens. *Avian Dis.* 25, 706–12. doi:10.2307/1590001.

Sohail, M. U., Ijaz, A., Younus, M., Shabbir, M. Z., Kamran, Z., Ahmad, S., Anwar, H., Yousaf, M. S., Ashraf, K., Shahzad, A. H. and Rehman, H. (2013). Effect of supplementation of mannan oligosaccharide and probiotic on growth performance, relative weights of viscera, and population of selected intestinal bacteria in cyclic heat-stressed broilers. *J. Appl. Poult. Res.* 22, 485–91. doi:10.3382/japr.2012-00682.

Stanley, D., Geier, M. S., Hughes, R. J., Denman, S. E. and Moore, R. J. (2013). Highly variable microbiota development in the chicken gastrointestinal tract. *PLoS ONE* 8, e84290. doi:10.1371/journal.pone.0084290.

Stanley, D., Hughes, R. J., Geier, M. S. and Moore, R. J. (2016). Bacteria within the gastrointestinal tract microbiota correlated with improved growth and feed conversion: Challenges presented for the identification of performance enhancing probiotic bacteria. *Front. Microbiol.* 7, 187. doi:10.3389/fmicb.2016.00187.

Stanley, D., Hughes, R. J. and Moore, R. J. (2014). Microbiota of the chicken gastrointestinal tract: influence on health, productivity and disease. *Appl. Microbiol. Biotechnol.* 98, 4301–10. doi:10.1007/s00253-014-5646-2.

Stringfellow, K., Caldwell, D., Lee, J., Mohnl, M., Beltran, R., Schatzmayr, G., Fitz-Coy, S., Broussard, C. and Farnell, M. (2011). Evaluation of probiotic administration on the immune response of coccidiosis-vaccinated broilers. *Poult. Sci.* 90, 1652–8. doi:10.3382/ps.2010-01026.

Sunkara, L. T., Achanta, M., Schreiber, N. B., Bommineni, Y. R., Dai, G., Jiang, W., Lamont, S., Lillehoj, H. S., Beker, A., Teeter, R. G. and Zhang, G. (2011). Butyrate enhances disease resistance of chickens by inducing antimicrobial host defense peptide gene expression. *PLoS ONE* 6, e27225. doi:10.1371/journal.pone.0027225.

Taheri, H. R., Moravej, H., Tabandeh, F., Zaghari, M. and Shivazad, M. (2009). Screening of lactic acid bacteria toward their selection as a source of chicken probiotic. *Poult. Sci.* 88, 1586–93. doi:10.3382/ps.2009-00041.

Talebi, A., Amani, A., Pourmahmod, M., Saghaei, P. and Rezaie, R. (2015). Synbiotic enhances immune responses against infectious bronchitis, infectious bursal disease, Newcastle disease and avian influenza in broiler chickens. *Vet. Res. Forum* 6, 191–7.

Talebi, A., Amirzadeh, B., Mokhtari, B. and Gahri, H. (2008). Effects of a multi-strain probiotic (PrimaLac) on performance and antibody responses to Newcastle disease virus and infectious bursal disease virus vaccination in broiler chickens. *Avian Pathol.* 37, 509–12. doi:10.1080/03079450802356995.

Teo, A. Y. and Tan, H.-M. (2007). Evaluation of the performance and intestinal gut microflora of broilers fed on corn-soy diets supplemented with *Bacillus subtilis* PB6 (CloSTAT). *J. Appl. Poult. Res.* 16, 296–303. doi:10.1093/japr/16.3.296.

Terada, A., Hara, H., Sakamoto, J., Sato, N., Takagi, S., Mitsuoka, T., Mino, R., Hara, K., Fujimori, I. and Yamada, T. (1994). Effects of dietary supplementation with lactosucrose (4G-beta-D-galactosylsucrose) on cecal flora, cecal metabolites, and performance in broiler chickens. *Poult. Sci.* 73, 1663–72.

Thorburn, A. N., McKenzie, C. I., Shen, S., Stanley, D., Macia, L., Mason, L. J., Roberts, L. K., Wong, C. H. Y., Shim, R., Robert, R., Chevalier, N., Tan, J. K., Mariño, E., Moore, R. J., Wong, L., McConville, M. J., Tull, D. L., Wood, L. G., Murphy, V. E., Mattes, J., Gibson, P. G. and Mackay, C. R. (2015). Evidence that asthma is a developmental origin disease influenced by maternal diet and bacterial metabolites. *Nat. Commun.* 6, 7320. doi:10.1038/ncomms8320.

Totton, S. C., Farrar, A. M., Wilkins, W., Bucher, O., Waddell, L. A., Wilhelm, B. J., McEwen, S. A. and Rajić, A. (2012) The effectiveness of selected feed and water additives for reducing *Salmonella* spp. of public health importance in broiler chickens: A systematic review, meta-analysis, and meta-regression approach. *Prev. Vet. Med.* 106, 197–213. doi:10.1016/j.prevetmed.2012.07.007.

Tsirtsikos, P., Fegeros, K., Balaskas, C., Kominakis, A. and Mountzouris, K. C. (2012). Dietary probiotic inclusion level modulates intestinal mucin composition and mucosal morphology in broilers. *Poult. Sci.* 91, 1860–8. doi:10.3382/ps.2011-02005.

Tuntufye, H. N., Lebeer, S., Gwakisa, P. S. and Goddeeris, B. M. (2012). Identification of avian pathogenic *Escherichia coli* genes that are induced *in vivo* during infection in chickens. *Appl. Environ. Microbiol.* 78, 3343–51. doi:10.1128/AEM.07677-11.

van den Borne, J. J. G. C., Heetkamp, M. J. W., Buyse, J. and Niewold, T. A. (2015). Fat coating of Ca butyrate results in extended butyrate release in the gastrointestinal tract of broilers. *Livest. Sci.* 175, 96–100. doi:10.1016/j.livsci.2015.02.016.

Wagenaar, J. A., Newell, D. G., Kalupahana, R. S. and Mughini-Gras, L. (2015). *Campylobacter*: Animal reservoirs, human infections, and options for control, in: Sing, A. (Ed.), *Zoonoses – Infections Affecting Humans and Animals*. Springer Netherlands, pp. 159–77.

Wang, X., Farnell, Y. Z., Peebles, E. D., Kiess, A. S., Wamsley, K. G. S. and Zhai, W. (2016). Effects of prebiotics, probiotics, and their combination on growth performance, small intestine morphology, and resident *Lactobacillus* of male broilers. *Poult. Sci.* 95, 1332–40. doi:10.3382/ps/pew030.

Wang, Y. and Gu, Q. (2010). Effect of probiotic on growth performance and digestive enzyme activity of Arbor Acres broilers. *Res. Vet. Sci.* 89, 163–7. doi:10.1016/j.rvsc.2010.03.009.

Wolfenden, A. D., Pixley, C. M., Higgins, J. P., Higgins, S. E., Vicente, J. L., Torres-Rodriguez, A., Hargis, B. M. and Tellez, G. (2007). Evaluation of spray application of a *Lactobacillus*-based probiotic on *Salmonella* Enteritidis colonization in broiler chickens. *Int. J. Poult. Sci.* 6, 493–6.

Wolfenden, R. E., Pumford, N. R., Morgan, M. J., Shivaramaiah, S., Wolfenden, A. D., Pixley, C. M., Green, J., Tellez, G. and Hargis, B. M. (2011). Evaluation of selected direct-fed microbial candidates on live performance and *Salmonella* reduction in commercial turkey brooding houses. *Poult. Sci.* 90, 2627–31. doi:10.3382/ps.2011-01360.

Wu, Y., Zhou, Y., Lu, C., Ahmad, H., Zhang, H., He, J., Zhang, L. and Wang, T. (2016). Influence of butyrate loaded clinoptilolite dietary supplementation on growth performance, development of intestine and antioxidant capacity in broiler chickens. *PLoS ONE* 11, e0154410. doi:10.1371/journal.pone.0154410.

Xu, Z. R., Hu, C. H., Xia, M. S., Zhan, X. A. and Wang, M. Q. (2003). Effects of dietary fructooligosaccharide on digestive enzyme activities, intestinal microflora and morphology of male broilers. *Poult. Sci.* 82, 1030–6. doi:10.1093/ps/82.6.1030.

Yitbarek, A., Echeverry, H., Munyaka, P. and Rodriguez-Lecompte, J. C. (2015). Innate immune response of pullets fed diets supplemented with prebiotics and synbiotics. *Poult. Sci.* 94, 1802–11. doi:10.3382/ps/pev147.

Yurong, Y., Ruiping, S., ShiMin, Z. and Yibao, J. (2005). Effect of probiotics on intestinal mucosal immunity and ultrastructure of cecal tonsils of chickens. *Arch. Anim. Nutr.* 59, 237–46. doi:10.1080/17450390500216928.

Zhang, G., Ma, L., and Doyle, M. P. (2007). Salmonellae reduction in poultry by competitive exclusion bacteria *Lactobacillus salivarius* and *Streptococcus cristatus*. *J. Food Prot.* 70, 874–8.

Zheng, A., Luo, J., Meng, K., Li, J., Zhang, S., Li, K., Liu, G., Cai, H., Bryden, W. L. and Yao, B. (2014). Proteome changes underpin improved meat quality and yield of chickens (*Gallus gallus*) fed the probiotic *Enterococcus faecium*. *BMC Genomics* 15. doi:10.1186/1471-2164-15-1167.

Zuidhof, M. J., Schneider, B. L., Carney, V. L., Korver, D. R. and Robinson, F. E. (2014). Growth, efficiency, and yield of commercial broilers from 1957, 1978, and 2005. *Poult. Sci.* 93, 2970–82. doi:10.3382/ps.2014-04291.

Chapter 11

Using models to optimize poultry nutrition

R. M. Gous and C. Fisher, University of KwaZulu-Natal and EFG Software, South Africa

1 Introduction

Commercial poultry production is all about making decisions and then implementing these decisions. The objective in most cases is to maximize profit for the enterprise. A decision is a choice between alternative courses of action, and it is made by weighing the consequences of these courses of action. The process of decision making is one that everyone practices every day: identifying the problem, evaluating alternative courses of action, choosing the most appropriate course of action on some or other basis, implementing the decision, evaluating the consequences and repeating the cycle. In order to evaluate the consequences effectively, these first need to be predicted. There is a common belief in poultry feeding that experience and/or experiments are an accurate means of predicting the consequences of different courses of action, but this is a slow and uncertain way of achieving a consensus. It is more likely that an accurate theory will lead to the correct decision being made in a fraction of the time required by any of the alternative methods.

Animal nutritionists continually face problems when formulating feeds for poultry: increases in input costs, problems in acquiring raw materials, changes in the growth or laying potential of the stock used and the occasional reduction in demand for product all lead to technical challenges that need an immediate response. This situation is very different from the one in which the producer controls the market and demands a closer examination of all aspects of the production process. The nutritionist, mill manager, farm manager and marketing department need to work together to maximize profitability of

http://dx.doi.org/10.19103/AS.2016.0011.11

the entire enterprise rather than regarding each department as an independent cost centre. Feed formulation, the tool at the disposal of the company nutritionist, is only a small part of the total process, and although in the past it was not possible to take account of all aspects of production when formulating feeds for animals, advances in simulation modelling have now improved the prospects for achieving this goal.

Poultry nutritionists base their nutritional strategies on the concept of a 'nutrient requirement'. This is seen as a characteristic of the bird and is the nutrient content required to support 'maximum' or 'optimum' production. Such requirements are published as guidelines by breeding companies, as tables by learned committees (e.g. NRC, 1994), by national extension services in many European countries and by some universities and research institutes. Within a company, the nutritionist will attempt to adapt these guidelines to local circumstances taking into account some of the objectives of the business and also of economic circumstances. The bases on which such adaptations are made are often tenuous and lacking in critical scientific rigour.

The application of systems thinking and modelling to the problem of feed formulation leads to the replacement of the above approach with one in which nutritional decisions are made entirely in terms of the objectives of the business. Nutrient specifications are chosen that will maximize a desired objective function, usually an economic objective such as margin/m^2 per annum. Feeding animals to achieve an overall company objective is not the same as feeding them to meet a 'requirement'. Economic circumstances will change from time to time and different nutritional strategies will be needed to maximize margins. Also, nutritional decisions will depend on the stage in the production process at which margin is to be assessed. For example, if nutrition is optimized for margin at the farm gate, with live bird weight (and perhaps downgrading) affecting revenue then nutritional responses in growth, feed conversion ratio and mortality will need to be considered. If, however, margin is measured on the basis of the production of processed portions or meat, then nutritional responses in these characteristics, as well as those operating at the farm gate, will affect the outcome. These are real differences, each requiring specific nutritional decisions. Modelling enables these differences to be accommodated.

The process followed is not dissimilar to that proposed by Deming (1986) to improve product quality, except that in this discussion it is profitability and not quality that is being targeted, which is not to say that quality should not be considered. The nutritionist's role at present is to complete the Plan/Do/Study/Act (PDSA) cycle by changing the formulation specifications and measuring the resultant response that is obtained in the field over a number of cycles. Decisions about the dietary energy content to be used, which nutrients to manipulate and by how much are generally made with reference to the shadow price of each limiting nutrient without regard to the value of the end product. Progress, if any, is at best slow, but may be elusive. Opportunities arising from changes in input costs or product value need to be grasped immediately if maximum benefit is to be gained, and this is possible only if performance can be predicted rather than measured in the field. The PDSA cycle could then be completed in seconds rather than months and the company would benefit in good and in bad times. The role of nutritional modelling in this process is thus to predict nutritional responses so that the PDSA cycle can be automated through the process of optimization. It should be made clear that the 'optimum' feeding strategy referred to throughout this discussion is that which will maximize or minimize the objective function defined by management.

These general points about decision making and feed formulation in poultry nutrition are pertinent to the industry as it is today. Looking ahead to the future suggests that they are likely to become ever more important. In most discussions about the sustainability of the poultry industry, maintaining the supply of feed ingredients is seen as a major issue. Increasing human populations and changing consumption patterns will create a momentum for rapid growth of poultry production. However, there will be a growing competition from the human population for feedstuffs and a growing stress on cultivable land availability and of water supplies. Thus, efficiency of feed use in the poultry industry will attract increasing attention and create an additional emphasis on accurate feed formulation.

In addition, markets are likely to become more variable and the accurate adjustment of feeding to these variations will have a higher economic value. Price volatility in commodities, including grains, is a growing feature of the modern world and the tools required to react quickly and optimally to these changes will be of growing importance. Systems of production may also become more variable in response to factors like changing consumer demands and increased energy costs of controlling the animal environment. For example, in Europe, the animal welfare concerns of the public are already creating a demand for slower growing breeds of broilers, with considerable diversity. Being able to predict the optimum nutrition of birds grown under these variable conditions will increasingly be seen as an economic necessity.

2 Predicting responses of poultry to nutrients

Predicting responses of poultry to nutrients has been the goal of nutritionists and modellers for a long time. The controlled feeding model of a growing pig (The Edinburgh Model Pig) was the first serious and successful attempt to integrate information about an animal, its feed and the environment in which it was kept, with a view to simulating its performance (Whittemore, 1976; Whittemore and Fawcett, 1976). This provided the impetus for the development of further models, for the modification of existing models and for research targeted at filling the gaps in our knowledge of critical aspects of the theory incorporated into these models. The most important subsequent contribution to response modelling was the theory proposed by Emmans (1981) to predict voluntary food intake in poultry and pigs, a formal description of which is given in Emmans (1997). This theory raised the value of prediction models inestimably by making food intake an output from, as opposed to an input to, the growth model. Models incorporating this theory are thus more realistic and useful, providing the nutritionist with a tool for making decisions about the most appropriate course of action to take under different circumstances. Advances continue to be made, and it is now possible to optimize the feeds and feeding programmes of broilers through the integration of a feed formulation programme, a simulation model and an optimization routine (Gous and Berhe, 2006). However, because models require a complete statement about each step in the chain of events, some interpolations must necessarily be used where appropriate data are missing (Whittemore, 1981). All models contain some such conjectures, so none can claim to be absolutely accurate. Also, as models become more sophisticated, the list of variables that may be predicted increases.

Until recently, mechanistic models developed for poultry have dealt with the simulation of responses in a single bird. Such responses are usually linear to the point where the genetic potential is reached (Fisher et al., 1973). Poultry nutritionists are interested in responses to nutrients in economically important outputs such as body weight (or protein) gain, breast meat yield, egg output, food intake and conversion efficiency, numbers of chicks produced per hen and so on. Because such responses are usually measured by using groups of birds, they are invariably curvilinear, being the result of integrating the responses of individuals making up that population. Populations of birds, therefore, cannot have 'requirements' for nutrients: what nutritionists seek are the optimum economic dietary contents of each nutrient and for this, they need to know how populations respond to increasing dietary contents of the essential nutrients. Descriptions of such responses, whilst taking account of the marginal costs and revenues, are, therefore, invaluable in determining how to maximize or minimize the objective function chosen for any given commercial operation. Clearly, being able to predict these nutrient responses may be seen as the foundation of successful nutrition. To date, the most successful method of simulating the response of a population to changes in food composition is to simulate the responses of the individuals making up that population.

Many empirical models have been developed and these attempt to optimize performance (Kenny et al., 2004; Eits et al., 2005a,b; and for a summary of others, see France and Kebreab, 2008), but these are not discussed further in this paper as they are not designed to predict performance but simply to predict the composition of the feed that will maximize performance or profitability. Being empirical in nature they are limited in their ability to respond to changes in the important variables such as genotype and environment. The greatest limitation of these empirical models is that none of them can predict food intake, which is the basis for being able to predict performance.

3 Predicting food intake

To be of any real value, models that attempt to optimize the feeding of animals must be capable of predicting voluntary food intake. Where this variable is an input to a model, as is most often the case, it is naïve to believe that feeding programmes can be successfully optimized, when the composition of the food offered has such important effects on voluntary food intake. Food intake must, therefore, be an output from a model and not an input. The theory of food intake and growth proposed by Emmans (1981, 1989) is based on the premise that birds attempt to grow or reproduce at their genetic potential, which implies that they attempt to eat as much of a given feed as would be necessary to achieve these goals. This 'desired' food intake (DFI) is defined as the amount of the nutrient required divided by the content of that nutrient in the feed and can thus be determined for each of the essential nutrients, and energy, required by the bird or animal. The nutrient resulting in the highest DFI is by definition the first limiting nutrient in the feed on offer. The process of calculating the DFI for each nutrient is relatively straightforward. Where this DFI cannot be achieved through constraints of gut capacity or environmental heat demand, the food intake is said to be constrained (CFI) and the actual food intake, being the lower of DFI and CFI, would in this case equal the constrained intake.

This theory has been shown to predict food intake and hence growth and carcass composition with considerable accuracy (Ferguson and Gous, 1997, Ferguson et al., 1997). Broilers (Burnham et al., 1992) and laying hens (Gous et al., 1987) have been shown to increase food intake as the limiting nutrient in the feed is reduced, attempting thereby to obtain more of the limiting nutrient, until a dietary concentration is reached where performance is so constrained that food intake falls. The common misconception that 'birds eat to satisfy their energy requirements' is clearly naïve and of no value in predicting voluntary food intake.

A growing or reproducing animal needs to be supplied with nutrients in order to meet its requirements for maintenance of the body and feathers (in the case of poultry), for the growth of all other components of the body, including feathers, and for reproduction. In order to predict voluntary food intake, it is necessary to predict the amount of each of these essential nutrients required by the bird or animal every day. This requires a description of the genotype (its body protein weight and potential growth rate or egg output on a day), the food being offered and the environment to which the animal is being subjected. Each of these provides challenges to the modeller, many of which have been described previously (Emmans and Fisher, 1986). The approach suggested by Emmans (1989) to describe and evaluate broiler genotypes, for example, begins with a definition of potential protein growth, and the live weight of the animal is built up from this, using the allometric relationships that exist between protein, water, ash and lipid, that is, a bottom-up approach. He has shown that a few, simple, assumptions can lead to a description of a growing animal that is sufficient for predicting its performance in non-limiting conditions and for calculating what these conditions are. It seems sensible to be able to predict performance in non-limiting conditions before the more difficult question is tackled, namely, that of defining growth in limiting conditions. Values for the genetic parameters that define a growing animal can be measured by rearing animals in environmental conditions that are as near to ideal as possible. Under these conditions, growth curves that represent the genetic potential for a particular genotype are obtained. The curves obtained in this way allow comparisons to be made between breeds and strains. Examples of such investigations can be found in Hancock et al. (1995) and Gous et al. (1999).

The resources needed to meet the requirements can be determined from knowledge of the growth rate and composition of the various components of the body and/or eggs being produced. The resources available for supplying the requirements, which are present in various feedstuffs, need to be described in the same terms as are used to describe the nutrient requirements. The requirement for protein (for maintenance, body or feather protein growth or egg production) depends on the amino acid composition of that protein and the rate at which it is being produced. The sum of each amino acid required for the maintenance and the growth of feather and of body and egg protein constitutes the daily requirement for each of the amino acids. The retention of lipid, water and ash has no protein requirement. The scale on which the amino acids required by the animal are measured and the scale on which the amino acids in the feed are described must be the same. The conventional wisdom is to express this in terms of digestibility. The marginal efficiency with which the first limiting amino acid is used for protein retention above maintenance is not necessarily constant, but can be modified by the supply of other amino acids and by the supply of energy (Kyriazakis and Emmans, 1992). Values for these marginal efficiencies need to be measured by using response experiments, but the methodologies used for measuring and interpreting such responses have not been completely resolved. A discussion on this topic is not appropriate here, but is important when interpreting results of published trials and when designing future response experiments.

4 Predicting potential laying performance

Describing the potential rate of lay of a laying hen is complex as this differs between individuals and over time within individuals. The mathematical model of Etches and Schoch (1984), based on the theory proposed by Fraps (1955), demonstrated that two functions, representing two independent but interacting systems of the hen's asynchronous ovulatory cycle, were able to predict realistic ovulation times and intra-sequence ovulation intervals. Johnston and Gous (2003) enhanced the value of this approach by producing a more generalized model that predicts mean rate of lay in a flock of hens at a particular age using the individual patterns of sequential laying at that time. The mean age at sexual maturity and the slope of the initial rise in flock egg production to peak rate of lay are influenced by the distribution of ages at sexual maturity and by the lengths of the individual prime sequences. The distribution of ages at sexual maturity can be predicted by using the model described for laying hens by Lewis and Morris (2004) and for broiler breeders by Lewis et al. (2007). The incidence of internal laying at the onset of maturity plays a role in modifying rate of lay but not ovulation rate. The persistency of lay after peak will be determined by the rate at which sequence lengths of individual hens decay over time, as well as by the number of pause days. Hence, the prediction of sequence length is a logical step in predicting the performance of a flock of laying hens over an entire laying cycle.

The methods described by Johnston and Gous (2006) to predict the potential laying performance of a flock of commercial laying hens appear to work satisfactorily for broiler breeders (Gous and Nonis, 2010) as long as appropriate functions are used to describe the relationships between age and yolk weight, albumen and yolk weight, and shell and egg content weight, which differ between strains. Differences in age at sexual maturity, maximum rates of lay, rates of decay in ovulation rate over time and the variation that exists between individuals in all these respects can be taken into account. Rules must be applied to account for minimum egg weights when essential nutrient intake is severely constrained, for the size of amino acid pools for potential albumen formation and for the rates at which lipid can be either deposited in, or withdrawn from, body reserves as a means of accounting for differences in energy balance.

One of the advantages of the method used by Johnston and Gous (2003) to model egg production is that it lends itself to stochasticity. Within a population of birds, individuals of the same age show considerable variation about a mean sequence length, which may be due to variation in the length of the open period for luteinizing hormone release, or variation in follicular dynamics. This variance may be accounted for by using mean values and standard errors for each of the parameters in the model. Such a population of birds would generate a range of ovulation times, the distribution of which would be unimodal and positively skewed in young hens, becoming bimodal with age. Reproductive senescence in hens manifests as an increase in the intra-sequence ovulation and oviposition intervals with time, as well as an increase in the number of pause days. With this information it is possible to determine the nutrients required daily by a flock of laying hens for reproductive purposes.

5 Modelling environmental factors affecting desired feed intake

High temperatures are the most common reason for birds and animals not achieving their desired feed intake. Whereas birds benefit in cold weather from the insulative properties of their feather cover, this thermal barrier constrains the amount of heat that may be lost to the environment in hot weather. As the potential growth rate of broilers is increased by genetic selection (Table 1), their inability to lose sufficient heat to the environment is becoming a major constraint in commercial broiler operations worldwide. In a review of nutritional interventions that may be used to alleviate the effects of high temperatures on

Table 1 A comparison of the predicted[1] daily heat outputs of male broilers[2], on a weekly basis, over six decades, and the highest temperature that would allow the potential protein growth to be achieved in each case

Age, d	Heat production, kJ/bird d					
	1970	1980	1990	2000	2010	2020
7	209	227	248	270	295	317
14	409	464	528	598	664	727
21	700	806	923	1038	1129	1193
28	945	1082	1221	1341	1505	1500
35	1184	1330	1470	1583	1774	1836
42	1400	1536	1671	1815	2057	2175
49	1642	1770	1915	2056	2212	2267
	Highest temperature possible for achieving potential protein growth, °C					
	1970	1980	1990	2000	2010	2020
7	33.7	32.9	31.8	30.7	29.4	27.9
14	32.6	31.3	29.8	28.1	26.0	23.2
21	31.4	29.8	27.7	25.2	22.0	17.4
28	30.2	28.1	25.4	22.1	17.7	11.9
35	29.0	26.4	23.0	19.1	14.4	9.0
42	27.6	24.6	21.0	17.0	13.1	10.7
49	26.2	23.0	19.6	16.5	14.9	13.8

[1]Predicted using the EFG Broiler Growth Model.
[2]The equivalent temperature for female broilers is lower than for males from 14 d onwards.

broiler performance, Gous and Morris (2005), using simulation modelling, showed that most nutritional strategies that have been proposed as a means of reducing the heat of digestion in the broiler result in a maximum theoretical saving in metabolic heat production equal to the effect of lowering dry bulb temperature in the broiler house by about 1°C. None of these strategies is as effective in terms of growth rate, feed conversion, liveability or carcass quality as reducing the radiant heat load on the birds by making appropriate modifications to the structure of the broiler house and to the husbandry practices employed.

Accounting for all the factors that contribute to the environmental heat demand placed on the birds, such as temperature, humidity, wind speed and thermal radiation, and then accounting for the response of birds to this 'effective' temperature are major challenges when modelling the response of broilers to nutrient supply.

Other constraints, such as social stress, microbial load, disease, vaccine reaction and so on, may be modelled relatively successfully by introducing a genetic parameter to describe the ability of each bird in a population to cope when exposed to such stressors (Wellock et al., 2003, 2004). This parameter will adjust appropriately the rate of maturing parameter in the Gompertz growth equation which reduces the potential according to the level and duration of stress and the susceptibility of the individual to the stressor (Kyriazakis and Sandberg, 2006).

In most models food intake needs to be input in some or other way, that is, it is not an output from the model. In such cases, it is difficult to imagine that the effects of the environment on food intake can be successfully modelled. But even where food intake is an output, such models presently take a relatively naïve approach when describing the environment, usually describing only the environmental temperature and the relative humidity. Yet, wind speed and radiation are important elements in determining the environmental heat demand on the animal, as is the fact that birds are capable of differential blood flow redistribution (McArthur, 1981) to the bare appendages of the body (wattles, comb and legs). By managing the vasoconstriction and vasodilation of the arterio-venous shunts of the skin in those anatomical regions, the bird is able to control sensible heat dissipation (Hillman et al., 1982; Hillman and Scott, 1989; Willmer et al., 2000). Accounting for these additional factors impacting on the response of the bird to its environment implies that a dynamic approach is necessary.

The micro-environmental conditions at which the bird needs to expend the least thermoregulatory effort have not yet been adequately defined. Such a definition would seem to be a prerequisite for predicting the environmental effects on birds when they are not in a resting state. Many of the experiments involving the effect of thermal stress on birds have been conducted by using constant temperatures applied over long periods of time, which implies that the physiological and productive responses of the bird are in a steady state [such as the model suggested by Mount (1979) for a homoeothermic animal], thus ignoring the important point that the environment has a dynamic, cumulative effect on chickens (Blanco and Gous, 2006). Responses of chickens to environmental conditions are dynamic and depend not only on the thermoregulatory abilities of the birds and the conditions of the environmental variables to which they are exposed, but also on the time of exposure to such conditions. Birds varying in body weight will achieve thermal equilibrium with the environment following different lengths of exposure, depending on the environmental conditions as well as on their thermal properties (feather cover, comb and wattle size, acclimatization, etc.). Blanco and Gous (2006) have modelled these thermal responses, considering the animal characteristics as well as the environmental conditions.

It appears that there is still much to be done in explaining these thermal responses in a dynamic, time-dependent manner before being able to take an accurate account of environmental heat demand when predicting voluntary food intake. It would seem that research aimed at addressing these issues would be of great value, given the changes in global environmental conditions that are predicted.

6 Using models to optimize feeding programmes

Once food intake has been predicted, the performance of the growing or reproducing bird can also be predicted. This leads to the possibility of determining the feeds and feeding programme that will maximize or minimize the objective function defined by management. Normally, the objective function will be some measure of productivity such as margin/m^2 per annum, or breast meat yield at a given body weight, but any output parameter from the growth model may be used for this purpose. To be of economic relevance, objective functions should include revenue, space and time, the most obvious example being margin/m^2 per annum. This takes account of both the fixed and variable costs of production including the cost and logistics of manufacturing, transporting and storing feed on the farm, and the income derived from the sale of product. In broiler production, fixed costs are invariably high, so throughput is particularly important. Reducing the age at slaughter by 1 or 2 days results in considerable savings in fixed costs. Such objective functions are more sensible than attempting to minimize feed conversion ratio.

Determining the optimum concentrations of amino acids relative to energy in each feed, the optimum nutrient density and the optimum length of time (or amount) that each feed should be fed is, therefore, both a nutritional and an economic decision. The information required for optimization consists of feed costs at different levels of amino acid provision, a description of all the relevant responses, both fixed and variable costs affecting the production system, and details of revenue. The complexity of the information required would depend on the level of organization at which the optimization is to be made. If profit of the broiler grower is to be maximized at the farm gate, then responses in liveability, growth and feed conversion ratio will probably suffice. However, and more realistically, a wider view will be required, and the effect of broiler nutrition on slaughterhouse variables (eviscerated yield, rejects, etc.) and further processing (carcass composition) will need to be defined. Mack et al. (2000) emphasized the importance of broiler companies considering all aspects of the production cycle when making nutritional decisions.

6.1 Optimizing amino acid content in each feed

The optimum relationships between the essential amino acids and energy change during the growing period because of changes in the potential growth rates of body and feather protein, which differ in amino acid composition, as well as the requirement for maintenance, which increases as a proportion of the total daily requirement over time. It is unreasonable to consider changing the feed composition daily and so feeding periods have to be defined either in terms of amounts fed (preferably) or in terms of time periods, and the optimum relationship between the amino acids and energy within each specified feeding period that maximizes (usually) or minimizes the objective function needs to be found. The overall objective is to determine the optimum amino acid to energy ratio in each of the feeds in the feeding programme in such a way that the overall performance is maximized. The objective is not to determine independently the optimum ratio in each of the feeds on offer. Because the performance on one feed impacts on the performance on subsequent feeds (Gous et al., 2012), this is an essential prerequisite in optimizing the feeding of broilers.

6.2 Optimizing nutrient density

Once the optimum ratio between the essential amino acids and energy in each feed has been established, the density at which these nutrients are included in each feed that will maximize or minimize the objective function needs to be determined. The cost of feed will increase as nutrient density is increased, but food intake will be reduced, so the cost of feeding needs to be considered together with changes in growth rate and composition when optimizing nutrient density. The nutrient density in each of the feeds in the feeding programme is that which maximizes the objective function over the entire growth period. As Fisher and Wilson (1974) have shown, the optimum nutrient density depends on such factors as sex, the ratio between input and output costs and mixing and transport costs. These factors, and others, should be considered by the user in determining the optimum nutrient density of each of the feeds in the programme.

6.3 Optimizing the feeding schedule

Many broiler producers do not have the opportunity of having feeds mixed according to their specifications, but are constrained to make use of proprietary feeds. An almost infinite number of options is open to such producers in designing their feeding schedule, which can be based on amounts fed in each phase or on fixed feeding periods for each feed. The optimum feeding schedule is dependent on the composition of the feeds, their respective prices, the revenue to be derived from the sale of the product and many other biological and economic considerations. The optimization process should be sufficiently versatile to account for multiple harvesting (cropping) from single- or mixed sex flocks and for revenue to be calculated from any mixture of whole-bird sales and processing.

6.4 Optimizing feeds for laying hens and broiler breeders

The situation with broiler breeder hens differs from that of broilers and full-fed laying hens in that a daily allowance of feed is allocated, this being less than would normally be consumed if the birds were given *ad libitum* access to feed. Yet, the principles applied to voluntary intake prediction, described above, remain: the difference is that the desired food intake of broiler breeder hens is hardly ever achieved, and thus, the actual food intake is that constrained by the farm manager. Consequently, egg output will be a function of the amount of limiting nutrient remaining after the maintenance requirement of the hen has been met. Nonis and Gous (2015) have shown that it is doubtful that broiler breeders have a requirement for growth once sexual maturity has been achieved; thus, it could be argued that body protein and lipid deposition, or utilization, leading to a change in body weight, should be regarded as being a consequence of the nutrients consumed and not as an obligatory daily process. This being the case, the balance of metabolizable energy intake remaining after accounting for maintenance and egg production would be converted into body lipid with varying efficiencies depending on whether the dietary lipid was deposited directly as body lipid or first converted to CO_2 and H_2O (Emmans, 1994). Similarly, any excess protein would be deaminated and converted into body lipid. This, we regard, as a more sensible approach than assuming that a broiler breeder hen has a need to grow body protein or lipid during the reproductive phase.

Optimizing the feed or feeding programme for a flock of laying hens or broiler breeders would involve determining the ratio between the amino acids and energy that would maximize or minimize a defined objective function such as margin over feed cost or number of chicks hatched. As with broilers, this optimum ratio could be fed at different nutrient densities, and because food intake can be predicted, the optimum nutrient density can be determined by simulation. This is an advance on the technique suggested by Morris (1968, 1969), which was one of the earliest attempts at optimizing the nutrient density of feeds for laying hens. In the case of broiler breeders, where a daily amount of feed is allocated, as opposed to allowing the birds to eat *ad libitum*, an additional question regarding optimization is introduced: not only is it possible to optimize the amino acid to energy ratio and the nutrient density, but the amount allocated on each day of the laying period can also be determined, which will maximize the number of chicks produced per hen. It is, of course, possible to consider changing the feed composition during the laying period in the case of both laying and broiler breeder flocks and to optimize the composition of the feeds during each of these different phases, as long as the objective is applied to overall performance and not to the performance within each phase.

Because the egg production periods of laying hens and broiler breeders are so long, experiments designed to determine the optimum feeds and feeding programmes for these strains frequently fail because of the almost infinite combination of variables that could be applied during this period. Furthermore, it has been demonstrated that the response of a bird or animal to a feed depends on what was being fed previously (Kyriazakis et al., 1991; Gous and Nonis, 2009), which adds a further dimension to the possible feed treatments that could be explored over the laying period and points to the limitation of attempting to determine the optimum feeds and feeding programmes by experimentation. Short-term trials are certainly not the answer. The approach described above, of accurately predicting the effects of daily allocations of a given feed on performance, is far more likely to result in optimum solutions under the varying circumstances likely to be encountered in practice.

7 Summary

For practical reasons, only energy and amino acid levels need to be considered in the optimization process, thereby reducing the magnitude of the problem. The range of responses that need to be predicted is not fixed, although it will have to be quite extensive if the models are to be useful: the actual range will, of course, depend on the particular markets involved. As with all optimization routines, it is possible to find a 'local' minimum or maximum that is not the best option available in the nutrient space being investigated. To overcome this problem, different starting points are often used, which increases the time taken to arrive at the optimal solution. Other methods are available, and these explore the entire space and then produce three-dimensional plots of the response surface, which assist in sensitivity analysis, determining which factors are of importance and under which circumstances.

It has been 40 years since the Edinburgh Model Pig entered the scientific arena and since then the progress that has been made in predicting performance of broilers and pigs has been enormous. The Edinburgh Broiler Model (Emmans, 1987) was an improvement on the Pig Model, mainly because it predicted voluntary food intake, as opposed to using a controlled feeding approach. The theory used to predict food intake (Emmans, 1981) has had major advantages for modellers, as it has been successfully applied in simulating the effect of, among others, changes in dietary amino acid and protein content, environmental temperature, infection and social stress. It has led to food intake being an output from models instead of being an input, which has enabled models to be used to optimize feeds and feeding programmes, a process not possible unless food intake is accurately predicted. It has spawned many useful scientific studies that have been designed to test the theory, and it has led to a simplified method of accounting for the heat produced by an animal when consuming a given feed, known as the effective energy system (Emmans, 1994). And because the effective (or net) energy value of a feed is a function of both the feed and the animal being fed, what would be the advantage of describing feeds in these terms if a model were not available to determine the value of this feed to the animal itself? These early models stimulated useful and purposeful research targeted at filling the gaps in our knowledge of critical aspects of the theory incorporated into these models, this in itself leading to useful improvements in the scientific value of research.

In spite of the progress made in the past decades, there are still challenges that lie ahead for those wishing to predict responses to nutrients in poultry. Many of these have been raised through the development of existing models. For example, the need to understand how to describe the environment and the way it impacts on broiler performance has greater meaning when this information can be linked to a prediction of the constraining effect of high temperatures on voluntary food intake. It has been demonstrated that significant changes have taken place in broiler genotypes over time, and that the genotypes available today differ substantially in their composition and in the way they deal with marginally deficient feeds. Yet, updating the description of these genotypes would be of little value if this information could not in some way be used to simulate the performance of these birds under varied feeding and environmental conditions, from which optimum feeds and feeding programmes could be predicted. Also, it only really matters whether the efficiency of utilization of an amino acid by a broiler for growth is 0.75 or 0.80 if this is to be used to predict the requirement by the broiler for that amino acid: this information is of no value if the requirement for the amino acid is being derived purely from the results of a growth trial.

Sadly, the world does not seem to be as enthusiastic about models as are the modellers themselves. This is partially the result of scepticism brought about through bad experiences with (bad) models. Many sets of empirical equations have been termed 'models', some of which are robust and useful, such as the model to predict age at first egg of laying pullets for any specified pattern of photoperiod used during the rearing period (Lewis et al., 2003), whilst many others are simply equations representing the result of a single experiment, with little or no predictive value outside of the experiment itself. It is the latter that have justifiably caused this scepticism. One of the challenges faced by those predicting responses to nutrients is to convince the poultry industry that good models have the potential to be of immense benefit to nutritionists, geneticists and other decision makers in the industry. The important principle is that models should be 'open': the user must know on what ideas and assumptions the model is based. Models that are 'black boxes' and closed to the user are not likely to contribute much to the improved management of nutrition. It is also imperative that the methods used by scientists to measure the numbers that make theories work are robust and unambiguous such that models based on the results of such experiments can be used to assist the poultry industry to become more efficient. It is perhaps apposite to end with a quote from Morris (2006), who said:

> Empirical modelling is not to be despised if it is the best available tool for solving a particular problem. What is not acceptable is the use of empirical modelling where others have already developed a mechanistic model capable of resolving the particular question being approached empirically. I believe that the modelling of growth and development in pigs and poultry has now reached a sufficiently advanced state that empirical models in this area can no longer be justified.

8 Where to look for further information

Undoubtedly the publications of Emmans (1981, 1987, 1989 and 1997) listed in the References are the best source of information on the prediction of food intake in poultry. The paper by Emmans and Fisher (1986) is an excellent review of the problems encountered in nutritional theory that can be resolved through modelling, and that by Emmans (1994) should be obligatory reading for those working with nutrient energy systems. Many of the issues seen as being important in modelling broilers and laying hens have been researched more extensively since these earlier publications, and are expounded in two recent books: *Mechanistic Modelling in Pig and Poultry Production*, R. M. Gous, T. R. Morris and C. Fisher (Eds), 2006, CABI, Wallingford, U.K. and the follow-up publication *Nutritional Modelling for Pigs and Poultry*, N. K. Sakomura, R. M. Gous, I. Kyriazakis and L. Hauschild (Eds), 2015, CABI, Wallingford, U.K.

9 References

Blanco, O. A. and Gous, R. M. 2006. Considerations for representing the micro-environmental conditions in simulation models for non-sweating livestock. In R. M. Gous, T. R. Morris and C. Fisher (Eds), *Mechanistic Modelling in Pig and Poultry Production*, pp. 76–96. CAB International, Wallingford, U.K.

Burnham, D., Emmans, G. C. and Gous, R. M. 1992. Isoleucine responses in broiler chickens. Interactions with leucine and valine. *Br. Poult. Sci.* 33: 71–87.

Deming, W. Edwards. 1986. *Out of the Crisis*. MIT Press. ISBN 0-911379-01-0. OCLC 13126265.

Eits, R. M., Kwakkel, R. P., Verstegen, M. W. A. and den Hartog, L. A. 2005a. Dietary balanced protein in broiler chickens. 1. A flexible and practical tool to predict dose-response curves. *Br. Poult. Sci.* 46: 300–9.

Eits, R. M., Giesen, G. W. J., Kwakkel, R. P., Verstegen, M. W. A. and den Hartog, L. A. 2005b. Dietary balanced protein in broiler chickens. 2. An economic analysis. *Br. Poult. Sci.* 46: 310–17.

Emmans, G. C. 1981. A model of the growth and feed intake of *ad libitum* fed animals, particularly poultry. In *Computers in Animal Production*, pp. 103–10. Occasional Publication No. 5. British Society of Animal Production.

Emmans, G. C. 1987. Growth, body composition and feed intake. *Worlds Poult. Sci. J.* 43: 208–27.

Emmans, G. C. 1989. The growth of turkeys. In C. Nixey and T. C. Grey (Eds), *Recent Advances in Turkey Science*, pp. 135–66. Butterworths, London.

Emmans, G. C. 1994. Effective energy – a concept of energy utilisation applied across species. *Br. J. Nutr.* 71: 801–21.

Emmans, G. C. 1997. A method to predict the food intake of domestic animals from birth to maturity as a function of time. *J. Theor. Biol.* 186: 189–200.

Emmans, G. C. and Fisher, C. 1986. Problems in nutritional theory. In C. Fisher and K. N. Boorman (Eds), *Nutrient Requirements of Poultry and Nutritional Research*, pp. 9–39. Butterworths, Oxford.

Etches, R. J. and Schoch, J. P. 1984. A mathematical representation of the ovulatory cycle of the domestic hen. *Br. Poult. Sci.* 25: 65–76.

Ferguson, N. S. and Gous, R. M. 1997. The influence of heat production on voluntary food intake in growing pigs given protein-deficient diets. *Anim. Sci.* 64: 365–78.

Ferguson, N. S., Gous, R. M. and Emmans, G. C. 1997. Predicting the effects of animal variation on growth and food intake in growing pigs using simulation modelling. *Anim. Sci.* 64: 513–22.

Fisher, C., Morris, T. R. and Jennings, R. J. 1973. A model for the description and prediction of the response of laying hens to amino acid intake. *Br. Poult. Sci.* 14: 469–84.

Fisher, C. and Wilson, B. J. 1974. Response to dietary energy concentration by growing chickens. In T. R. Morris and B. M. Freeman (Eds), *Energy Requirements of Poultry*, British Poultry Science Ltd. Edinburgh, British Poultry Science Symposium 9: 151–84.

France, J. and Kebreab, E. 2008. *Mathematical Modelling in Animal Nutrition.* CAB International, Wallingford, U.K.

Fraps, R. M. 1955. Egg production and fertility in poultry. In J. Hammond (Ed.), *Progress in the Physiology of Farm Animals*, Vol. 2, pp. 661–740. Butterworths, London.

Gous, R. M. and Berhe, E. T. 2006. Modelling populations for the purposes of optimization. In R. M. Gous, T. R. Morris and C. Fisher (Eds), *Modelling in Pig and Production*, pp. 188–208. CAB International, Wallingford, U.K.

Gous, R. M. and Morris, T. R. 2005. Nutritional interventions in alleviating the effects of high temperatures in broiler production. *Worlds Poult. Sci. J.* 61: 463–6.

Gous, R. M. and Nonis, M. A. 2010. Modelling egg production and nutrient responses in broiler breeder hens. *J. Agric. Sci.* 148: 287–301.

Gous, R. M., Emmans, G. C. and Fisher, C. 2012. The performance of broilers on a feed depends on the feed protein content given previously. *S. Afr. J. Anim. Sci.* 42: 63–73.

Gous, R. M., Griessel, M. and Morris, T. R. 1987. Effect of dietary energy concentration on the response of laying hens to amino acids. *Br. Poult. Sci.* 28: 427–36.

Gous, R. M., Moran, E. T., Stilborn, H. R., Bradford, G. D. and Emmans, G. C. 1999. Evaluation of the parameters needed to describe the overall growth, the chemical growth, and the growth of feathers and breast muscles of broilers. *Poult. Sci.* 78: 812–21.

Hancock, C. E., Bradford, G. D., Emmans, G. C. and Gous, R. M. 1995. The evaluation of the growth parameters of six breeds of commercial broiler chickens. *Br. Poult. Sci.* 36: 247–64.

Hillman, P. and Scott, N. R. 1989. Energy budget of the chicken foot. *J. Therm. Biol.* 14: 205–17.

Hillman, P., Scott, N. R. and Van Tienhoven, A. 1982. Vasomotion in chicken foot: a dual innervation of arteriovenous anastomoses. *Am. J. Physiol.* 242: R582–90.

Johnston, S. A. and Gous, R. M. 2003. An improved mathematical model of the ovulatory cycle of the laying hen. *Br. Poult. Sci.* 44: 752–60.

Johnston, S. A. and Gous, R. M. 2006. Modelling egg production in laying hens. In R. M. Gous, T. R. Morris and C. Fisher (Eds), *Mechanistic Modelling in Pig and Poultry Production*, pp. 229–59. CABI, Wallingford, U.K.

Kenny, M., Kemp, C. and Fisher, C. 2004. What protein levels will maximise margin? *Int. Poult. Prod.* 12: 11–17.

Kyriazakis, I. and Emmans, G. C. 1992. The effects of varying protein and energy intakes on the growth and body composition of pigs. 1. The effects of energy intake at constant, high protein intake. *Br. J. Nutr.* 68: 603–13.

Kyriazakis, I. and Sandberg, F. B. 2006. The problem of predicting the partitioning of scarce resources during sickness and health in pigs. In R. M. Gous, T. R. Morris and C. Fisher (Eds), *Mechanistic Modelling in Pig and Poultry Production*, pp. 117–42. CAB International, Wallingford, U.K.

Kyriazakis, I., Stamataris, C., Emmans, G. C. and Whittemore, C. T. 1991. The effects of food protein content on the performance of pigs previously given foods with low or moderate protein contents. *Anim. Prod.* 52: 165–73.

Lewis, P. D. and Morris, T. R. 2004. Research note: amendments to the model for predicting age at sexual maturity for growing pullets of layer strains following changes in photoperiod. *J. Agric. Sci.* 142: 613–14.

Lewis, P. D., Gous, R. M. and Morris, T. R. 2007. A model to predict age at sexual maturity in broiler breeders given a single increment in photoperiod. *Br. Poult. Sci.* 48: 625–34.

Lewis, P. D., Morris, T. R. and Perry, G. C. 2003. Effect of two opposing changes in photoperiod upon age at first egg in layer-hybrid pullets. *J. Agric. Sci. (Cambridge)* 140: 373–9.

McArthur, A. J. 1981. Thermal insulation and heat loss from animals. In J. A. Clark (Ed.), *Environmental Aspects of Housing for Animal Production*, pp. 37–60. Butterworths, London.

Mack, S., Hohler, D. and Pack, M. 2000. Evaluation of dose-response data and implications for commercial formulation of broiler diets. *Proc. 21st Aust. Poult. Sci. Symp.* 82–7.

Morris, T. R. 1968. The effect of dietary energy level on the voluntary calorie intake of laying birds. *Br. Poult. Sci.* 9: 285–95.

Morris, T. R. 1969. Nutrient density and the laying hen. In H. Swan and D. Lewis (Eds), *Proceedings of the Third Nutrition Conference for Feed Manufacturers*, pp. 103–14. University of Nottingham, Nottingham, U.K.

Morris, T. R. 2006. Introduction to modelling in Animal Science. In R. M. Gous, T. R. Morris and C. Fisher (Eds), *Mechanistic Modelling in Pig and Poultry Production*, pp. 1–5. CAB International, Wallingford, U.K.

Mount, L. E. 1979. *Adaptation to Thermal Environment: Man and his Productive Animals*. Edward Arnold, London.

National Research Council (NRC) (1994). Nutrient Requirements of Poultry. Ninth Revised Edition, National Academy Press, Washington, DC.

Nonis, M. and Gous, R. M. 2015. Changes in the feather-free body of broiler breeder hens after sexual maturity. *Anim. Prod. Sci.* http://dx.doi.org/10.1071/AN14713.

Wellock, I. J., Emmans, G. C. and Kyriazakis, I. 2003. Predicting the consequences of social stressors on pig food intake and performance. *J. Anim. Sci.* 81: 2995–3007.

Wellock, I. J., Emmans, G. C. and Kyriazakis, I. 2004. Modeling the effects of stressors on the performance of populations of pigs. *J. Anim. Sci.* 82: 2442–50.

Whittemore, C. T. 1976. A study of growth responses to nutrient inputs by modelling. *Proc. Nutr. Soc.* 35: 383–91.

Whittemore, C. T. and Fawcett, R. 1976. Theoretical aspects of a flexible model to simulate protein and lipid growth in pigs. *Anim. Prod.* 22: 87–96.

Whittemore, C. T. 1981. Animal production response prediction. In *Computers in Animal Production*, pp. 47–64. Occasional Publication No. 5. British Society of Animal Production.

Willmer, P., Stone, G. and Johnston, L. 2000. *Environmental Physiology of Animals*. Blackwell Science, Oxford.

Chapter 12

Developments in feed technology to improve poultry nutrition

Charles Stark, Kansas State University, USA; and Adam Fahrenholz, North Carolina State University, USA

1 Introduction

Feed accounts for over 70% of the cost of poultry production. Ingredients make up the greatest proportion of the cost, while feed manufacturing and delivery account for approximately five percent of the total diet cost. Properly leveraging the feed manufacturing process, combined with least-cost formulation and the use of alternative ingredients, can significantly lower the diet cost. However, variations in the quality of ingredients and finished feed, as well as in the manufacturing process, can significantly affect broiler performance. In addition, advancements in poultry nutrition, changes in broiler genetics, increased biosecurity, and fluctuations in ingredient quality require greater flexibility in the manufacturing process. For these reasons, and to remain competitive, poultry producers

http://dx.doi.org/10.19103/AS.2016.0011.12

must routinely evaluate the quality of their ingredients and the manufacturing process in the feed mill. A systematic evaluation of each process within the feed manufacturing facility will identify opportunities for improvement in manufacturing efficiency and reduced nutrient variation of finished feed, which ultimately results in a lower cost of poultry production.

2 Purchasing and formulating feed ingredients

Purchasing high-quality ingredients from approved suppliers is the first step in manufacturing safe high-quality feed. Ingredient specification sheets provide guidance to purchasing agents, suppliers, transporters, and receiving personnel and are the cornerstone of producing high-quality finished feed, limiting product liability and lowering the cost of feed (Stark and Jones, 2012). Ingredient specification sheets should include a product description, expected nutrient content, analytical methods, physical characteristics and the basis for rejection. Constraints in feed mill design and capacity can impact ingredient-purchasing decisions. Purchasing decisions should take into account the truck driver's wait time, railcar demurrage, unexpected formulation changes due to the lack of ingredients and the potential negative impact that constant formula changes can have on bird performance as a result of the lack of ingredients (Stark, 2012a).

Alternative ingredients are often included in corn–soyabean meal formulas to lower the cost of diets. However, alternative ingredients require added analytical costs to develop and maintain least-cost formulation matrix values, increase receiving time at the feed mill, reduce batching times due to the addition of more ingredients to the scales and reduce pellet mill throughput. Alternative ingredients can also change pellet quality, feed density and palatability. These issues should be taken into account in a cost–benefit analysis to determine if the reduced diet cost truly represents a system-wide savings.

Precision formulation allows the broiler production system to meet the goals for animal nutrient utilization, while minimizing costly and wasteful excretion of valuable nutrients (Stark and Jones, 2015). Pomar and Pomar (2012) stated that the essential elements for precision feeding include evaluating the nutritional content of feed ingredients, precise determination of nutrient requirements of animals, formulating diets that limit the amount of excess nutrients and the adjustment of nutrients to match the requirements of animals fed. Companies that combine purchasing of quality ingredients with precision formulation based on genetics, housing and management systems can reduce the cost of raising broilers.

3 Automation technology

Poultry feed mills have become highly automated over the last 20 years. The large control room with push-button panels has been replaced with programmable logic controllers (PLC) and server-based networks that can be used to control the automation system from any place in the world. In addition to the system controlling the equipment, there is significantly more data captured during the manufacturing process. Automation packages have detailed reporting systems that can track both individual equipment operating parameters and system-wide efficiencies through real-time data management protocols. Automation systems apply statistical process control (SPC) software to monitor the time required to unload an ingredient truck, the accuracy of ingredient additions to a batch of feed and the time required to load finished feed trucks. SPC can be used to improve the precision of the batching process, which reduces variation in the nutrient content of the finished feed. Managers can review both real-time and historical data and make changes in parameter settings and equipment set-up that improve the accuracy of each batch of feed (Stark, 2016).

The next generation of automation systems will not only control equipment and collect data; they will also be integrated into the company's animal food safety systems. Consumers around the world are not only demanding more meat but also demanding that the meat is produced in a safe, sustainable and responsible production system. Automation systems will track ingredients from the time of purchase through receiving, processing and delivery to the animal producer (Stark, 2016). Next-generation automation systems will track ingredients from farm to fork through the use of global positioning system (GPS) equipment mounted on ingredient delivery trucks and finished feed trailers to ensure that the right feed is delivered to the right bin. Similarly, radio frequency identification (RFID) systems can be used to verify specific location and position information. Such systems are currently used to automatically load feed trucks and control the delivery of finished feed using sensors on finished feed bins and the end of the boom on the feed trailer.

The seamless exchange of data between different software packages is the key to improving the efficiency of the manufacturing process. The adoption of new technologies in the feed mill is occurring at a faster pace as software companies improve data exchanges between processing automation software and business enterprise software (Stark, 2016). Companies are now monitoring the nutrient content of ingredients with in-line near-infrared (NIR) technology, sending that information to the feed formulation software to make adjustments in the formulation and then adjusting the batching software with minimal human interaction. In such a system, the formula for the next batch of feed can be altered based on least-cost principles and the composition of ingredients used in the current batch. This allows the feed manufacturer to take real-time advantage of nutrient composition information versus evaluating formulations weekly, bi-weekly or monthly based on information gathered from composite samples. While monitoring would continue to be necessary, technology such as this could greatly reduce human effort required to develop formulas, input them into the automation system and conduct quality tests.

4 Feed ingredient receiving

Biosecurity in the receiving area has become a concern due to disease outbreaks and increased movement of ingredients into feed mills. The greatest risks for pathogens entering a feed mill are from vehicles, people and ingredients. Biosecurity programmes reduce the risk of disease outbreaks by inspecting and disinfecting delivery vehicles, limiting the movement of visitors and purchasing from approved suppliers.

Feed mills are designed to manage a wide variety of dry ingredients delivered in bulk, totes and bags as well as liquid ingredients delivered in bulk or totes. Bulk dry ingredients arrive at feed mills by covered hopper railcars, hopper bottom and hydraulic lift trailers and pneumatic trucks. Receiving systems must quickly unload unit trains of 65 cars or more in less than 15 hours and trucks within minutes of arrival to avoid demurrage charges. The government and consumers are also demanding near-complete traceability of feed ingredients as they move from storage to processing, and finally to finished feed.

Simple, inexpensive quality assurance tests at the point of receiving can help segregate ingredients based on moisture, protein, fat and starch content. Equipment such as the grain moisture analyser, NIR or moisture balance can be used to rapidly determine the nutrient content of an ingredient prior to its receipt; typically these tests take less than ten minutes (Stark, 2012b). Tabletop NIR technology ranges from basic machines that measure moisture, protein and fat to models that can measure moisture, protein, fat, fibre, starch and amino acids. While tabletop models can provide useful points in time analysis, the greatest opportunity moving forward is in the use of in-line technology that has the potential to change how ingredients are received and characterized. In-line NIR can quantify the moisture, protein, fat and fibre content of ingredients as they pass across in-line sensors, which allows for continuous monitoring and segregation of ingredients during the receiving process (Stark, 2013). While understanding the proximate nutrient composition of received ingredients is important, the economic advantages in diet formulation are based on the company's ability to develop prediction equations to estimate energy (ME or NE) and amino acid composition of ingredients and then separate ingredients based on their nutrient values within least-cost formulation (Stark and Jones, 2015).

Feed mills that have the ability to segregate and manage ingredients based on supplier and/or plant location can capture savings through least-cost formulation if they can develop matrix values specific to those suppliers and/or locations. Segregation of ingredients based on nutrient content requires communication between the purchasing group, nutritionists and the feed mill. In order for such a system to be effective, formulas must be regularly updated based on the nutrient content of the ingredients currently in the feed mill's inventory.

5 Particle size reduction in feed

Cereal grains are ground at the feed mill, whereas ingredients such as soyabean meal, rendered products, distillers dried grains and wheat by-products are ground by the supplier and can be used without further processing by the feed mill. Owens and Heimann (1994) indicate the major reasons for particle size reduction are:

- Expose greater surface for digestion
- Improve ease of handling of some ingredients
- Improve mixing characteristics of ingredients
- Improve pelleting efficiency and pellet quality

Particle size reduction of cereal grains is a small fraction of the overall cost of manufacturing poultry feed, but can significantly affect bird performance and gut health (Fahrenholz and Brake, 2015). The cost of grinding an ingredient is inversely related to particle size; the cost will increase as the target particle size decreases. Researchers continue to evaluate the optimal particle size for poultry and how to manipulate the grinding process to most effectively balance bird performance, gut health and processing costs (Fahrenholz, 2014).

Fine grinding cereal grains for broilers has led to poor nutrient absorption and gut health issues due to the lack of reverse peristalsis in the bird's GI tract (Chewning et al. 2012; Xu et al. 2015). Large particles help stimulate the gizzard and reverse peristalsis. Studies have shown a correlation between the particle size of corn and gizzard development and feed utilization. Nir et al. (1994) stated that greater coarseness of feed increased relative gizzard weight while Amerah et al.

(2008) suggested gizzard stimulation was due to the length of time that the coarse particles resided in the gizzard. Larger gizzards relative to body weight (BW) have resulted in improved feed utilization (Xu et al. 2015) and gastric intestinal tract health (Ferket, 2000).

The particle size of ground grain is most commonly expressed as the geometric mean diameter by weight (d_{gw}) of the sample as determined by the Method of Determining and Expressing Fineness of Feed Materials by Sieving (ANSI/ASAE S319.4 FEB 2009 R2012). The standard method involves the analysis of the material utilizing wire-cloth sieves and a sieve shaker. Although a standard method exists, researchers, laboratories and feed mills may use modified methods, which can significantly change the results. The method states, 'For industrial applications, the end-point determination process can be omitted, and the end-point is set to be the sieving time of 15 min.' Additionally, the use of dispersion agents and sieve agitators will reduce the estimated particle size result of the sample. Stark and Chewning (2012) reported that the addition of a dispersion agent to hammermill ground corn reduced the estimated particle size of the sample from 411 to 332 microns, whereas the addition of the sieve agitators only resulted in a 41-micron decrease in particle size (392 vs 351 microns). These results were in general agreement with the results by Fahrenholz et al. (2010). Kalivoda (2016) reported no difference in particle size results between a 10- and 15-minute sieving time when sieve agitators and sieving agent were used to determine particle size. With the above in mind, it is important to document and understand how particle size results are generated and reported.

6 Feed batching

The number of ingredients added to broiler diets has increased over the last five to ten years. The use of more alternative ingredients other than corn and soyabean meal as well as increased numbers of micro-ingredients has created challenges for proportioning and mixing systems. The proportioning systems in poultry feed mills may be designed with multiple major scales, a minor scale, a tote system and multiple micro-system bins that facilitate fast and accurate batching of feed. The growth of the feed additive industry, in addition to increased precision in formulation with the use of synthetic amino acids, has added further complexity to automated batching systems. Older batching systems may not be capable of weighing ingredients to within the desired upper and lower specification limits (1–2%) of the required quantities (Stark and Jones, 2015). Unfortunately, this has resulted in lower manufacturing productivity and less accuracy, especially in the batching process. Feed mills often must choose between accuracy of inclusion and speed of the batching process due to the increased number of ingredients added to a batch of feed. In addition to the increased number of micro-ingredients, older micro-systems were not designed for small inclusion ingredients such as concentrated enzymes or small inclusions of the third and fourth limiting amino acids in poultry diets. In the past, ingredients were typically included at rates of 0.5 to 1.0 lb per ton. However, due to concentration of enzymes, vitamins and trace minerals in premixes, the inclusion levels have decreased to less than 0.5 lb per ton in many instances. Smaller inclusion levels require greater scale resolution, finer control of equipment and a higher degree of accuracy during weighing, which is a challenge when trying to weigh ingredients to the nearest 0.01 lb. The relatively new innovation of loss-in-weight micro-systems have become popular, both for their improved accuracy and for inventory tracking of medications. Loss-in-weight micro-systems have greater accuracy than conventional scaling systems since product free-fall determinations are eliminated during the batching process (Stark, 2016).

The evolution of automation systems has significantly changed the batching process. The need for traceability and enhanced data collection is being driven by new government regulations and consumer demands. Automation systems have moved from merely controlling equipment and routing feed to precision feed manufacturing, ingredient lot tracing, using barcode readers during hand additions and summarizing process data (Stark, 2014). The latest technology adds variable frequency drive (VFD) controllers to the PLC-based systems to increase the accuracy of ingredient additions during the batching process. The accuracy of the batching system is a function of time, screw conveyor diameters and VFD setting (when available) for the micro-system, as well as for the minor and major scales (Stark and Jones, 2015). The addition of the VFD allows for a screw feeder to be operated at multiple speeds: a fast speed at the beginning of the ingredient draw and a slower speed when nearing the target weight. This arrangement can be both faster and more accurate than traditional systems. The installation of a single VFD to control multiple screw conveyors is becoming common practice in new feed mill construction and is being retrofit in older feed mills (Stark, 2016).

Precision formulation requires greater monitoring and control of the batching process to ensure the correct amount of each ingredient is added to each batch of feed. The cost of not adding the correct amount of each ingredient will result in poor animal performance, especially in diets that have small nutritional safety factors built into the formulation. Conversely, the continuous over-addition of ingredients to feed results in ingredient shrink. Small differences between the amount listed on the master formula and the actual amount of ingredients weighed on the scales can create a significant inventory deviation over time (Stark, 2015).

7 Feed mixing

Changes in mixer design and size have led to increased productivity in feed mills, especially where new mixer design has decreased the amount of time required to mix ingredients. Double shaft ribbon and paddle mixers have dramatically decreased mixing time from three to four minutes to two to three minutes. While a one-minute decrease may not seem significant, it results in an additional five batches of feed each hour. The discharge order and location in which ingredients are added to the mixer can affect mixing efficiency. The discharge order should be major scale, minor scale and finally micro-scales with a small delay after each discharge. The size of the batch should also be taken into consideration when mixing feed. A mixer is designed and constructed for a specific volume of feed. Therefore, as the density of the mixture decreases due to the addition of more alternative by-products, the batch size must also be decreased in order to not exceed the maximum volume of the mixer. A general rule of thumb is that the ribbons should be visible when the mixer is operating and all the feed is in motion (Stark, 2016). Routine maintenance and inspection of the ribbons and paddles for material build-up will ensure that the mixer is functioning properly.

$$\text{CV\%} = \frac{\text{standard deviation X 100\%}}{\text{mean}}$$

Table 1 Interpretation of mixer uniformity test

CV	Rating	Corrective action
<10%	Excellent	None
10–15%	Good	Increase mixing time by 25–30%
15–20%	Fair	Increase mixing time by 50%, look for worn equipment, overfilling or sequence of ingredient addition
20% +	Poor	Possible combination of all the above
		Consult extension personnel or feed equipment manufacturer

Feed mills should conduct a mixer uniformity test when a mixer is installed and then at least annually to ensure the mixer is producing a homogeneous feed. The concentration of one single-source nutrient or tracer is determined for 10 individual samples taken from the mixer, the mixer surge hopper, or most commonly from the discharge conveyor at equally spaced intervals during the discharge cycle. The tracer levels from each sample are then used to calculate the CV% of a batch of feed.

The CV% should be less than 10%. Once the mix time (dry and wet mix) has been established it should not be changed without validating the new mixing times (Stark, 2016). Herrman and Behnke (1994) provided guidelines for interpretation of mixer uniformity results and possible corrective actions (Table 1). Mix uniformity can have a significant impact on animal performance. McCoy et al. (1994) reported that a lower CV% (10% vs 40%) resulted in improved BW and feed conversion in chicks raised to 28 d of age. The feed mill should also have established sequencing and flushing procedures that minimize the cross-contamination of medicated feed additives. Martinez-Kawas (2008) demonstrated that increasing the size of the flush after mixing a medicated feed from 2.5% to 20% of the mixer's capacity decreased the concentration of a medication in the flush material and subsequent batches of non-medicated feed. The amount of medication found in the subsequent feeds was very low, with the highest concentrations observed in the bucket elevator and finished feed bin.

8 Feed pelleting

There are many factors that impact the efficient production of high-quality pellets. These include ingredient composition, equipment design and manufacturing parameters. There are entire reference works devoted to the topic of pelleting (Jones et al. 2014). Therefore, in the following paragraphs, pelleting is discussed specifically as it relates to the production of feed for poultry. The benefits of pelleting have been well documented in poultry. Broiler diets are typically pelleted to increase feed consumption and BW gain, which improves feed conversion. In addition to feeding pellets, there is added value in feeding a greater percentage of whole pellets (Stark, 2014). Behnke (1994) lists the following benefits to feeding pelleted diets to animals:

- Decreased feed wastage
- Reduced selective feeding
- Decreased ingredient segregation

- Less time and energy expended in prehension
- Destruction of pathogens
- Improved palatability.

Pellets are produced by conditioning dry mash with steam and then extruding the material through a pellet die. The steam adds both heat and moisture, which helps soften the brittle particles and makes them pliable. The heat and moisture also activate the natural adhesive properties of both starch and protein present in the grains and protein meals. The mash is conditioned to 180°–195°F and retained in single pass conditioners for 30–45 seconds or double pass conditioners for 75–90 seconds. While higher retention times tend to improve pellet quality and pellet mill throughput, some research has suggested that long-term heat processing will negatively affect nutritional properties of the feed, such as reducing amino acid digestibility (Wamsley and Moritz, 2013). In addition to long-term conditioning, the use of high compression pellet dies, which impart greater amounts of frictional heat, has been shown to reduce performance.

The benefits of pelleting tend to diminish as pellet quality decreases and the amount of fines in the feeder increases. Reducing the amount of fines at the feeder may require a change in formulation, addition of pellet binders and/or reduced pellet mill throughput. These changes tend to increase the cost of the diet, due to both ingredient and manufacturing costs. However, most poultry feed mills do not remove fines due to capacity constraints, and ship product that has both pellets and fines. The feed may contain anywhere from 20% to 60% fines when delivered to the birds. Moritz et al. (2001), McKinney and Teeter (2004), Lemme et al. (2006) and Amerah et al. (2007) reported an improvement in feed conversion when birds were fed pellets versus mash feed. However, Greenwood et al. (2004) reported no difference in feed conversion in diets containing 20% to 60% fines. McKinney and Teeter (2004) indicated there was no advantage to producing more than 40% pellets unless pellet quality was in excess of 60%; the results of their study were used to develop an effective caloric value model. The model estimates the effects that pellet fines (0% to 80%) have on the caloric value of the diet. The model suggests an 80% change in fines results in a caloric difference of 111 ME/kg of diet, attributed primarily to a higher eating frequency associated with poor-quality pellets. Hu et al. (2012) reported that birds fed crumbles in any quantity versus mash exhibited a higher feed intake, which resulted in a greater BW when fed in cages; however, most of the positive effect on BW that occurred from feeding high-quality crumbles was no longer apparent at 35 d of age after the broilers were moved to the floor pens. The results were similar to those of Nir et al. (1995) who reported higher BW (2298 g) with crumbled feed as compared to mash diets (2236 g) at 35 d of age in male broilers. Additionally, Hu et al. (2012) demonstrated that 49-d-old birds fed screened pellets (83% grower, 97% finisher pellets) versus pelleted feed that contained fines (46% grower, 51% finisher pellets) were 4.5% and 6.0% heavier in both males and females, respectively. Lemons and Moritz (2016) reported that feeding a higher percentage of crumbles and pellets to fines ratio improved feed intake, especially in pens with a high bird density. These findings were similar to other researchers who reported that the addition of fines to broiler diets reduced performance in all diet phases (Nir et al., 1995; Engberg et al., 2002; Svihus et al., 2004; Corzo et al., 2011, Lemons and Moritz, 2016). Amerah et al. (2007) suggested that these improvements could be attributed to increased nutrient density of the pellets, improved starch digestibility due to conditioning and pelleting the mash, increased nutrient intake due to the physical form of the feed, reduced feed wastage and decreased energy expenditure during feeding. Abdollahi and Ravindran (2013) reported improved feed intake and weight gain in starter chicks fed 3-mm-length pellets as compared to 5- and 7-mm-length pellets.

Additionally, Buchanan et al. (2010) reported that diet formulation and manufacturing techniques affected broiler performance and intestinal morphology.

De Jong (2015) reported a difference in the nutrients composition of fines and pellets loaded onto a bulk delivery truck. The percentage of crude protein was higher and crude fat was lower in the pellets as compared to fines in feed taken after the fat coater and bulk load-out. However, there was no difference in the mineral content of the fines and pellets. The variation in nutrient content of fines and pellets may create greater variation within the house especially as the percentage of fines tend to be delivered to the feeders closest to the bin, whereas the pellets end up in the feeder pans the farthest away from the feed bin on the farm.

Research studies clearly indicate that the percentage of fines in the feed can have a significant impact on animal performance. However, higher quality pellets cost more to produce, and there is generally a negative impact on feed mill production capacity. Therefore, economic models should be used to determine the value of producing more feed for more birds as compared to feeding less fines to a smaller number of birds. While either producing more durable pellets and/or screening all the fines prior to delivery may result in the best bird performance, it may not result in the lowest cost of production due to the additional cost of manufacturing, the cost of adding pellet binders or the cost to screen and re-pellet the fines (Stark, 2015).

Cooling is an important part of the pelleting process that is often overlooked by feed mill management. The goal of the cooling process should be to reduce the moisture content of the pelleted feed to be less than or equal to the original moisture content of the mash feed prior to entering the conditioner. Residual moisture left in the pellets can lead to the growth of mould, degradation of pellets during handling and dilution of the energy content of the feed (Stark, 2012a). A pelleted broiler diet that contains as little as 1% additional moisture after cooling will have a lower calculated energy content, which will negatively affect feed conversion. Moisture levels greater than target are typically due to uneven bed depths in the cooler, the incorrect volume of air or too many fines that block the air flow through the bed of pellets in the cooler. The speed of the cooler fan may need to be adjusted based on the ambient temperature, which affects the water holding capacity of the air. Monitoring the mash before pelleting and the pellets after cooling will help the manager make the appropriate cooler and fan adjustments (Stark, 2012b).

9 Post-pellet liquid application

The addition of liquids, such as fat or heat-sensitive enzymes post pelleting, has become popular over the last 20 years. Post-pellet liquid application (PPLA) of fat can occur at the pellet die while the pellets are hot or after the pellets have been cooled. The addition of fat at the die is a simple process that does not require a significant capital investment in equipment as compared to PPLA after the pellets have been cooled. However, applying fat at the die reduces the efficiency of the cooling process and requires more routine cleaning and maintenance of the cooling equipment. The application of fat on cooled pellets requires equipment that can accurately measure the dry flow of feed and then metre on the correct amount of liquid fat and/or enzymes (Stark, 2016). A dry flow system determines the amount of incoming dry feed either by volume and density or by force using a device such as a CentriFlow® meter. The automation system then metres on liquid ingredients in proportion to the amount of dry flow. The liquid can be applied with spray nozzles or spinning discs. The liquid-coated pellets are mixed in cut-section screws before being conveyed to the finished feed bin. Feed mills that use PPLA systems should frequently calibrate the liquid metres and dry flow devices, implement a liquid inventory reconciliation process and develop a sampling and monitoring programme that validates uniform and accurate addition of the liquid ingredients (Froetschner, 2007).

10 Feed delivery

The delivery of feed is the final and most important step in getting the right feed to the right bin at the right time. The delivery of the wrong feed to a house of birds can significantly impact the performance of the flock. Automation of the process can help ensure that the correct feed is loaded on the truck in the correct order, and GPS-based software can track the delivery of the feed to the house. RFID systems can be used to check what was hauled on the delivery truck prior to loading, automatically load the compartments of the truck and track the delivery of the feed. Once on the farm, RFID technology installed on each compartment of the truck, the end of the delivery boom and each farm bin can be used to verify that the right feed goes to the right bin (Stark, 2016). The design and tare weight of the feed delivery equipment will determine the total weight of feed that can be hauled on a load. Selecting the correct type and design of equipment should be based on the desired unloading speed, longevity of the equipment and pellet quality. Advances in equipment design have decreased the unloading time and reduced the degradation of pellets during the unloading process. De Jong (2015) reported a 43% reduction in unloading time by increasing the speed on the discharge augers from 900 to 1400 RPM with no change in the percentage of fines created during the unloading process.

11 Finished feed quality

The quality of finished feed is often defined by nutrient content and physical characteristics or percentage of fines in the feed, also known as pellet quality. At the feed mill, the durability of pellets can be measured, giving some indication of pellet quality once the feed reaches the birds. The most common method used in the poultry industry to determine pellet durability is the pellet durability index (PDI) (ASAE, 1997; S269.4), which was developed at Kansas State University. The PDI method is often modified to create a more aggressive test by adding hex nuts or ball bearings to the tumble chamber to model a feed mill's manufacturing and delivery processes. Another method commonly used by the industry is the Holmen NHP 100, which uses air to create abrasion of the pellets versus the tumbling action that occurs in the metal box of the PDI tester. The latest development in pellet durability testing is the Holmen NHP 300, a fully automated in-line tester that obtains the sample directly after the pellet mill to determine pellet durability. The Holmen NHP 300 collects, cools and sieves the sample; performs the tests; calculates the durability; and exports the data to a computer. The operator can then use the results to make the appropriate adjustments to the pellet mill (Stark, 2014).

The pellet durability test should be a predictive model of the manufacturing and handling processes that occur as feed is handled throughout the feed mill and delivery system.

Each poultry production system will have unique characteristics that must be taken into account when developing the model. Feed mills should develop a method to estimate pellet quality at the feeder and then make the appropriate adjustments to processes within the feed mill to achieve the desired pellet durability. Additionally, the model can provide feedback to the nutritionist and purchasing agent as to the effect an ingredient or formulation change had on pellet durability and resultant quality (Stark, 2012a).

NIR technology has been used in feed laboratories for over 30 years, but recently feed mills have started to embrace the use of tabletop NIR that can be located in different process areas of the feed mill. NIR technology has evolved from the need to grind and load sample cells to simply placing the product on a plate in the unit. The user interface has also improved with touchscreen technology (Stark, 2016). Additional advantages of NIR include minimal sample preparation, production line or on-line analysis, high precision, low sample cost, and no chemicals or waste (Eubanks, 2013). All of the above means that NIR can now be as valuable in process and finished feed analysis as it has historically been in analysing incoming ingredients.

While NIR technology is an exciting advancement, companies must recognize that NIR units are only as good as the calibrations used to predict nutrient content and the wet chemistry analyses used to bias the results. Companies must be willing to invest in both the time and resources required to maintain robust calibrations (Stark, 2016). New network-based software solutions allow an expert to precisely configure, manage and monitor NIR instruments from a remote location, which reduces the need for a NIR specialist at the facility (Tollecback and Mills, 2009).

12 Conclusion and future trends

Feed manufacturing continues to evolve at a rapid pace, with improvements to automation systems, updates in analytical methods and greater understanding of how feed processing impacts animal performance each playing a significant role. There will continue to be advancements in these areas and others. With this in mind, it is of utmost importance for feed manufacturers to be aware of the technologies available and for there to be good communication between decision makers when it comes to deciding how a company can best impact the bottom line. Depending on the company, decisions on innovation can be from the bottom up (e.g. a mill manager requesting funds to purchase a new technology) or from the top down (e.g. management being sold on a new idea and driving its implementation). In both cases, someone has to be doing the groundwork to see these innovations coming, and then detailed discussions must occur as to how a new technology can be implemented, what it will mean to ongoing operations and what the cost–benefit will be.

Moving forward, much of the focus on feed manufacturing will relate to compliance with increasing food safety regulations, and the adaptation of formulations and processes based on the inclusion of alternative ingredients as consumers drive a demand for medication and by-product-free animal feeding. On the regulatory side, much of the attention will be on generating records and streamlining compliance-related processes. Accordingly, automation systems will need to include increasingly robust records packages and/or have an increased ability to communicate with other management programmes. The changes to animal feeding strategies is a double-edged sword for feed manufacturers. On the one side, reducing the use of medications limits regulatory exposure and pulling out by-products frees up bin space and in many cases helps with pellet quality issues. On the other side, some alternative ingredients are very low inclusion, necessitating more precise batching, and may be very sensitive to thermal processing, requiring strict control of temperature profiles or post-pellet application. While these issues are not necessarily new or particularly difficult to address, it is the wide range of alternative ingredients entering the market that creates a challenge. In short, the future will likely require feed manufacturers to be more flexible in both how they conduct and how they document practices in the feed mill. And technology will certainly have a significant role to play.

13 Where to look for further information

The *Feed Manufacturing Technology V* book published by the American Feed Industry Association provides a comprehensive overview of the feed industry. The most current information related to quality assurance, manufacturing processes and management are available through extension bulletins at Kansas State University and North Carolina State University.

14 References

Abdollahi, M. R., and V. Ravindran. 2013. Influence of pellet length on pellet quality and performance of broiler starters. *J. Appl. Poult. Res.* 22: 516–22.

Amerah, A. M., V. Ravindran, R. G. Lentle and D. G. Thomas. 2007. Influence of feed particle size and feed form on the performance, energy utilization, digestive tract development, and digesta parameters of broiler starters. *Poult. Sci.* 86: 2615–23.

Amerah, A. M., V. Ravindran, R. G. Lentle and D. G. Thomas. 2008. Influence of feed particle size on the performance, energy utilization, digestive tract development, and digesta parameters of broiler starters fed wheat- and corn-based diets. *Poult. Sci.* 87: 2320–8.

ASAE. 1997. Cubes, Pellets, and Crumbles – Definitions and Methods for Determining Density, Durability, and Moisture Content. ASAE Standard S269.4. American Society of Agricultural and Biological Engineers, St. Joseph, MI.

ASAE. 2012. Method of Determining and Expressing Fineness of Feed Materials by Sieving. ANSI/ASAE Standard S319.4 R2012. American Society of Agricultural and Biological Engineers, St. Joseph, MI.

Behnke, K. C. 1994. Factors affecting pellet quality. Proceedings of Maryland Nutrition Conference, 20–25 March 1994. Department of Poultry Science, College of Agriculture, University of Maryland, College Park.

Buchanan, N. P., K. G. S. Lilly and J. S. Moritz. 2010. The effect of diet formulation, manufacturing technique, and antibiotic inclusion on broiler performance and intestinal morphology. *J. Appl. Poult. Res.* 19: 121–31.

Corzo, A., L. Mejia and R. E. Loar II. 2011. Effect of pellet quality on various broiler production parameters. *J. Appl. Poult. Res.* 20: 68–74.

Chewning, C. G., C. R. Stark and J. Brake. 2012. Effects of particle size and feed form on broiler performance. *J. Appl. Poult. Res.* 21: 830–7.

Deaton, J. W., B. D. Lott and S. L. Branton. 1995. Corn grind size and broilers reared under two temperature conditions. *J. Appl. Poult. Res.* 4: 402–6.

De Jong, J. 2015. Feed processing challenges facing the swine industry. Ph.D. Dissertation. Kansas State University, Manhattan, KS.

Engberg, R. M., M. S. Hedemann and B. B. Jensen 2002. The influence of grinding and pelleting of feed on the microbial composition and activity in the digestive tract of broiler chickens. *Br. Poult. Sci.* 43: 569–79.

Eubanks, H. 2013. Expanding the uses of rapid assessment technology for feed production. 34th Western Nutrition Conference. Saskatoon, SK.

Fahrenholz, A. C., L. J. McKinney, C. E. Wurth and K. C. Behnke. 2010. The Importance of Defining the Method in Particle Size Analysis by Sieving. Kansas State University Agricultural Experiment Station Report of Progress 1038 (Swine Day 2010), p. 261.

Fahrenholz, A. C. 2014. Manipulating Particle Size to Impact Animal Nutrition. Proceedings of 75th Minnesota Nutrition Conference, p. 75.

Fahrenholz, A. C., and J. Brake. 2015. Feed Particle Size and Poultry Performance. Proceedings of Midwest Poultry Federation Convention, St. Paul, Minnesota.

Ferket, P. 2000. Feeding whole grains to poultry improves gut health. *Feedstuffs* 72: 12–14.

Froetschner, J. R. 2007. Micro-ingredient application and equipment: issues and advances. Animal Feed Manufacturer Assn. Forum Proceedings. Sun City, South Africa.

Greenwood, M. W., P. M. Clark and R. S. Beyer. 2004. Effect of fines level on broilers fed two concentrations of dietary lysine from 14 to 30 days of age. *Int. J. Poult. Sci.* 3(7): 446–9.

Herman, T., and K. C. Behnke. 1994. Testing mixer performance. MF-1172. Department of Grain Science and Industry. Kansas State University, Manhattan.

Hu, B., C. R. Stark and J. Brake. 2012. Evaluation of crumble and pellet quality on broiler performance and gizzard weight. *J. Anim. Vet. Adv.* 11: 2453–8.

Jones, C., A. Fahrenholz and C. Stark, eds. 2014/ Feed Pelleting Reference Guide. WATT Global Media, http://www.wattagnet.com/feedpelleting.html.

Kalivoda, J. 2016. Effect of sieving methodology on determining particle size of ground corn, sorghum, and wheat by sieving. M.S. Thesis, Kansas State University, Manhattan, KS.

Lemme, A., P. J. A. Wijtten, J. van Wichen, A. Petri and D. J. Langhout. 2006. Responses of male growing broilers to increasing levels of balanced protein offered as coarse mash or pellets of varying quality. *Poult. Sci.* 85: 721–30.

Lemons, M. E. and J. S. Moritz. 2016. The effect of feeder space access and crumble- or pellet-to-fine ratio on 38-day-old broiler performance. *J. Appl. Poult. Res.* 25: 12–20.

Martinez-Kawas, A. 2008. Evaluating flushing procedures to prevent drug carryover during medicated feed manufacturing. Master's Thesis. Kansas State University, Manhattan, KS.

McCoy, R. A., K. C. Behnke, J. D. Hancock and R. R. McEllhiney. 1994. Effect of mixing uniformity on broiler chick performance. *Poult. Sci.* 73: 443–51.

McKinney, L. J. and R. G. Teeter. 2004. Predicting effective caloric value of nonnutritive factors: I. Pellet quality and II. Prediction of consequential formulation dead zones. *Poult. Sci.* 83: 1165–74.

Moritz, J. S., R. S. Beyer, K. J. Wilson and K. R. Cramer. 2001. Effect of moisture addition at the mixer to a corn-soybean-based diet on broiler performance. *J. Appl. Poult. Res.* 10: 347–53.

Nir, I., Twina, E. Grossman and Z. Nitsan. 1994. Quantitative effects of pelleting on performance, gastrointestinal tract and behaviour of meat-type chickens. *Br. Poult. Sci.* 35: 589–602.

Nir, I., R. Hillel, I. Ptichi and G. Shefet, 1995. Effect of particle size on performance. 3. Grinding pelleting interactions. *Poult. Sci.* 74: 771–83.

Owens, J. M., and M. Heimann. 1994. Material processing cost center. In R. R. McElhinney (ed.), *Feed Technology IV.* American Feed Industry Assn. Arlington, VA, p. 81.

Pomar, C., and J. Pomar. 2012. Sustainable precision Livestock farming: a vision for the future of the Canadian Swine industry. In *Advanced in Pork Production*, pp. 207–13.

Stark, C. R. 2012a. Feed processing to maximize feed efficiency. In J. F. Patience (ed.), *Feed Efficiency in Swine.* Enfield Pub. Enfield, NH.

Stark, C. R. 2012b. Feed Processing to Improve Poultry Performance. Arkansas Nutrition Conference, Rogers, AR.

Stark, C. R. 2013. Feed processing to increase performance and profit. 34th Western Nutrition Conference. Saskatoon, SK.

Stark, C. R. 2014. Feed processing technology to improve feed efficiency in pigs and poultry. Recent Advances in Animal Nutrition 2014. In P. C. Garnsworthy and J. Wiseman (eds), Context Products Ltd.

Stark, C. R. 2015. Feed processing to improve swine and poultry performance. PHILSAN 28th Annual Convention, Manila, Philippines.

Stark, C. R. 2016. Feed processing to improve swine and poultry performance. The 5th Global Feed and Food Congress, Antalya, Turkey.

Stark, C. R., and C. G. Chewning. 2012. The effect of sieve agitators and dispersing agent on the method of determining and expressing fineness of feed materials by sieving. *Anim. Prod. Sci.* 52(1): 69–72.

Stark, C. R., and F. T. Jones. 2012. Quality Assurance Programs in Feed Manufacturing. Feedstuffs Reference Issue 2013 Miller Publishing, pp. 84–9.

Stark, C. R., and C. K. Jones. 2015. Feed processing to improve animal performance. Canada Eastern Animal Nutrition Conference. Quebec, Canada.

Svihus, B., K. H. Klovstd, V. Perez, O. Zimonja, S. Sahlstrom and R. B. Schuller. 2004. Physical and nutritional effects of pelleting of broiler chicken diets made from wheat ground to different coarsenesses by the use of roller mill and hammer mill. *Anim. Feed Sci. Technol.* 117: 281–93.

Tollecback, C., and R. Mills. 2009. In shape for feed analysis. *Focus* 33 (2): 16–17.

Wamsley, K. G. S., and J. S. Moritz. 2013. Resolving poor pellet quality and maintaining amino acid digestibility in commercial turkey diet feed manufacturing. *J. Appl. Poult. Res.* 22: 439–46.

Xu, Y., C. R. Stark, P. R. Ferket, C. M. Williams and J. Brake. 2015. Effects of feed form and dietary coarse ground corn on broiler live performance, body weight uniformity, relative gizzard weight, excreta nitrogen, and particle size preference behaviors. *Poult. Sci.* 94 (7): 1549–56.

Chapter 13

Alternative sources of protein for poultry nutrition

Paul A. Iji, Mehdi Toghyani, Emmanuel U. Ahiwe and Apeh A. Omede, University of New England, Australia

1 Introduction

Protein sources are the second largest component of practical poultry diets. A limited number of ingredients are used by the mainstream commercial poultry industry to supply protein; these are limited in distribution and are also generally more expensive than energy sources. Soybean seed is the premier protein source used by the poultry industry. Rapeseed or canola seed is probably the second most important protein source. Both seeds are rarely fed as whole seed meals but rather the residue left over after oil extraction is the main ingredient used by the poultry industry. Both soybean and rapeseed are cultivated in only a few places in the world. Soybeans are predominantly

http://dx.doi.org/10.19103/AS.2016.0011.13

produced in the United States, Brazil and Argentina, while the leading producers of rapeseed/canola are Canada, some parts of Europe and China. The key producers of soybean meal export the product to several countries around the world, to the extent that it would seem that it is produced all over the world. A large amount of canola seed meal is now also exported but not to the same extent as soybean meal (SBM). Owing to the limited number of producers and demand, the prices of SBM and canola meal are high and tend to fluctuate with changes in climatic conditions and social situations in the countries where they are produced. These are the drivers of change, and many countries which do not produce soybeans or canola explore alternative sources of protein to support their industries.

The range of ingredients that can supply protein for poultry is wide. Many animals contain high levels of protein, and if these are not used for human food, they can be directly processed into poultry feed. The animal industries, including the poultry industry, yield by-products that are also useful sources of protein. A few examples include blood, feathers and meat-on-bones. Among plant sources, different ingredients command different levels of importance. Some alternative sources are of such local importance that the poultry industry in those areas relies almost entirely on them rather than SBM or canola meal. Good examples of such ingredients are peanut meal or groundnut cake and sunflower seed meal, with substantial production outputs globally. Many other plant protein sources are truly marginal in scope and their potential to replace soybean is negligible. Regardless of the volume of production of an alternative protein source, any shift in use of ingredients other than SBM and canola will greatly reduce the pressure on these key protein sources. Such a shift would also help promote the development of the poultry industry in many areas of the world. This chapter examines the range of alternative ingredients that are available to the poultry industry, their sustainability and means to develop them to prominence as protein sources.

2 Regional supply of conventional protein sources

North and South America are the predominant producers of the two leading crops for oil production: soybean and rapeseed (canola) (Table 1). The United States, Brazil and Argentina together produce more than 80% of all soybeans, while Canada produces about 22% of rapeseed. These are the two most important vegetable protein sources in the world. Although the focus of soybean and canola producers is not animal feed but oil, the residue following oil extraction is an important source of protein for the animal industry. Other important crops that yield protein for animal feeding are groundnuts (peanuts) and sunflower, with total tonnage of 42.4 and 41.3 million tonnes, respectively, in 2014 (FAO, 2015). About 77 million tonnes of cottonseed were produced during the same period, although the meal from cottonseed is not a common ingredient in poultry diets due to a number of reasons.

The poultry industry around the world relies on supply of meals or cakes from the few countries that dominate in the production of soybean, rapeseed and sunflower, and to a lesser extent, peanuts, cottonseed and palm kernels. The prices of oilseed meals are therefore dependent on location, demand, fluctuation in currency exchange rates and transportation costs.

Table 1 Important ingredient sources of vegetable proteins for poultry feeding

Ingredient (intact)	Global output, 2014 (million tonnes)	Leading producers (in order of production volumes)
Soybean	308.01	United States, Brazil, Argentina, China
Rapeseed	15.56	Canada, China, India, Germany
Cottonseed	76.87	India, China, United States, Pakistan
Groundnuts	42.32	China, India, Nigeria, United States
Sunflower	41.33	Ukraine, Russia, China, Romania
Sesame	5.46	India, Sudan, China, Myanmar
Linseed	2.56	Canada, Russia, China, Kazakhstan
Castor oil seed	1.94	India, Mozambique, China, Brazil
Palm kernels	15.91	Indonesia, Malaysia, Nigeria, Thailand
Chickpeas	14.24	India, Australia, Pakistan, Myanmar
Lupins	0.98	Australia, Poland, Russia, Germany
Safflower seed	0.87	Kazakhstan, Mexico, India, United States

Source: FAO (2015).

3 Finding alternative sources of protein for poultry

In intensive poultry production systems, feed is the most important input and accounts for 60–80% of total production costs. Protein is one of the main compartments of poultry feed and is one of the major contributors to the finished feed cost. However, the rapidly growing poultry industry and the increasing demand for poultry feeds have led to a considerable increase in feedstuff prices. It has been predicted that traditional sources of protein for poultry will become scarce and more expensive in the near future. Furthermore, beside the rising prices of feedstuff, the traditional and conventional protein sources in poultry feed such as oil seed meals (SBM and canola meal) and fishmeal are failing to meet the increasing demand due to both sector and human population growth. With these present trends of rising prices and shortage of supply, it seems inevitable to consider alternative protein supplements to fully or partially replace the conventional protein sources in poultry feed.

Generally, a product can qualify as a competent alternative protein substitute in poultry diets provided it is in good supply, reasonably priced and has proper nutrient levels. However, there are some nutritional and technological considerations, which determine the feasibility of an alternative protein source to be introduced into the diet (Van der Poel et al., 2013). These include nutritional and technical aspects such as the variability in nutrient level and quality particularly the essential amino acid balance of the ingredient, presence of naturally occurring anti-nutritional and/or toxic factors such as tannins and enzyme inhibitors, presence of pathogenic microorganisms and need for supplementation. The technical aspects worthy of consideration are availability and supply throughout the year; bulkiness, wetness and/or powdery texture; processing requirements; predicted availability of ingredient in the long term (beyond 2020); and consistency in research and development efforts.

The potential unconventional/alternative protein resources can be categorized as animal or plant protein sources. Most of the sources require much further processing before adding it into the diet.

4 Alternative plant protein sources: grain by-products

4.1 Brewers' and distillers' dried gains and solubles

Many of the cereal grains used as animal feed are also used for human consumption or the development of industrial products. Brewers' grain (BG) and distiller's dried grains with solubles (DDGS) are the major by-products derived from grain during the industrial fermentation and distillation of starch-rich grains for various alcoholic and non-alcoholic drinks, and ethanol production. BGs are the solid residue left after the processing of germinated and dried cereal grains (malt) for the production of beer and other malt products (malt extracts and malt vinegar). Although barley is the main grain used for brewing, beers are also made from wheat, maize, rice, sorghum and millet. BGs are collected at the end of the mashing process, once all sugars have been removed from the grain. The remaining product is a concentrate of proteins and fibre that is suitable for animal feeding (Crawshaw, 2004). BGs are a highly variable by-product and their nutritional value depend on the grain used, on the industrial process (temperature, fermentation etc.) and on the method of preservation.

DDGS are a co-product of the ethanol production process, which is rich in protein, fat, minerals and vitamins (Table 2). Like BG the nutrient composition of DDGS is a function of the starting grain and the specific methods used to make ethanol and other products. Distillers' grains and solubles have very low concentrations of starch because most of the starch in the starting grains is converted to ethanol. The concentrations of protein, fibre, fat and minerals are increased, depending on the concentration of starch in the grain. The bioavailability of phosphorus in DDGS and brewers product is fairly high at around 60%, but the digestibility of lysine in these by-products is lower than in the starting grain (Swaitkiewwicz and Koreleski, 2008). The main anti-nutritional factor (ANF) limiting high inclusion of DDGS and BG in poultry feed is the high crude fibre content, ranging between 7.0 and 13.0%, with the major part existing in the form of non-starch polysaccharides (NSPs). Like many by-products, the nutrient composition and nutritive value of DDGS vary widely. Some American DDGS samples are generally low in ether extract, and this will affect the AME value of such meals (Meloche et al., 2013; Kerr et al., 2014).

4.2 Gluten feed and meal

Gluten feed and gluten meal are by-products of the manufacture of starch by wet milling process (RFA, 2008). Gluten is the substance remaining after removal of the germ and the starchy endosperm. Gluten feed and meal are considered high-protein sources (Table 3). The most common cereals used for starch production are corn and sorghum. Corn gluten meal (CGM) is a protein-rich feed, containing about 65% crude protein, making it one of the richest potential protein sources. However, CGM is deficient in lysine

Table 2 Chemical composition of Brewers and DDGS of different grains

	DDGS				Brewers	
% of DM	Barley	Corn	Sorghum	Wheat	Barley	Sorghum
Dry matter	91.3	89.0	89.9	90.6	90.7	94.2
Crude protein	28.2	29.5	33.5	37.3	27.8	26.0
Crude fibre	13.8	7.9	8.1	7.7	11.6	8.2
NDF	60.1	34.2	38.5	34.0	39.7	40.0
ADF	23.8	13.6	19.8	14.5	15.5	24.6
Lignin	5.1	4.3	19.5	4.6	3.8	–
Ether extract	7.5	11.1	9.4	5.0	8.5	6.5
Crude ash	5.4	5.4	4.5	5.9	5.8	7.3
GE (kcal/kg)	5060	5100	4990	4890	5080	4600
Minerals (g/kg)						
Calcium	2.0	1.6	0.8	2.1	1.6	7.7
Phosphorus	6.5	7.9	7.4	9.1	9.7	8.1
Potassium	7.6	10.3	3.5	10.9	10.5	9.8
Sodium	2.4	2.4	0.4	4.9	0.5	1.9
AA (% of CP)						
Arginine	5.1	4.3	4.3	4.0	4.2	6.5
Cysteine	1.6	2.0	3.4	1.9	1.8	2.1
Isoleucine	3.7	3.8	4.3	3.5	3.9	5.2
Leucine	7.5	11.6	13.2	6.6	7.1	8.2
Lysine	2.5	3.0	2.9	2.3	4.3	3.8
Methionine	1.4	2.0	1.8	1.5	1.4	2.3
Threonine	3.6	3.7	3.6	3.0	3.7	4.2
Tryptophan	–	0.8	0.8	1.0	–	2.5
Valine	5.3	5.1	5.5	4.3	5.2	5.4

Source: Heuzé et al., 2015.

with 1.7%, compared with SBM, which has 6.2%. Corn gluten feed (CGF) consists mainly of corn bran and steep liquor (liquid separated after steeping) but may also contain distillers' solubles, germ meal, cracked maize screenings as well as minor quantities of end products from other microbial fermentations. The chemical composition of CGF varies hugely, as it depends on the milling process and on the relative proportions of bran, steep liquor and other components. In particular, the energy and protein content of CGF are positively correlated to the proportion of steep liquor in the blend (Stock et al., 1999).

Sorghum gluten feed (SGF) is part of the grain that remains after the extraction of the larger part of the starch and germ, but further processing and separation of the bran results in sorghum gluten meal (SGM). SGF contains at least 26.0% protein, but the protein content of SGM is approximately two times more than that of SGF. As the SGF includes the seed coat, it has higher tannin content than SGM, depending on sorghum variety.

Table 3 Chemical composition of gluten meal and feed from corn and sorghum

	Gluten feed		Gluten meal	
% of DM	Corn	Sorghum	Corn	Sorghum
Dry matter	88.3	91.2	90.0	91.0
Crude protein	21.7	25.3	67.2	47.9
Crude fibre	8.3	8.0	1.2	5.4
NDF	39.6	–	4.1	–
ADF	10.6	–	1.6	–
Lignin	1.2	–	0.3	–
Ether extract	3.4	5.8	2.9	7.3
Crude ash	6.9	6.4	2.1	3.6
Starch	21.5	–	17.6	–
Total sugar	1.8	–	0.5	–
GE (kcal/kg)	4490	4680	5520	5110
Minerals (g/kg)				
Calcium	1.6	0.3	0.3	0.1
Phosphorus	10.2	1.7	4.0	1.7
Potassium	15.4	0.6	1.0	0.2
Sodium	3.4	0.2	0.8	0.2
AA (% of CP)				
Arginine	4.4	1.7	3.0	2.1
Cysteine	1.9	1.6	1.8	1.2
Isoleucine	3.1	1.9	4.0	4.1
Leucine	8.2	8.0	15.9	12.8
Lysine	2.9	1.0	1.7	1.1
Methionine	1.7	1.0	2.4	1.4
Threonine	3.4	3.2	3.3	2.5
Tryptophan	0.6	0.4	0.5	0.6
Valine	4.6	5.6	4.5	4.4

Source: Heuzé et al., 2015.

5 Alternative plant protein sources: oil seed and fruit by-products

Many oilseed crops produce a by-product meal or cake, which generally is a good-quality protein source for poultry feeding. Reference has been made to the two most important crops, soybean and canola, on which extensive research has been carried out. However, some other less common oilseed meals such as groundnut, cottonseed, linseed, sesame and sunflower, and oil fruits such as coconut and palm kernel also have considerable amount of protein and could be good sources of protein in poultry diets. Oilseeds and fruits are crushed to extract oil and the residue is used as a feed ingredient. Different kinds of oil are extracted from the fruit (pericarp) and seed of oil palm fruit. The pulp resulting from crushing and extraction of the pericarp is less useful as a feed material but the residue obtained after oil is extracted from the seed (kernel) itself is a useful feed ingredient. Coconut pericarp contains no oil but oil is extracted from the seed, leaving a residue that is similar to palm kernel meal. The following are the three different methods employed by crushing plants to extract oil from oil-bearing seeds or fruits:

1 Cold-press extraction: the oil and meal are physically separated without heat. This method is less efficient and results in a meal with high residual oil content of up to 15%.
2 Expeller-press extraction: the oil is physically extracted with added heat resulting in a meal with a residual oil content of around 8–10%.
3 Solvent extraction: the oil is extracted with the combined physical extraction followed by solvent washing. The solvent extraction method is the commonest and the most efficient method of extracting the oil, usually resulting in a meal with less than 3% residual oil.

Each method of extraction results in different meal characteristics. High temperatures and pressure during processing will decrease the concentration of some ANF in the meal, but will result in reduction of meal quality, as well.

5.1 Cottonseed meal

Cottonseed meal (CSM) is valued as a protein feed, but the protein content is highly variable as it depends on the extent of dehulling and on the efficiency of oil extraction. The protein content ranges from 30% DM for non-dehulled CSM up to 50% DM for fully dehulled meals. Lower and higher values than these extremes have also been recorded. The crude fibre content varies accordingly, from 25% (non-dehulled) to 5% (fully dehulled). The various methods used for oil extraction also explain the large range of residual oil present in CSM. Some solvent-extracted meals contain less than 2% oil, like the other major oilseed meals, but many CSM contain higher oil values, often in the 5–10% range, and over 20% is possible. The CSM protein is less rich in lysine than SBM (4% vs 6% of the protein) and since the protein content is generally lower, the total content of lysine and essential amino acids is lower for CSM. The main constraint of CSM is the presence of gossypol, which limits its use in non-ruminant animals such as pig and poultry. CSM usually contains gossypol, unless it has been obtained from glandless (gossypol-free) seeds. Gossypol toxicity can be alleviated through the addition of iron salts (Henry et al., 2001).

5.2 Linseed meal

Linseed meal is the by-product of oil production from linseeds (*Linum usitatissimum* L.). Linseeds are primarily used for the production of linseed oil, which is used in paints and in other industries, such as the manufacture of linoleum. Linseeds and linseed meal have attracted considerable attention since the 1990s due to the presence in the oil of polyunsaturated fatty acids (PUFAs), notably alpha-linolenic acid (ALA, an omega-3 fatty acid) and conjugated linoleic acid (CLA). These fatty acids are being used in diets of livestock to alter the fatty acid profile of meat, milk and eggs in order to provide health benefits to human consumers (Newkirk, 2008). Linseed meal is deficient in lysine and has a lower content of methionine compared to other oilseed meals. The dietary fibre content of linseed meal is relatively high, which limits its high inclusion in poultry diet.

5.3 Sesame meal

Sesame (*Sesamum indicum L.*) is an annual plant 0.5–2.5 metre high, which matures in 70–150 days. The elliptical pod consists of two to four chambers containing about 20 seeds each. Solvent extraction of the oil from sesame seeds oil results in a meal with 45–50% protein and less than 3% fat. Tannins, NSP and phytic acid are the main ANFs present in sesame meal. Since phytate (up to 60 mg/g meal) binds to approximately 80% of total phosphorus of the meal, high inclusion of sesame meal in diet without using a phytase supplement can negatively compromise Ca and P bioavailability (Jansman, 1993).

5.4 Sunflower meal

Sunflower (*Helianthus annuus* L.) is the fifth most important oilseed crop in the world and accounts for 8% of world oilseed production (FAO, 2011). Sunflower meals can be made from whole or decorticated seeds, and can be extracted mechanically and/or using solvents. The quality of sunflower meal depends on the plant characteristics (seed composition, hulls/kernel ratio, dehulling process, growth and storage conditions) and on the processing (dehulling, mechanical and/or solvent extraction) (Swetman, et al., 2002) . The major ANFs found in sunflower meal are its high crude fibre and NSP, and low available lysine. Sunflower meal also contains chlorogenic acid (1.2%), which can interfere with the biological function of hydrolytic enzymes and consequently retard growth performance. The negative effect of chlorogenic acid can be reduced by adding methionine and choline to poultry diets. Its protein content ranges from 23% for some non-dehulled, mechanically extracted meals, to more than 40% for highly decorticated, solvent-extracted meals. However, usual ranges for protein are 29–33% for non-dehulled meals and 35–39% for dehulled and partially dehulled meals. The fibre content is directly linked to the presence of hulls: crude fibre ranges from 27 to 31% for non-dehulled meals and from 20 to 26% for dehulled and partially dehulled sunflower meals.

5.5 Coconut meal

Coconut meal, also known as copra meal, is the by-product of the oil extraction from dried coconut kernels. The residual oil in coconut meal is highly variable ranging from 1 to 22%, depending on oil extraction method. The meal protein content is fairly low compared to other oilseed meals with low digestibility of lysine and methionine. Also, the high crude fibre in coconut meal decreases its metabolizable energy content for poultry. The high

Table 4 Chemical composition of less conventional oilseed and fruit meals[1] used in poultry feeds

% of DM	Coconut	Cottonseed	Groundnut	Linseed	Palm kernel	Sesame	Sunflower
Dry matter	91.5	92.2	88.8	90.6	90.9	92.2	89.8
Crude protein	22.4	45.0	55.9	34.1	18.7	44.4	37.7
Crude fibre	14.2	10.6	8.5	10.5	20.2	7.8	22.8
NDF	54.7	23.7	22.4	25.4	73.2	21.3	38.7
ADF	28.7	15.0	9.8	15.3	45	10.2	26.6
Lignin	6.7	5.4	3.6	6.2	12.8	1.7	8.6
Ether extract	9.8	8.9	5.3	10.2	2.8	11.1	1.8
Crude ash	6.8	7.0	7.1	6.3	4.6	12.4	7.7
GE (kcal/kg DM)	4800	5060	4990	4950	4490	4890	4600
Minerals (g/kg)							
Calcium	1.2	2.0	2.0	4.3	2.7	19.8	4.6
Phosphorus	5.8	12.4	6.4	9.0	6.6	12.7	13.0
Potassium	21.1	16.6	13.5	11.8	7.4	10.3	17.7
Sodium	0.6	0.3	0.3	0.8	0.2	0.1	0.1
AA (% of CP)							
Arginine	10.7	11.1	9.5	9.6	12.6	12.6	8.5
Cysteine	1.2	1.6	0.5	1.8	1.4	2.2	1.7
Isoleucine	0.3	3.2	2.9	4.4	3.7	3.7	4.1
Leucine	5.9	5.9	5.5	6.0	6.5	6.7	6.2
Lysine	2.6	4.2	2.6	4.0	3.2	2.5	3.5
Methionine	1.3	1.4	0.9	1.9	1.7	2.7	2.3
Threonine	3.0	3.3	2.4	3.9	3.3	3.4	3.6
Tryptophan	1.3	1.1	0.9	1.6	0.7	1.3	1.2
Valine	4.7	4.2	3.4	5.2	5.4	4.6	4.9

[1] Coconut meal: expeller-extracted; cottonseed meal: low fibre, low oil; groundnut meal: decorticated, solvent-extracted; linseed: cold-press, palm kernel meal: solvent-extracted; sesame meal: mechanically extracted; sunflower meal: dehulled, solvent-extracted.

oil content of copra meal renders it susceptible to rancidity and the product should not be used after prolonged storage. A peculiarity of copra meal is its high NSP content, and notably its high levels of mannan and galactomannan (25–30%), which have well-documented anti-nutritional properties in non-ruminant animal species (Kempton, 2006).

5.6 Groundnut meal

The residual substance after oil removal from groundnut seed is known as groundnut meal or cake. The seed can be processed either with or without the hull resulting in different protein and fibre levels in the meal. The meal produced from decorticated fruit is low in fibre and high in protein contents, which makes it a valuable ingredient for poultry feeding. Groundnut meal is still deficient in lysine and methionine, but the main constraint to its utilization is its easy contamination by aflatoxin (Green et al., 1987), particularly with long storage under high temperature and humidity.

5.7 Palm kernel meal

Palm kernel meal is a common feed ingredient, especially in ruminant feeding. It is the least nutritionally valuable of the major oil meals due to its low protein content (14–20%, lower than copra meal) and its large quantities of cell wall constituents (crude fibre 14–28%; NDF 60–80%; ADF 35–50%; lignin 10–18%). The proportion of lignin in the cell walls of palm kernel meal (18% of NDF) is higher than in copra meal (13% of NDF). For these reasons, its nutritive value is inferior to that of the main oil meals, notably SBM, groundnut meal and CSM. Unlike these products, palm kernel meal is often obtained from mechanical extraction (expeller) and its oil content is quite high (6–15%). The less common solvent-extracted palm kernel meal contains less oil (about 3%) and slightly more protein (19 vs17% on average) than the expeller meal, but the cell wall constituents and minerals are only slightly affected by the extraction process (Sundu et al., 2006).

6 Alternative plant protein sources: grain legumes or pulses

Grain legumes, also called pulses, are cultivated for their seeds. The seeds are used for human consumption and animal feeding or for the production of oils for industrial use. Grain legumes are important food crops due to their high protein and essential amino acid content, although the methionine content of legumes is relatively low. The FAO has distinguished eleven groups of pulses that belong to the family of Leguminosae, including beans (*Phaseolus* spp. and some *Vigna* species, such as kidney bean, Lima bean and mung bean); broad bean (*Vicia faba*; such as horse bean and field bean); peas (*Pisum* spp.); chickpea (*Cicer arietinum*); cowpea (*Vigna unguiculata*; blackeye bean); pigeon pea (*Cajanus cajan*; cajan pea); lentils (*Lens culinaris*); Bambara groundnut (*Vigna subterranea*; earth pea); vetch (*Vicia sativa*); lupines (*Lupinus* spp.) and minor pulses such as jack bean, winged bean and yam bean.

6.1 Beans

Beans (*Vicia faba*) are basically cultivated for human consumption. Bean seeds are rich in protein (25–33%) and starch (40–48%) and are therefore a valuable source of protein and energy for livestock. They have a moderate content of fibre (crude fibre 7–11%). Their composition is quite similar to peas, although beans are richer in protein. The amino acid profile of beans has a high lysine concentration in the protein (5.4–6.8%) and a relative deficiency in sulphur amino acids (0.6–1.0% methionine). Faba beans contain about 1% lipids, with a high proportion of linoleic and linolenic acids, which makes them susceptible to rancidity if ground and stored for more than about a week (Blair, 2007). The ANFs in beans are protease inhibitors, tannins and lectins. Beans contain glycosides, vicine and convicine, which are responsible for favism, an acute haemolytic disease characterized by oxidative damage in red blood cells that affects human populations suffering from glucose-6-phosphate dehydrogenase deficiency (Enneking and Maxted, 1995).

6.2 Chickpea

Chickpea (*Cicer arietinum*) is a legume that also belongs to the family Fabaceae, like *Vicia faba*. Its seeds are high in protein and it is one of the oldest cultivated pulses. Chickpeas are particularly rich in lysine (6–7% of the protein) but sulphur-containing amino acids and threonine may be deficient for non-ruminant animal species. Chickpeas contain negligible amounts of lipids (sometimes less than 5%). Chickpeas contain a variety of secondary compounds that can impair nutrient absorption from the gastrointestinal tract (Bampidis et al., 2011). Depending on the variety, chickpea seeds contain variable amounts of trypsin and chymotrypsin inhibitors that may decrease their feed value in poultry. Tannins and oxalic acid are the other ANFs found in chickpea.

6.3 Lupine

Lupine (*Lupinus* spp.) (white lupines, yellow lupines and blue lupines) is cultivated to harvest the beans for food and feed, and as forage. Lupine is an interesting crop because of the high protein content in its beans (35%). All three lupine varieties contain various types of alkaloids of which the quinolizidine alkaloids are the most relevant ANF. Alkaloids have a bitter taste; this may result in not only inhibition of feed intake, but also neurophysiological effects, for example, tremors, convulsions and pulmonary arrest have been described. Low-alkaloid varieties, also known as sweet lupines, are generally available. In contrast to *Vicia faba* and peas, lupines contain hardly any trypsin inhibitor activity and only low levels of saponins (Helsper et al., 2006).

6.4 Pea

Pea (*Pisum sativum*) is a grain legume mostly cultivated for human consumption. Peas are regarded as a highly valuable protein source due to their high protein content (usually about 22–24%, ranging between 16 and 32%), which is intermediate between cereals and oil meals (Castell et al., 1996). Peas can be an alternative to SBM. Their amino acid profile is well balanced in lysine (similar to that of SBM and higher than those of cereal grains, particularly maize grain) so that they can be a protein supplement in cereal-based diets (Duranti et al., 1997). However, they are deficient in tryptophan and sulphur-containing amino acids (notably methionine) (Vander Pol et al., 2008). Peas can be a particularly valuable protein source in organic livestock farming when usual sources such as SBM and industrial amino acids are prohibited (Schumacher et al., 2011). The ANFs present in pea are protease inhibitors, tannins, lectins and phytate (Stegeman et al., 2010).

6.5 Guar meal

Guar (*Cyamopsis tetragonoloba* (L.) Taub.) is an erect, bushy annual herbaceous legume up to 3 m high, with trifoliate leaves, about 10 cm long, and white or rose coloured flowers. Guar is a multi-purpose plant, mostly used today as a source of galactomannan gum, which is used as a thickener and stabilizer in foods such as salad dressings, ice cream and yoghurt. Guar meal is the main by-product of guar gum production. It is a mixture of

Table 5 Chemical composition of grain legumes as protein sources in poultry diets

Analysis (% of DM)	Bean	Chickpea	Lupine	Pea
Dry matter	86.6	89.0	90.2	86.5
Crude protein	29.0	22.1	33.8	23.9
Crude fibre	9.1	10.5	16.1	6.0
NDF	15.9	22.8	25.6	14.2
ADF	10.7	13.8	20.9	7.0
Lignin	1.0	0.7	1.5	0.4
Ether extract	1.4	5.0	6.1	1.2
Crude ash	3.9	3.3	3.5	3.5
Starch	44.7	35.6	4.7	51.3
Total sugar	3.6	3.6	5.8	4.9
Tannins	4.8	4.9	2.1	0.1
GE (kcal/kg)	4460	4680	4850	4370
Minerals (g/kg)				
Calcium	1.5	1.7	2.7	1.2
Phosphorus	5.5	3.9	3.5	4.5
Potassium	11.5	11.9	9.3	11.3
Sodium	0.1	0.2	0.5	0.1
AA (% of CP)				
Arginine	9.0	8.6	11.0	8.4
Cysteine	1.2	1.2	1.5	1.4
Isoleucine	4.1	3.8	4.2	4.2
Leucine	7.1	7.1	6.9	7.1
Lysine	6.2	6.6	4.7	7.2
Methionine	0.8	1.2	0.7	1.0
Threonine	3.5	3.4	3.4	3.8
Tryptophan	0.8	0.9	0.8	0.9
Valine	4.6	3.9	3.9	4.8

Source: Heuze et al., 2015.

germs and hulls at an approximate ratio of 25% germs to 75% hulls (Lee et al., 2004). It is a protein-rich material containing about 40% protein and is used as a feed ingredient, but may require processing to improve palatability and remove ANFs. Its lysine (1.72%) and sulphur amino acids (methionine + cysteine 0.96%) contents are comparable to those of groundnut meal but much lower than those of SBM. The main constraints for all species are palatability and content of ANF. Guar meal contains about 12% gum residue (7% in the germ fraction and 13% in the hulls) (Lee et al., 2005), which increases viscosity in the intestine and reduces digestibility and growth (Lee et al., 2009). In addition, guar meal contains other types of ANF, including trypsin inhibitors, saponin, haemagglutinins, hydrocyanic acid and polyphenols (Gutierrez et al., 2007). The large saponin content of guar seed (up to 13% DM) could have both anti-nutritional effect and a positive anti-microbial activity (Hassan et al., 2010).

7 Alternative plant protein sources: algae and duckweed

7.1 Algae

Algae are a heterogeneous group of plants with a complex and often controversial taxonomy. There are two main types of algae: the macroalgae (seaweeds), which occupy the littoral zone and can be of very large size, and the small-sized microalgae, which are found in benthic and littoral habitats as well as throughout the ocean as phytoplankton (Hasan et al., 2009; El Gamal, 2012). There are about 10 000 species of seaweeds (Guiry, 2014), but only a few of them are of interest in animal feeding. *Aphanizomenon flos-aquae* (blue-green alga), *Spirulina* (blue-green alga) and *Chlorella* (green alga) are the most prominent protein-rich algae, which are commercially produced. Blue-green algae (also called cyanobacteria) are microorganisms, because of their simple cellular structure. Some *Aphanizomenon* and *Spirulina* are toxic. However, the strains cultured for consumption do not contain toxins. In contrast to *Spirulina*, *Aphanizomenon flos-aquae* are able to fix nitrogen.

Seaweeds have a highly variable composition, which depends on the species, time of collection and habitat, together with external conditions such as water temperature, light intensity and nutrient concentration in water. All of these factors markedly influence the content of protein, amino acids, mineral, lipid and fibre in seaweed (Mišurcová, 2012). One common feature of fresh seaweeds is that they contain very large amounts of water (70–90%) and need to be dried or consumed quickly. The total protein content varies between different seaweed strains and is rather low in brown seaweed (10–24% of dry weight), whereas higher protein contents are observed in green and red seaweed species (up to 44% of dry weight) (Holdt and Kraan, 2011). In a recent review on the importance of seaweeds for poultry feeding, Makkar et al. (2016) surmised that when compared to SBM, seaweeds are deficient in most indispensable amino acids except the sulphur-containing amino acids (Table 6). However, as already highlighted, there is great variability even within the same species of seaweeds.

7.2 Duckweed

Duckweeds are tiny free-floating vascular plants found throughout the world on fresh (or sometimes brackish) waters. The entire duckweed plant is composed of non-structural, metabolically active tissue. Most photosynthesis is devoted to the production of protein and nucleic acids, making duckweeds very high in nutritional value, typically rich in protein and minerals and poor in fibre. However, the chemical composition of duckweeds varies

Table 6 Concentrations of indispensable amino acids (g/16 g N) of seaweeds versus soybean meal

Amino acid	*Ascophyllum nodosum*	*Undaria pinnatifida*	*Saccharina japonica*	*Macrocystis pyrifera*	*Ulva* sp.	SBM
Methionine	0.7–1.9	1.7–2.2	0.9–2.4	1.9 0.4	1.3–1.9	1.32
Lysine	4.3–4.9	5.6–6.8	3.9–7.7	4.7 0.7	3.7–3.9	6.18
Valine	3.7–4.4	5.2–10.3	3.8–9.7	5.2 1.7	4.2–4.5	4.5
Isoleucine	2.8–3.4	4.3–4.9	2.7–4.2	3.4 0.2	2.2–2.9	4.16
Leucine	4.6–6.0	7.4–8.4	4.9–7.2	5.8 0.4	5.0–5.3	7.58
Phenylalanine	2.3–4.0	4.7–4.8	3.2–4.5	3.8 0.4	3.4–3.8	5.16
Histidine	1.3–1.5	2.5–3.2	2.2–3.8	1.3 0.3	1.6–2.3	3.06
Threonine	2.8–4.3	4.4–4.5	3.5–5.5	4.6 0.6	3.8	3.78
Tryptophan	NA	0.3–0.7	0.3–0.5	0.9 0.1	NA	1.36
Alanine	5.3–5.4	4.7–16.7	5.7–7.3	10.9 + −1.8	5.5–6.2	4.54

Source: Makkar et al. (2016). NA – Not available.

considerably due to the age of the plant, environmental temperature and the nutrient content of the water. On a dry matter basis, duckweed has high protein content with a valuable amino acid composition (Rusoff et al., 1980). The crude protein content of duckweeds ranges from 7 to 45% DM, depending on nitrogen availability (Culley et al., 1981). Under optimal conditions, duckweed contains considerable protein, fat, starch and minerals. Duckweeds grown in enriched waters containing minerals or effluents from agricultural and municipal waste lagoons can have a protein content as high as 30–40% DM.

8 Alternative animal protein sources

8.1 Silkworm pupae meal

Silkworms are the caterpillars of moth species raised for silk. Ninety percent of the world production results from the cocoons of the domesticated mulberry silk moth *Bombyx mori*, a Bombycidae moth (Longvah et al., 2011). Spent silkworm pupae are a waste material often discarded in the open environment or used as fertilizer (Wei ZhaoJun et al., 2009). It can be extracted to yield valuable oil used in industrial products such as paints, varnish, pharmaceuticals, soaps, candles, plastic and bio-fuels (Trivedy et al., 2008). The extracted meal is sometimes used for the production of chitin, the long-chain polymer of *N*-acetylglucosamine, which is the main component of the exoskeleton (Suresh et al., 2012). Silkworm pupae meal is a protein-rich feed ingredient with a high nutritional value. Its crude protein content ranges from 50% DM to more than 80% DM (for defatted meal). The lysine (6–7% of the protein) and methionine (2–3% of the protein) contents are particularly high. However, the true protein (calculated as the sum of amino acids) in silkworms was found to correspond to only 73% of the crude protein content (Finke, 2002), which is explained by the presence of chitin, since this component contains nitrogen.

However, the chitin content of pupae meal is relatively low, about 3–4% DM (Finke, 2002; Suresh et al., 2012). The presence of chitin and insoluble protein may also explain the level of fibre, and values of 6–12% DM of ADF have been reported (Finke, 2002; Ioselevich et al., 2004). Undefatted pupae meal is rich in fat, typically in the 20–40% DM range. The defatted meal contains less than 10% oil in the DM. Silkworm oil contains a high percentage of PUFAs, notably linolenic acid (18:3), with reported values ranging from 11 to 45% of the total fatty acids (Ioselevich et al., 2004; Usub et al., 2008). Compared to other animal by-products, silkworm pupae meal is relatively poor in minerals (3–10% DM). Silkworm litter appears to have an extremely variable composition, with crude protein values reported to be between 15 and 58% (Patil et al., 2013).

8.2 Hatchery by-products

Hatchery by-product meal results from the processing of poultry hatchery waste, such as shells of hatched eggs, infertile eggs, dead embryos and dead or culled chicks. Hatchery waste contains protein and minerals and can be rendered by cooking, drying and grinding into a meal (hatchery by-product meal or hatchery waste meal) suitable to feed livestock. Hatchery by-product meal contains 22–33% protein, with 1.1–1.8% lysine and 0.5–0.8% methionine. As can be expected, crude fat (11–30%) and ash (about 60% but values as low as 22% have been reported) are extremely variable (Glatz and Miao, 2009). In addition to supplying energy and protein, hatchery by-product meal is an important source of calcium (17.2–24.6%), although it contains little phosphorus (0.3–0.6%) (El Boushy et al., 2000). Due to the fat content, the gross energy can be as high as 6900 kcal/kg.

Like other animal by-products, the use of hatchery by-products for animal feeding is often regulated and even prohibited in certain countries due to concerns over the transfer of pathogens (Al-Harthi et al., 2010). Therefore, hatchery wastes have to be properly heat-treated during processing to assure that the resulting hatchery by-product meal is pathogen-free. Acid fermentation of hatchery wastes can prevent undesirable bacterial development (Deshmukh et al., 1997).

Raw eggs contain avidin, a biotin-binding protein responsible for biotin deficiency in animals (Göhl, 1970). Biotin deficiency causes skin and hair abnormalities (alopecia and loss of hair colour) as well as locomotor and reproductive problems. It reduces feed intake and subsequent growth. In poultry, biotin deficiency may cause sudden death (Whitehead, 1985). Thermal destruction of avidin occurs between 73.3°C and 125.6°C and properly heat-processed hatchery by-product meal should contain little or no avidin (Göhl, 1970).

8.3 Insects

Insects can be considered as a sustainable and high-quality alternative protein source in poultry diets. Some of the more important groups of insects, which have been used as animal food include grasshoppers, caterpillars, beetle grubs and sometimes adults, winged termites, bee, wasp and ant brood (larvae and pupae), as well as winged ants, cicadas, and a variety of aquatic insects.

The protein content of edible insects ranges from 30% (woodworms) to 80% (some wasp species). Insects generally have a comparable, if not higher, amount of calories per unit weight compared to cereals, vegetables, legumes and meats. In addition, edible insects have a diverse range of mineral salts (e.g. Na, K, Ca, Zn, Fe and Mg) similar or comparable to or in higher amounts than conventional meat products (e.g. beef, fish, turkey, milk and eggs). Insects also have good feed conversion efficiency due to their poikilothermic nature not needing to maintain their body temperature. For example, crickets convert plants into biomass five times faster than cows. If reared on waste organic matter and waste materials, insects may act as bio-transformers, converting organic bio-waste into animal biomass rich in proteins and suitable for use in animal feeding. Worm meal is another growing alternative source of protein for poultry. It may be easier to grow, harvest and process worms than flying species. The reader is referred to the review by Tiroesele and Moreki (2012) on the potential role of termites and earthworms as protein sources for poultry.

9 Poultry responses to diets containing alternative protein sources

The biological and hence, nutritive value of dietary proteins is determined by the pattern and quantity of essential amino acids present. The presence of one or more of the essential amino acids in adequate amounts increases the nutritive value of the protein (Rangel et al., 2004). However, the biological value of alternative protein sources extends beyond its amino acid composition and digestibility, and can be influenced by additional factors in a tissue-specific manner. The advantage of understanding and using biological values of alternative protein sources, for example, the digestibility of amino acids, in diet formulation and poultry nutrition in general is that it makes it possible to increase the inclusion level of these alternative ingredients in poultry diets, in particular, low-quality protein sources. In effect, it will increase the range of ingredients that can be incorporated, improve the precision of formulation and ensure more predictable bird performance (Bryden et al., 2009).

Dietary sources of protein are heterogeneous mixtures of different proteins. It would be anticipated, therefore, that different proteins would be different in their biological values, and this in turn would result in variation in their potential to meet animal protein and amino acid requirements. This is also further affected by processing methods applied in preparing different protein sources for use in diets for poultry. In the following sections, we will review the response of poultry on diets containing various alternative sources of protein, vis-à-vis the nutrient profiles. In most such studies, it is common practice to compare SBM, although in some regions these alternative ingredients may be used as regular sources of protein.

Meals from groundnut and fully dehusked sunflower can often be used to completely replace SBM in poultry diets. CSM is used when a low nutrient density diet is required (Lordelo et al., 2004), since it has only 58% of the total lysine and 43% of the digestible lysine of SBM (NRC, 1994). CSM in combination with lysine-rich supplements has been shown to have the potential to replace up to 40% of the SBM protein in diets for broilers without any negative effects (Ravindran and Blair, 1992; Henry et al., 2001). In layer diets, the use of CSM is restricted due to the effects of cyclopropenoid fatty acids on internal egg quality.

Although groundnut meal is comparable to SBM (Elkin, 2002), it is not balanced in amino acids for poultry and has sub-optimal amounts of cysteine and methionine (McDonald et al., 1995). Carew et al. (1988) concluded from their studies that the performance of groundnut meal-fed chicks was inferior to SBM-fed chicks. They concluded that groundnut meal diets were first limiting in methionine and then in lysine, because supplementation with methionine alone improved performance, whereas lysine improved performance only in the presence of supplemental methionine. Pesti et al. (2003) fed groundnut meal to laying hens in comparison to SBM and concluded that maximum performance levels (hen-day egg production) by feeding 16% protein from groundnut meal can be achieved. However, the choice of an appropriate protein level will be dependent on the cost of feeding higher protein levels and the increased value of the larger eggs that may be produced. Existing studies suggest that groundnut meal can be successfully used in poultry nutrition if adequate levels of dietary lysine and methionine are provided (Zhang and Parsons, 1996; Costa et al., 2001). The protein quality of groundnut meal is further compromised or reduced by excess heating during processing, just like in most oilseed meals (Zhang and Parsons, 1996).

In a trial on broiler chicks, Senkoylu and Dale (2006) reported that 20–30% of high-oil sunflower seed meal was as effective as SBM in the starter diets of young chicks, suggesting that this level might even meet the insoluble fibre requirement of broiler chicks for enhanced gizzard function and broiler livability when given in mash form. With high-oleic-acid sunflower varieties, an inclusion of 10% sunflower seed meal can be used to increase the oleic acid (a monounsaturated fatty acid) of chicken meat with no adverse effects on broiler performance (Rebolé et al., 2006).

Occasionally, some minor ingredients may be fed to improve animal product quality. A case in point is linseed meal. While linseed meal is very similar to canola meal in protein and energy content, it tends to be limiting in lysine. The amino acid digestibility is less than that of SBM, so it is important to formulate on a digestible amino acid basis (Newkirk, 2008). González-Esquerra and Leeson (2000) fed 10% linseed meal to broiler chickens and demonstrated an increase in ALA (C_{18}:3n-3), an ω-3 fatty acid, from 11 to 54 mg/100g cooked skinless breast meat and from 43 to 183 mg/100 g cooked skinless thigh meat. The long-chain ω-3 fatty acid content also increased from 17 and 0 to 89 and 23 mg/100g of cooked skinless breast meat and thigh meat, respectively. However, bird performance was typically reduced by feeding this level of linseed meal. Therefore, the maximum recommended inclusion of linseed or meal in broiler diets is 3%. If more than 10% linseed or flax seed meal is incorporated into the diet, supplementation with additional vitamin B_6 is recommended to overcome the negative effects of linatine, provided sufficient premium for the product can be achieved. While feeding up to 15% flaxseed meal increases the ω-3 fatty acid level of eggs, at dietary levels higher than 10%, there is a significant drop in egg production (Najib and Al-Yousef, 2010), with long-term use of flaxseed increasing the incidence of liver haemorrhages in supplemented laying hens (Bean and Leeson, 2003). In addition to the presence of linatine, an anti-pyridoxine factor, moderate amounts of cyanogenic glycosides and mucilage limit the use of this meal in poultry diets (Ravindran and Blair, 1992; McDonald et al., 1995). The amino acid profiles of differently processed linseed meals are shown in Table 7.

In recent years, raw and processed seaweeds have been fed to poultry and other animals, as sources of protein (Gongnet et al., 2001) or to enhance the product quality, particularly the level of PUFA and pigmentation (Zheng et al., 2012; Swiatkiewicz et al., 2015; Ao et al., 2015). Several studies have been conducted to examine the effects of feeding seaweed products on the performance of layers and broiler chickens. There are conflicting reports as to the benefit of these supplements on the gross response of the birds. In a recent trial on a *Chlorella* by-product in diets for layers, Kim and Kang (2015) reported a linear improvement in feed intake and hen-day egg production when the product was fed at up to 75 g/kg diet. Eggshell thickness and strength were not affected. In another trial involving a commercial *Spirulina* algae product in broiler chicken diets, Evans et al. (2015) obtained satisfactory growth when the product was used at up to 16% of the diet but body weight gain was reduced at 21% of the product. The digestibility of methionine was also improved in the diets containing the test product. Other seaweed products have been employed largely as prebiotics, with some health benefits to poultry. The abundance of beneficial bacteria, including *Bifidobacterium longum* and *Streptococcus salivarius*, was increased while the prevalence of *Clostridium perfringens* was reduced in response to dietary supplementation with a combination of red seaweed products for layers (Kulshereshtha et al., 2014).

As highlighted in a previous section of this chapter, insects have a great potential as sources of protein for poultry and other non-ruminant animals. According to Khusro et al. (2012), the safety and economic viability of breeding and rearing insects on organic waste need to be assessed. The process will be appealing if the overall cost of rearing and feeding insects to chickens is lower than the cost of feeding conventional protein sources. Insect meals are generally high in lipids and can therefore supply energy along with protein. However, like most alternative feed ingredients, there is great variability between products, not only because of differences in insect species but also because of the stage of growth of insect from which the meal is made. Insect meals can be made from larvae, pupae or adult insects. In a study in which SBM was replaced by the larvae of yellow mealworm (*Tenebrio molitor*), Bovera et al. (2015) found that the feed intake and growth rate of broiler chickens were not affected by the level of the insect meal in the diet. In later growth, however, the feed conversion ratio of birds on the larva-supplemented diets was

Table 7 Protein and amino acid composition of expeller linseed meal and solvent-extracted linseed meal

	Kratzer and Vohra (1996)		Batal and Dale (2010)	
Protein and amino acids	Expeller linseed meal	Solvent-extracted linseed meal	Expeller linseed meal	Solvent-extracted linseed meal
Crude protein, %	34.3	34.6	–	–
Methionine, %	0.58	0.54	0.47	0.48
Cysteine, %	0.61	0.61	0.56	0.58
Lysine, %	1.18	1.16	1.10	1.10
Tryptophan, %	0.50	0.51	0.47	0.48
Threonine, %	1.14	1.22	1.10	1.20
Isoleucine, %	1.69	1.68	1.70	1.80
Histidine, %	0.65	0.69	0.60	0.70
Valine, %	1.61	1.74	1.50	1.60
Leucine, %	1.92	2.02	1.90	2.00
Arginine, %	2.81	2.94	2.60	2.70
Phenylalanine, %	1.38	1.46	1.40	1.50
Tyrosine, %	0.96	1.09	–	–
Glycine, %	1.63	1.74	–	–
Serine, %	1.89	1.92	–	–

better than that of birds on the control (SBM) diet. In another study, broiler chickens on diets supplemented with 10% housefly larva meal grew equally as well as those on a 10% fishmeal diet and productivity on these two diets was superior to that on a control diet, with SBM alone as the main source of protein (Pieterse et al., 2014). In another study, Pieterse et al. (2014) investigated the eating quality of meat from birds that had been fed on diets supplemented with housefly larvae meal. Meat from the larvae-fed group was judged to have a prominent chicken aroma but less prominent chicken flavour. The meat from the birds on the test diets also had a higher metallic aroma and aftertaste value, although it was concluded that these values were low and less likely to be detected by consumers. Similar results have been obtained with housefly maggots, replacing groundnut cake (Adeniji et al., 2007). The material was used to completely replace groundnut cake in one diet, with no adverse effect on productivity.

10 Constraints on the use of alternative protein sources

A major constraint to the use of most alternative protein sources is the presence of ANF in the diet. Although such factors are present in conventional ingredients, not much effort has been put into developing many of the alternative ingredients. Protease inhibitors, phytate, lectins, polyphenolic compounds, glucosinolates, saponins and NSP are examples of ANF that depress nutrient digestion and utilization (Bryden, 1996; Hughes and Choct, 1999). Ironically, those feedstuffs (grain legumes, oil seed meals) that are used extensively as sources of dietary protein also tend to contain the highest concentrations of ANF. For example, SBM contains a range of ANF, many of which are heat-labile and are destroyed during feedstuff manufacture (Dale, 1996). Unless destroyed or inactivated by heat or some other suitable treatment, these substances can exert adverse physiological effects when fed to animals (Liener, 1994a). However, heat treatment may reduce protein quality through protein denaturation and development of Maillard-type reaction products (Bryden et al., 2009).

Dietary trypsin inhibitors are often responsible for the poor digestibility of dietary protein by interference with the proper function of endogenous proteases, leading to growth retardation and pancreatic hypertrophy (Liener, 1994a). Trypsin inhibitors are rich in sulphur-containing amino acids, and thus can create stress and cause a deficiency of methionine, which is basically the first limiting amino acid in soybeans and many of the alternative feed ingredients.

There have been several research studies focused on the unique structure of phytate that gives it the ability to bind minerals, proteins and starch, to lower bioavailability of nutrients and to inhibit enzymatic digestion of both proteins and starch (Oatway et al., 2001). Phytate is present in the form of protein–phytate or protein–phytate–protein complexes, which impart resistance to digestion by proteolytic enzymes, thereby reducing the utilization of dietary proteins (Chen et al., 2013). Phytate also binds multivalent cations (Pallauf and Rimbach, 1997). Its binding with certain proteins and amino acids initiates a reduction in the activities of intestinal proteases and amylases (Ravindran et al., 2001).

Gossypol is a polyphenolic compound (pigment) found in every part of the cotton plant that binds to iron molecules in the diet, in the bloodstream and in the yolks of eggs, causing anaemia in the bird and discoloured brown yolks in the eggs (Davis et al., 2002; Perez-Maldonado, 2003). Gossypol may also bind with lysine during processing, thus reducing the nutritional value of the protein (Heidarinia and Malakian, 2011). Non-ruminant animals appear to be more prone to overall gossypol toxicity than ruminants, which apparently are able to detoxify gossypol in the rumen (Chenoweth et al., 1994). However, an effective processing strategy has been developed for reducing and detoxifying the gossypol content of CSM (Sterling, 2002; Heidarinia and Malakian, 2011).

Lectins are a class of proteins or glycoproteins characterized by their ability to bind particular sugar residues that belong to polysaccharide moieties of glycoproteins, glycolipids, polysaccharides or simple glycosides (Murray, 1984). Two of the main attributes of the different lectins that determine how much they affect the bird are their ability to survive intestinal proteolysis and the avidity and selectivity with which they bind to the brush-border glycosyl units (Smithard, 2002). Lectins have been reported to have the potential for agglutinating animal erythrocytes, stimulating mitosis in resting lymphocytes (Hankins and Shannon, 1978) and binding to the intestinal mucosa, thereby impairing digestion and absorption of nutrients (Liener, 1994b) and further reduce protein digestibility by inhibiting digestive enzymes (Thompson et al., 1986).

The effect of phenolic compounds is not restricted to only proteins. While they decrease the digestibility of proteins and carbohydrates, they also limit the availability of vitamins and minerals. They lower the activity of digestive enzymes such as α-amylase, trypsin, chymotrypsin and lipase and may cause damage to the mucosa of digestive tract and also reduce the absorption of nutrients such as vitamin B_{12} (Liener, 1994b). Velvet bean (*Mucuna pruriens* var. *utilis*), an underutilized tropical legume has a nutritional quality comparable to soybeans and other conventional legumes (Pugalenthi et al., 2005), but the total phenolic content of velvet bean has been reported to be between 3.1 and 4.9% (Vijayakumari et al., 2002; Vadivel and Janardhanan, 2000) and that this appears to be higher when compared to commonly cultivated pulses.

The exploitation of alternative protein ingredients is also constrained by under-production and supply. Many alternative feed ingredients are produced under subsistent agricultural systems or grow naturally. In such situations, the crops are not improved through breeding, resulting in low yield per unit area. This limits their use in commercial agriculture and animal feeding.

11 Improving the nutritive value of alternative protein sources for poultry

There have been various attempts at mitigating the constraints highlighted above, in order to re-position the key alternative feed ingredients in poultry feeding. Most of these interventions are on a very limited scale and while such efforts can support the needs of small-to-medium scale producers, they remain a long way to increasing the importance of the ingredients. All the same, the procedures outlined below are the ones that have been proven to be effective with many of the ingredients that have become conventional today, so they should be effective on alternative ingredients.

11.1 Crop breeding

In recent times, research interest has been ignited towards the possibilities of improving the nutritive value of different alternative leguminous crops through crop breeding. This

Table 8 Some alternative protein sources, anti-nutritional factors present and reported processing methods

Alternative protein source	Anti-nutritional factor present	Suitable processing methods
Cottonseed (*Gossypium hirsutum*)	Gossypol (polyphenolic compound),linatine, cyanogenic glycoside	Dehulling, oil extraction, pelleting, extrusion, cooking, $Ca(OH)_2$ treatment, roasting (Dinesh et al., 2003)
Linseed (*Linum usitatissimum*)	Deficient in lysine and methionine, high in fibre content, HCN	Lysine and methionine supplementation, soaking, autoclaving, pelleting (Alonso, 2000)
Sesame (*Sesamum indicum* L., Pedaliaceae)	Non-starch polysaccharides, phytic acid, high oil content, high phosphorus	Solvent extraction, use of phytase, dehulling (Carvalho,1997)
Sunflower meal (*Helianthus annuus* L.)	High in crude fibre, low in lysine, NSP, chlorogenic acid	Dehulling, mechanical and/or solvent extraction (Yust et al., 2003)
Groundnut meal (*Arachis hypogaea*)	Deficient in lysine and methionine, easily contaminated by aflatoxin	Roasting, mechanical and/or solvent extraction (Faezana, 2005)
Beans (*Vicia faba*)	Protease inhibitors, tannin, lectins, glycosides (vicine and convicine)	Dehulling, solvent extraction, germination (Alonso, 2000)
Chickpea (*Cicer arietinum*)	Trypsin, chymotrypsin inhibitor, tannin, oxalic acid, alkaloid, saponin	Dehulling, heating application (Rajni Mittal et al., 2012)
Velvet bean (*Mucuna pruriens*)	Phenol, L-3,4-dihydroxyphenylalanine, phytate, tannins, trypsin inhibitor activity	Dehulling, roasting, solvent extraction, dry heat, autoclaving, soaking in $NaHCO_3^+$ autoclaving (Yasmin et al., 2008; Gurumoorthi et al., 2008; Vadivel and Pugalenthi 2008)

approach seeks to identify and select desired traits in various alternative protein sources and also eliminate undesired traits, including the ANF in these crops or seeds (Bond and Duc, 1993). As all traits of a plant or their seeds are controlled by genes located on chromosomes, conventional plant or crop breeding can be considered as the way forward in eliminating undesirable traits through gene engineering by applying gene transfer, gene silencing and targeted gene mutation.Crop breeding is a long-term process and the results, for example, the removal of trypsin inhibitors in peas (*Pisum sativum*) are not always consistent. Some cultivars after cross-breeding had higher trypsin inhibitor activities than one of their parents, and cultivars derived from the same cross showed different trypsin inhibitor activities. These results suggest that the hereditary transmission of trypsin inhibitor activities is not systematic and that plant breeding does not produce fast or predictable results (Duc et al., 1999). It is noteworthy that crop breeding methods of improving the nutrient value or reducing the ANFs content should be done with care to prevent the total elimination of these ANFs, as some of the ANF serve as the plant's defence mechanism against pests and microorganisms during growth and storage. Hence, removal of ANFs prior to consumption is the better way of handling the problem (Khattab and Artntfield, 2009).

11.2 Processing

Most of the alternative protein sources are excluded from poultry feed formulations because they contain substantial amounts of ANFs, some of which have been mentioned in early parts of this chapter. It is noteworthy that some ANFs are plant- or seed-specific. Based on this fact, different alternative protein sources may require different processing methods or a combination of methods to effectively eliminate these ANFs (Soetan, 2009). Processing is an act of applying suitable method(s) to reduce or totally eliminate single or several anti-nutrients present in them.

11.2.1 Dehulling

Dehulling is a process of mechanically removing the outer coat/husk of a seed (Jansen, 1993). Most seeds of alternative protein source have seed coats/hulls, and tannins are mainly concentrated in the seed coat (hulls). If removed, there is a significant increase in protein digestibility and protein content in the legume seed meal. Petterson (2000) and Frias et al. (2000) reported that dehulling is effective in reducing the tannin contents in various leguminous seeds. Brene et al. (1993) concluded that there was an improvement in broiler performance for birds fed dehulled lupin (*L. albus*) diet for 7–21 days compared to birds fed a diet containing raw lupin diet. Alonso et al. (2000) also reported a decrease of 92% in tannin content found in faba beans after dehulling.

11.2.2 Heat treatment

Heat processing is widely accepted as an effective means of inactivating the thermo-labile ANF of legume grains. The nutritive quality of most tropical legume seeds, particularly cowpea, soybean, pigeon pea, lima bean and winged beans is notably improved by heat treatment (Akande and Fabiyi, 2010). Heat treatment has been shown to be adequate for reducing contents or activity of ANFs such as protease inhibitors and lectins (Carvalho and Sgarbieri, 1997; O´Doherty and Keady, 2000; Jiménez-Martínez et al., 2001). Steam heating or toasting at 100°C for 15, 30, 60 and 120 min reduced the total contents of glucosinolates in rapeseed meal by 24, 46, 70 and 95%, respectively (Jensen et al., 1995). Frias et al. (2000) reported a reduction of tannin content in dehulled and cooked chickpea.

11.2.3 Extrusion method

A further method to reduce the ANF contents is extrusion, where feeds are treated under varying conditions of high temperature and high pressure. Extrusion involves preparing and treating the given product using pressure and hot steam. Alonso et al. (1998) found that extrusion is the best method to eliminate trypsin, chymotrypsin and α-amylase inhibitors, as well as haemagglutinin activity, without modifying protein content. Extrusion of faba beans at 152–156°C significantly reduced the level of condensed tannins by about 54%, and led to a decrease in trypsin inhibitor activity (TIA) by about 53% (Alonso et al., 2000). It has been suggested that a reduction in TIA and condensed tannin contents may contribute to improvement in ileal protein and amino acid digestibilities in extruded peas (Mariscal-Landin et al., 2002).

11.2.4 Cooking

Omeje (1999) reported that cooking legume seeds for about 30 minutes resulted in the destruction of ANF such as trypsin inhibitors, haemagglutinins, phytic acid, lectin and goitrogen. In soybeans, there were 52.3% reduction in raffinose and 20.7% reduction in stachyose contents after cooking (Mulimani et al., 1997). Armour et al. (1998) reported complete inactivation of soya lectin and protease inhibitory activity by aqueous heat treatment of fully imbibed soya seeds at 100°C for 10 min. Hydration and heating at boiling temperature for 6 h in water and in a 0.5% sodium bicarbonate solution reduced alkaloid contents in *Lupinus campestris* (a wild lupin species) from 27 g/kg to 0.3 g/kg and to 0.02 g/kg, respectively (Jimenéz-Martínez et al., 2001). Autoclaving and limewater treatment of seeds for 10 minutes at 15 lb mm^{-2} destroyed the trypsin inhibitors. In contrast to the above findings, Rao and Belavady (1978) showed an increase in the level of oligosaccharides after cooking of pulses.

11.2.5 Soaking

Overnight soaking of pulses results in significant reduction in α-galactoside concentration (Okolie and Ugochukwu, 1988). Soaking of cowpea flour for 16 h reduced 26% and 28% of stachyose and raffinose concentrations, respectively. With chickpeas, Frias et al. (2000) reported that overnight soaking reduced the concentration of α-galactosides by 16–27%. The efficiency and effectiveness of the soaking process can be enhanced by the addition of sodium bicarbonate, longer soaking time, higher temperature or lower seed:water ratio (Jood et al., 1985; Vijayakumari et al., 1996; Ibrahim et al., 2002). In addition to these, differential solubilities of ANF and their diffusion rates are the two important parameters that influence the losses of ANFs.

11.2.6 Germination

Germination has been reported to reduce the α-galactoside concentration of pulses. During the germination process, complex sugars are converted into simple sugars. Ndonda (2011) showed that germination for a period of three days effectively improved the nutritional value of soybeans and can be considered as an alternative treatment of soybeans in situations where heat treatment is impossible or impractical. Three days of germinating lentils reduced the raffinose oligosaccharide (RFO) concentration by 18–40% (Frias et al., 1996). Eighty-three percent reduction in α-galactosides was also noticed in pigeon peas after four days of germination. Similarly, there were drastic reductions in RFO contents for several non-conventional legumes such as cowpeas, jack bean, mucuna and dolichos (*Lablab purpureus*) (Martin-Cabrejas et al., 2008) after germination. During the germination process phytate is degraded by the endogenous phytase.

11.2.7 Solvent extraction

The use of some solvents to extract oil from oil-bearing seeds has been reported to also enhance the nutritive value of such seeds. For instance, the ANF in cottonseed is soluble in organic solvents like ethyl alcohol, hexane, aqueous acetone and aliphatic amines. These substances have been reported to effectively reduce the gossypol content in CSM. The extraction of CSM with 95% ethanol resulted in reduction in total gossypol content by about 70% (Hron et al., 1994).

11.3 Nutrient supplementation

The objective of feed formulation is to derive a balance diet that will provide the appropriate quantities of biologically available nutrients required by poultry. In addition to providing energy and protein, formulations may contain supplements to provide additional minerals, vitamins and specific amino acids that may be lacking because of the use of non-conventional feed material that may be deficient in one or more nutrients. Adequate knowledge of feed material and nutrients to be supplemented for each feed material is important before a farmer or feed formulator can use alternative protein sources since most of these feed materials are high in ANF and are deficient in amino acids, vitamins and minerals. The use of alternative feed ingredients often necessitates dietary nutrient supplementation. Various authors have reported the enormous improvement in the nutritive value of diets formulated with alternative protein sources through supplementation. For instance, CSM in combination with lysine-rich supplements has the potential to replace up to 40% of the SBM protein in diets for broilers without any negative effects (Ravindran and Blair, 1992; Henry et al., 2001). Gossypol toxicity can be alleviated through the addition of iron salts (Henry et al., 2001). McDonald et al. (1995) reported that linseed meal can be incorporated at up to 10% in poultry diets when supplemented with vitamin B_6 to overcome the effect of linatine. Several researchers (Ryan et al., 1986; Zhang and Parsons, 1996; Elkin, 2002; Costa et al., 2001; Newkirk, 2008) have suggested that supplementation of poultry diets with the relevant deficient amino acid would help to solve the issue of levels of such amino acids in the alternative ingredient. Such research has been conducted on groundnut cake, CSM, linseed and many other alternative protein sources.

11.4 Non-nutrient supplementation

The application of feed enzymes to poultry diets to improve nutrient digestibility has gained relevance in the poultry industry over about three decades. It can be regarded as one of the most significant advances in feed management. Enzymes are added to diets to enable the bird to degrade anti-nutrient feed components, in particular, NSP and phytate. It has also been demonstrated that addition of feed enzymes improves amino acid digestibility and the metabolizable energy value of the diet. The response to feed enzymes is dependent on diet composition, source and level of enzyme addition (Ravindran et al., 1999, 2001; Hew et al., 1999; Selle et al., 2006) and may reflect improved dietary protein digestion per se and/or a reduction in endogenous amino acid losses.

In some studies, it has been shown that the application of xylanase and phytase alone and in combination improves amino acid digestibility to levels which can be quite significant in terms of overall feed formulation. The positive effect of enzymes on amino acid digestibility again demonstrates the impact of ANF on both protein digestion or endogenous amino acid loss (Bryden et al., 2009).

12 Conclusion

There are several alternative protein sources with the potential for use in poultry nutrition. To properly utilize these resources, it is important to have a good understanding of their properties and biological values. However rich these resources are in protein and amino acids contents, their use as replacements for SBM, which is the major protein source in poultry nutrition is limited either by the level of nutrient digestibility or their anti-nutrient contents. It is possible to improve the quality of these alternative protein sources through processing or supplementation with nutrient and non-nutrient additives such as enzymes in order to improve their utilization by poultry. A major limitation remains the low volume of production of many of the alternative ingredients.

13 Where to look for further information

Further information on this subject can be found in the following texts:

- Animal production and health, Agriculture and consumer protection department, FAO. http://www.fao.org/ag/againfo/themes/en/poultry/AP_nutrition.html
- Protein sources for the animal feed industry. Proceedings of expert consultation and workshop. FAO, 2004. http://www.fao.org/docrep/007/y5019e/y5019e00.htm
- Scientific Advances in Animal Nutrition: Promise for the New Century, Proceedings of a Symposium, Committee on Animal Nutrition, Board on Agriculture and Natural Resources, Division on Earth and Life Studies, National Research Council, 2002. National Academy Press, Washington, DC.
- Leson, S. and Summers, J. D. (2009), *Commercial Poultry Nutrition*. Nottingham University Press, Nothingham, UK.
- McNab, J. M. and Boorman, K. N. (2002), *Poultry Feedstuffs: Supply, Composition and Nutritive Value*. CAB International, London, UK.

14 References

Adeniji, A. A. (2007), Effect of replacing groundnut cake with maggot meal in the diet of broilers. *International Journal of Poultry Science* 6: 822–5.

Akande, K. E., Doma, U. D., Agu, H. O. and Adamu, H. M. (2010), Major anti-nutrients found in plant protein sources: Their effect on nutrition. *Pakistan Journal of Nutrition* 9(8): 827–32.

Al-Harthi, M. A., El-Deek, A. A., Salah El-Din, M. and Alabdeen, A. A. (2010), A nutritional evaluation of hatchery by-product in the diets for laying hens. Egypt. *Poult. Sci.* 30(1): 339–51.

Alonso, R., Aguirre, A. and Marzo, F. (2000), Effects of extrusion and traditional processing methods on antinutrients and *in vitro* digestibility of protein and starch in faba and kidney beans. *Food Chem.* 68: 159–65.

Alonso, R., Orue, E. and Marzo, F. (1998), Effect of extrusion and conventional processing methods on protein and antinutritional factor contents in pea seeds. *J. Food Chem.* 63: 505–12.

Ao, T., Macalintal, L. M., Paul, M. A., Pescatore, A. J., Cantor, A. H., Ford, M. J., Timmons, B. and Dawson, K. A. (2015), Effects of supplementing microalgae in laying hen diets on productive performance, fatty-acid profile, and oxidative stability of eggs. *J. Appl. Poult. Res.* 24: 394–400.

Armour, J. C., Perera, R. L. C., Buchan, W. C. and Grant, G. (1998), Protease inhibitors and lectins in soybeans and effects of aqueous heat-treatment. *J. Sci. Food Agric.* 78: 225–31.

Bampidis, V. A. and Christodoulou, V. (2011), Chickpeas *(Cicer arietinum* L.) in animal nutrition: A review. *Anim. Feed Sci. Technol.* 168: 1–20.

Batal, A. and Dale, N. (2010), Feedstuffs ingredient analysis tables: 2011 edition. [Online]. *Feedstuffs.* http://fdsmagissues.feedstuffs.com/fds/Reference_issue_2010 (Accessed 15 March 2016).

Bean, L. D. and Leeson, S. (2003), Long-term effects of feeding flaxseed on performance and egg fatty acid composition of brown and white hens. *Poult. Sci.* 82: 388–94.

Blair, R. (2007), *Nutrition and Feeding of Organic Pigs.* Cabi Series, CABI, Wallingford, UK.

Bond, D. A. and Duc, G. (1993), Plant breeding as a means of reducing antinutritional factors in grain legumes. In A. F. B. Van der Poel, J. Huisman and H. Saini (eds), *Recent Advances of Research in Antinutritional Factors in Grain Legume Seeds*, pp. 379–96. WageningenPers, Wageningen.

Bovera, F., Piccolo, G., Gasco, L., Marono, S., Loponte, R., Vassalotti, G., Mastellone, V., Lombardi, P., Attia, Y. A. and Nizza, A. (2015), Yellow mealworm larvae (Tenebrio molitor, L.) as a possible alternative to soybean meal in broiler diets. *Br. Poult. Sci.* 56: 569–75.

Brenes, A., Marquardt, R., Guenter, W. and Rotter B. A. (December 1993), Effect of enzyme supplementation on the nutritional value of raw, autoclaved, and dehulled lupins (Lupinusalbus) in chicken diets. *Poult Sci.* 72(12): 2281–93.

Bryden, W. L. (1996), Antinutritive factors in protein metabolism. In A. F. Nunes, A. V. Portugal, J. P. Costa and J. R. Ribeiro (eds), *Protein Metabolism and Nutrition*, pp. 517–18. Estacao Zootecnica Nacional: Vale de Santarem, Portugal.

Bryden, W. L., Li, X., Ravindran, G., Hew, L. I. and Ravindran, V. (2009), Ileal digestible amino acid values in feedstuffs for poultry. Rural Industries Research and Development Corporation Publication 09/071, p. 76.

Canola Council of Canada (2009), Canola meal feeding guide. 5th edition. https://cigi.ca/wp-content/uploads/2011/12/2009-Canola_Guide.pdf (Accessed 18 February 2016).

Canola Council of Canada (2015), Canola meal feeding guide. 5th Edition. http://www.canolacouncil.org/media/516716/2015_canola_meal_feed_industry_guide.pdf (Accessed 15 March 2016).

Carew, S. N., Olomu, J. M. and Offiong, S. A. (1988), Amino acid supplementation of groundnut meal protein in broiler diets. *J. Trop. Agric.* 65: 329–32.

Carvalho, M. R. B. and Sgarbieri, V. C. (1997), Heat treatment and inactivation of trypsin-chymotrypsin inhibitors and lectins from beans (*Phaseolus vulgaris* L.). *J. Food Biochem.* 21: 219–33.

Castell, A. G., Guenter, W. and Igbasan, F. A. (1996), Nutritive value of peas for non – ruminant diets. *Anim. Feed Sci. Technol.* 60: 209–27.

Chen, Y. P., Duan, W. G., Wang, L. L., Zhang, S. L. and Zhou, Y. M. (2013), Effects of thermostable phytase supplementation on the growth performance and nutrient digestibility of broilers. *Int. J. Poult. Sci.* 12: 441–4.

Chenoweth, P. J., Risco, C. A., Larsen, R. E., Velez, J., Tran, T. and Chase Jr, C. C. (1994), Effects of dietary gossypol on aspects of semen quality, sperm morphology and sperm production in young Brahman Bulls. *Theriogenology* 42: 1–13.

Costa, E. F., Miller, B. R., Pesti, G. M., Bakalli, R. I. and Ewing, H. P. (2001), Studies on feeding peanut meal as a protein source for broiler chickens. *Poult. Sci.* 80: 306–13.

Crawshaw, R. (2004), *Co-product Feeds: Animal Feeds from the Food and Drinks Industries.* Nottingham University Press, Nottingham, UK.

Culley Jr, D. D., Rejmankova, E., Kvet, J. and Frye, J. B. (1981), Production, chemical quality and use of duckweeds (Lemnaceae) in aquaculture, waste management, and animal feeds. *J. World Maricult. Soc.* 12: 27–49.

Dale, N. (1996), Variation in feed ingredient quality: oilseed meals. *Anim. Feed Sci. Tech.* 59: 129–35.

Davies, K. M. (2007), Genetic modification of plant metabolism for human health benefits. *Mutat. Res.* 622: 122–37.

Davis, A. J., Lordelo, M. M. and Dale, N. M. (2002), The use of cottonseed meal with or without added soapstock in laying hen diets. *J. Appl. Poult. Res.* 11: 127–33.

Deshmukh, A. C. and Patterson, P. H. (1997), Preservation of hatchery waste by lactic acid fermentation. 2. Large-scale fermentation and feeding trial to evaluate feeding value. *Poult. Sci.* 76 (9): 1220–6.

Dinesh, K., Phundan, S.,Mukta C., Shaikh, A. and Gayal, S. (2003), Cottonseed oil quality, utilization and processing. *Cicr. Technical Bulletin. No.* 25: 1–16.

Duc, G., Marget, P., Esnault, R., Le Guen, J. and Bastianelli, D. (1999), Genetic variability for feeding value of faba bean seeds (*Vicia faba*): Comparative chemical composition of isogenics involving zero-tannin and zero-vicine genes. *J. Agric. Sci.* 133: 185–96.

Duranti, M. and Gius, C. (1997), Legume seeds: protein content and nutritional value. *Field Crops Res.* 53 (1/3): 31–45.
El Boushy, A. R. Y. and Van der Poel, A. F. B. (2000), *Handbook of Poultry Feed from Waste: Processing and Use*. Springer-Verlag, New York, p. 428.
Elkin, R. G. (2002), Nutritional components of feedstuffs: A qualitative chemical appraisal of protein. In J. M. McNab and K. N. Boorman (eds), *Poultry Feedstuffs: Supply, Composition and Nutritive Value*, pp. 57–86. CAB International, London, UK.
Enami, H. R. (2011), A review of using canola meal/rapeseed meal in aquaculture feeding. *J. Fish. Aqua. Sci.* 6(1): 22–36.
Enneking, D. and Maxted, N. (1995), Narbon bean (Vicia narbonensis L.). In J. Smartt and N. W. Simmonds (eds), *Evolution of Crop Plants*, 2nd ed., pp. 316–21. Longman: London.
Evans, A. M., Smith, D. L. and Moritz, J. S. (2015), Effects of algae incorporation into broiler starter diet formulations on nutrient digestibility and 3 to 21 d bird performance. *J. Appl. Poult. Res.*, 24: 206–14.
Ewing, W. R. (1963), *Poultry Nutrition*. The W. R. Ewing Company, Pasadena, California.
Farzana, P. (2005), Anti-nutritional factors in oil seeds as aflatoxin in groundnut. Digitalverlag GmbH, Germany. www.chemlin.com, pp. 1–7.
Finke, M. D. (2002), Complete nutrient composition of commercially raised invertebrates used as food for insectivores. *Zoo Biology* 21 (3): 269–85.
Food and Agricultural Organization (2015), *Production – Crops Processed*. FAO, Rome, Italy.
Frias, J., Dıaz-Pollan, C., Hedley C. L. and VidalValverde C. (1996), Evolution and kinetics of monosaccharide, disaccharide and α-galactosides during germination of lentils. *Z. LebensmUntersForsch* 202: 35–9.
Frias, J., Vidal-Valverde, C., Sotomayor, C., DıazPollan, C. and Urbano, G. (2000), Influence of processing on available carbohydrate content and antinutritional factors of chickpeas. *Eur. Food Res. Technol.* 210: 340–5.
Friedman, M. (1996), Nutritional value of proteins from different food sources: A review. *J. Agric. Food Chem.* 44 (1): 6–29.
Glatz, P. and Miao, Z. (2009), High value products from hatchery waste. RIRDC Publication No. 09/061, RIRDC Project No PRJ-000078. Australian Government, Rural Industries Research and Development Corporation, Australia.
Göhl, B. (1970), *Animal Feed from Local Products and By-products in the British Caribbean*. Rome, FAO. AGA/Misc/70/25.
Gongnet, G. P., Niess, E., Rodehutscord, M. and Pfeffer, E. (2001), Algae-meal (Spirulina platensis) from lake Chad replacing soybean-meal in broiler diets. *Archiv fur Geflugelkunde* 65, 265–8.
González-Esquerra, R. and Leeson, S. (2000), Effects of menhaden oil and flaxseed in broilers on sensory quality and lipid composition of poultry meal. *British Poult. Sci.* 41: 481–8.
Green, S., Bertrand, S. L., Duron, M. J. C. and Maillard, R. (1987), Digestibilities of amino acids in soybean, sunflower and groundnut meals, determined with intact and caecectomised cockerels. *Br. Poult. Sci.* 28 (4): 643–52
Guiry, M. D. and Guiry, G. M. (2014), *Algaebase*. World-wide electronic publication, National University of Ireland, Galway.
Gurumoorthi, P., Janardhanan, K. and Myhrman, R. V. (2008), Effect of differential processing methods on L-Dopa and protein quality in velvet bean, an underutilized pulse. *LWT – Food Science and Technology* 41: 588–96.
Gutierrez, O., Zhang, C., Cartwright, A. L., Carey, J. B. and Bailey, C. A. (2007), Use of guar by-products in high-production laying hen diets. *Poult. Sci.* 86: 1115–20.
Hankins, C. N. and Shannon, L. M. (1978), The physiological and enzymatic properties of a phytoheamagglutinin from mung beans. *J. Bio. Chem.* 253: 7791–7.
Hassan, S. M., Haq, A. U., Byrd, J. A., Berhow, M. A., Cartwright, A. L. and Bailey, C. A. (2010), Haemolytic and antimicrobial activities of saponin-rich extracts from guar meal. *Food Chem.* 119: 600–5.
Heidarinia, A. and Malakian, M. (2011), Nutritional evaluation of cottonseed meal with and without ferrous sulfate for broiler chickens. *Res. J. Poult. Sci.* 4: 14–17.

Helsper, J. P. F. G., Balkema-Boomstra, A. G., Ribôt, S. A. and Van Loo, E. N. (2006), Novel protein crops as pig feed in organic farming. Edited by Wageningen Plant Research International, The Netherlands, Report 112.

Henry, M. H., Pesti, G. M., Bakalli,R., Lee, J. Toledo, R. T., Eitenmiller, R. D. and Phillips, R. D. (2001), The performance of broiler chicks fed diets containing extruded cottonseed meal supplemented with lysine. *Poult. Sci.* 80: 762–8.

Herber, S. M. and Elswyk, M. E. (1996), Dietary marine algae promotes efficient deposition of n-3 fatty acids for the production of enriched shell eggs. *Poultry Science* 75, 1501–7.

Hew, L. I., Ravindran, V., Ravindran, G., Pittolo, P. H. and Bryden, W. L. (1999), The apparent metabolisable energy and amino acid digestibility of wheat, triticale and wheat middlings for broiler chickens as affected by exogenous xylanase supplementation. *J. Sci. Food Agric.* 79: 1727–32.

Heuzé V., Tran G., Giger-Reverdin S., Renaudeau D., Bastianelli D. and Lebas F. (2015), Sorghum by-products. Feedipedia, a programme by INRA, CIRAD, AFZ and FAO. http://www.feedipedia.org/node/752. Last updated on 7 October 2015, 14:50.

Heuzé V., Tran G., Sauvant D. and Lebas F. (2015), Brewers grains. Feedipedia, a programme by INRA, CIRAD, AFZ and FAO. http://www.feedipedia.org/node/74. Last updated on 3 August 2015, 17: 01.

Higuchi, M., Suga, M. and Iwai, K. (1983), Participation of lectin in biological effects of raw winged bean seeds on rats. *Agric. and Bio. Chem.* 47: 1879–86.

Holdt, S. L. and Kraan, S. (2011), Bioactive compounds in seaweed: functional food applications and legislation. *J. Appl. Phycol.* 23 (3): 543–97.

Hron, R. J., Wan, P. J. and Keek, M. S. (1994), Ethanol vapour deactivation of gossypol in cotton seed meal. *J. Am. Oil Chem. Soc.* 73: 1337–9.

Huang, K. H., Li, X., Ravindran, V. and Bryden, W. L. (2006), Comparison of apparent ileal amino acid digestibility of feed ingredients measured with broilers, layers and roosters. *Poult. Sci.* 85: 625–34.

Hughes, R. J. and Choct, M. (1999), Chemical and physical characteristics of grains related to variability in energy and amino acid availability in poultry. *Aust. J. Agric. Res.* 50: 689-701.

Huisman, J. and Tolman, G. H. (2001), Antinutritional factors in the plant proteins of diets for non-ruminants. In P. C. Garnsworthy and J. Wiseman (eds), *Recent Developments in Pig Nutrition 3*, pp. 261–322. Nottingham University Press, Nottingham, UK.

Ibrahim, S. S., R. A. Habiba, A. A. Shatta and Embaby, H. E. (2002), Effect of soaking, germination, cooking and fermentation on antinutritional factors in cowpeas. *Nahrung* 46: 92.

Ioselevich, M., Steingaß, H., Rajamurodov, Z. and Drochner, W. (2004), Nutritive value of silkworm pupae for ruminants. VDLUFA Kongress, Qualitätssicherung in landwirtschaftlichen Produktionssystemen, Rostock, 13–17 September 2004, 116: 108.

Jansman, A. J. M. (1993), Tannins in feedstuffs for simple stomached animals. *Nutr. Res. Rev.* 6: 209–36.

Jensen, S. K., Liu, Y.-G. and Eggum, B. O. (1995), The effect of heat treatment on glucosinolates and nutritional value of rapeseed meal in rats. *Anim. Feed Sci. Technol.* 53: 17–28.

Jiménez-Martínez, C., Hernández-Sánchez, H., Alvárez-Manilla, G., Robledo-Quintos, N., Martínez-Herrera, J. and Dávila-Ortiz, G. (2001), Effect of aqueous and alkaline thermal treatments on chemical composition and oligosaccharide, alkaloid and tannin contents of *Lupinuscampestris*seeds. *J. Sci. Food Agric.* 81: 421–8.

Jood, S., Mehta, U., Singh, R. and Bhat, C. M. (1985), Effect of processing on flatus-producing factors in legumes. *J. Agri. Food Chem.* 33: 268–71.

Kamp, J. and van Reeuwijk, P. (2011), Naar een glutenvrije haverketen. Wageningen UR, Praktijkonderzoek Plant & Omgeving, Rapport in opdracht van ministerie van EL&I. PPO publicatienummer 409.

Kempton, T. J. (2006), Value-added coconut co-products. In S. W. Adkins, M. Foale and Y. M. S. Samosir (eds), *Coconut Revival—New Possibilities for the 'Tree of Life'*. Proc. Int. Coconut Forum, Cairns, Australia, 22–24 November 2005. ACIAR Proceedings No. 125.

Kerr, B. J., Dozier III, W. A. and Shurson, G. C. (2013), Effects of reduced-oil corn distillers dried grains with solubles composition on digestible and metabolizable energy value and prediction in growing pigs. *J. Anim. Sci.* 91: 3231–43.

Khajali, F. and Slominski, B. A. (2012), Factors that affect the nutritive value of canola meal for poultry. *Poult. Sci.* 91: 2564–75.

Khattab, R. Y. and Arntfield, S. D. (2009), Nutritional quality of legume seeds as affected by some physical treatments on anti-nutritional factors. *LWT – Food Sci. Technol.* 42: 1113–18.

Khusro, M., Andrew, N. R. and Nicholas, A. (2012), Insects as poultry feed: a scoping study for poultry production systems in Australia. *World's Poult. Sci. J.* 68: 435–46.

Kim, C. H. and Kang, H. K. (2015), Effect of dietary supplementation with a chlorella by-product on the performance, immune response and metabolic function in laying hens. *European Poul. Sci.* 79: 108.

Kulshreshtha, G., Rathgeber, B., Stratton, G., Thomas, N., Evans, F., Critchley, A., Hafting, J. and Prithiviraj, B. (2014), Feed supplementation with red seaweeds, Chondrus crispus and Sarcodiotheca gaudichaudii, affects performance, egg quality, and gut microbiota of layer hens. *Poult. Science* 93: 2991–3001.

Lee, J. T., Bailey, C. A. and Cartwright, A. L. (2009), In vitro viscosity as a function of guar meal and beta-mannanase content of feeds. *Int. J. Poult. Sci.* 8 (8): 715–19.

Lee, J. T., Connor-Appleton, S., Bailey, C. A. and Cartwright, A. L. (2005), Effects of guar meal by-product with and without beta-mannanase hemicell on broiler performance. *Poult. Sci.* 84 (8): 1261–7.

Lee, J. T., Connor-Appleton, S., Haq, A. U., Bailey, C. A. and Cartwright, A. L. (2004), Quantitative measurement of negligible trypsin inhibitor activity and nutrient analysis of guar meal fractions. *J. Agric. Food Chem.* 52: 6492–5.

Li, X., Mollet, P., Huang, K. and Bryden, W. L. (2002), Inclusion level of cottonseed meal in broiler diets. *Proc. of the Joint 7th WPSA Asian Pacific Federation Conference and 12th Australian Poultry and Feed Convention*, p. 256.

Liener, I. E. (1994a), Implications of antinutritional components in soybean foods. *Crit. Rev. Food Sci. Nutr.* 34: 31–67.

Liener I. E. (1994b), Antinutritional factors related to proteins and amino acids. In Y. H. Hul, J. R. Gorham, K. D. Murrel and D. O. Cliver (eds), *Food Borne Disease Hand Book*, Vol. 3, pp 261–309. New York: Dekker.

Liu, U, Singh S. and Green A. (2002), High – oleic and high stearic cotton seed oils: Nutritionally improved cooking oils developed using gene silencing. *J. Ani. Nutr.* 21(3 suppl): 205–11.

Longvah, T., Mangthya, K. and Ramulu, P. (2011), Nutrient composition and protein quality evaluation of eri silkworm (Samia ricinii) prepupae and pupae. *Food Chem.* 128 (2): 400–3.

Lordelo, M. M., Davis, A. J., Wilson, J. L. and Dale, N. M. (2004), Cottonseed meal diets improve body weight uniformity in broiler breeder pullets. *J. Applied Poult. Res.* 13: 191–9.

Mahajan, A. and Dua, S. (1997), Nonchemical approach for reducing antinutritional factors in rapeseed (*Brassica campestris*Var. Toria) and characterization of enzyme phytase. *J. Agric. Food Chem.* 45: 2504–8.

Makkar, H. P. S., Tran, G., Heuze, V., Giger-Reverdin, S., Lessire, M., Lebas, F. and Ankers, P. (2016), Seaweeds for livestock diets: a review. *Animal Feed Science and Technology* 212: 1–17.

Mariscal-Landin G., Lebreton Y. and Sève B. (2002), Apparent and standarised true ileal digestibility of protein and amino acids from faba bean, lupin and pea, provided as whole seeds, dehulled or extruded in pig diets. *Anim. Feed Sci. Tech.* 97: 183–98.

Martin-Cabrejas, M. A., Diaz, M. F., Aguilera, Y., Benitez, V., Molla, E. and Esteban, R. M. (2008), Influence of germination on the soluble carbohydrates and dietary fibre fractions in non-conventional legumes. *Food Chemistry* 107(3): 1045–52.

McDonald, P., Edwards, R. A., Greenhalgh, J. F. D. and Morgan, C. A. (1995), Animal Nutrition, 5th edition. John Wiley and Sons, New York, USA, p. 607.

McGinnis, J., Hsu, P. T. and Carver, J. S. (1948), Nutritional deficiencies of sunflower oil meal for chicks. *Poult. Sci.* 27: 389–93.

McLeod, A. and World Livestock (ed.) (2011), Livestock in Food Security. Rome: Food and Agriculture Organization of the United Nations.

Meloche, K. J., Kerr, B. J., Shurson, G. C. and Dozier, W. A., III (2013). Apparent metabolizable energy and prediction equations for reduced-oil corn distillers dried grains with solubles in broiler chicks from 10 to 18 days of age. *Poult. Sci.* 92: 3176–83.

Mišurcová, L. (2012), Chemical composition of seaweeds. In Se-Kwon Kim (ed.), *Handbook of Marine Macroalgae: Biotechnology and Applied Phycology*. John Wiley & Sons, p. 567.

Mulimani, V. H. and Ramalingam (1997), Enzymatic degradation of raffnose family sugars in chickpea flour. *World J. Microbiol. Biotechnol.* 13: 583–5.

Murray, D. R. (1984), Accumulation of seed reserves of nitrogen. In D. R. Murray (ed.), *Development: Vol. 1. Seed Physiology*, pp 83–137. New York, USA: Academic Press.

Nadeem, M. A., Gilani, A. H., Khan, A. G. and Mahr-Un-Nisa. (2005), Amino acids availability of poultry feedstuffs in Pakistan. *Int. J. Agric. Biol.* 7: 985–9.

Najib, H. and Al-Yousef, Y. M. (2011), Performance and essential fatty acids content of dark meat as affected by supplementing the broiler diet with different levels of flax seed. *Annual Rev. Res. Bio.* 1: 22–32.

National Research Council (1994), *Nutrient Requirements of Poultry*. 9th rev. ed. National Academy Press, Washington, DC.

Ndonda, C. K. (2011), Germination as a processing technique for soybeans in small-scale broiler farming. Degree thesis submitted to the Department of Animal and Wildlife Sciences Faculty of Natural and Agricultural Sciences University of Pretoria, Pretoria. September 2011.

Newkirk, R. W. and Classen, H. L. (2002), The effects of toasting canola meal on body weight, feed conversion efficiency and mortality in broiler chickens. *Poult. Sci.* 81: 815–25.

Newkirk, R. W. (2008), Flax feed industry guide. Flax Canada. http://flaxcouncil.ca/wp-content/uploads/2015/02/Flax-Feed-Industry-Guide-Final.pdf (Accessed 15 March 2016).

Newkirk, R. (2011), Wheat DDGS Feed Guide: Wheat dried distiller grains with solubles. Feed Opportunities from the Biofuels Industries (FOBI), 1st Edition. Canadian International Grains Institute (Cigi).

O'Doherty, J. V. and Keady, U. (2000), The nutritive value of extruded and raw peas for growing and finishing pigs. *Anim. Sci.* 70: 265–74.

Oatway, L., Vasanthan, T. and Helm, J. H. (2001), Phytic acid. *J. Food Rev. Int.* 17: 419–31.

Okolie, P. N. and E. N. Ugochukwu.(1988), Changes in activities of cell wall degrading enzymes during fermentation of cassava (Manihot esculentacCrantz) with Citrobacter freundii. *J. Sci. Food Agric.* 44: 51–61.

Omeje, S. K. (1999), *Issues in Animal Science*. Raykennedy Scientific Publication, Enugu Nigeria.

Oryschak, M. and Beltranena, E. (2013), Solvent-extracted vs. expeller-pressed *B. napus* and *B. juncea* fed to layers: Effects on feed intake, egg production and physical egg quality. *Poult. Sci.* 92 (Suppl. 1): 80.

Pallauf, J. and Rimbach, G. (1997), Nutritional significance of phytic acid and phytase. *Arch. Anim. Nutr.* 50: 301–19.

Patil, S. R., Amena, S., Vikas, A., Rahul, P., Jagadeesh, K. and Praveen, K. (2013), Utilization of silkworm litter and pupal waste-an eco-friendly approach for mass production of Bacillus thuringiensis. *Biores. Technol.* 131: 545–7.

Perez-Maldonado, R. A. (2003), Canola meal and cottonseed meal in broiler and layer diets. A report for the Australian-sourced feed ingredients for pigs and poultry. AECL Publication No. 03/10.

Pesti, G. M., Bakalli, R. I., Driver, J. P., Sterling, K. G., Hall, L. E. and Bell, E. M. (2003), Comparison of peanut meal and soybean meal as protein supplements for laying hens. *Poult. Sci.* 82: 1274–80

Petterson, D. S. (2000), The use of lupins in feeding systems – review. *Asian-Aust. J. Anim. Sci.* 13: 861–82.

Phelps R. (1966), Cottonseed meal for poultry: from research to practical application. *World's Poult. Sci. J.* 22: 86–112.

Pieterse, E., Pretorius, Q.; Hoffman, L. C. and Drew, D. W. (2014), The carcass quality, meat quality and sensory characteristics of broilers raised on diets containing either Musca domestica larvae meal, fish meal or soybean meal as the main protein source. *Anim. Prod. Sci.* 54: 622–8.

Pugalenthi, M., Vadivel, V. and Siddhuraju, P. (2005), Alternative Food/Feed Perspectives of an Underutilized Legume *Mucuna pruriens* var. *Utilis* – A Review. *Plant Foods Hum. Nutr.* 60: 201–18, 2005.

Ramesh, K. R., Devegowda, G. and Khosravinia, H. (2006), Effects of enzyme addition to broiler diets containing varying levels of double-zero rapeseed meal. *Asian-Aust. J. Anim. Sci.* 19: 1354–60.

Rangel, A., Saraiva, K., Schwengber, P., Narciso, M. S., Domont, G. B., Ferreira, S. T. and Pedrosa, C. (2004), Biological evaluation of a protein isolate from cowpea (*Vigna unguiculata*) seeds. *Food Chem.* 87 (4): 491–9.

Rao, B. B. (1978), Oligosaccharides in pulses: varietal differences and effect of cooking and germination. *J. Agric. Food Chem.* 26: 316–19.

Ravindran, G., Ravindran, V. and Bryden, W. L. (2006), Total and ileal digestible tryptophan contents of feedstuffs for broiler chickens. *J. Sci. Food Agric.* 86: 1132–7.

Ravindran, V. and Blair, R. (1992), Feed resources for poultry production in Asia and the Pacific region. II. Plant protein sources. *World's Poult. Sci. J.* 48: 205–31.

Ravindran, V., Cabahug, S., Bryden, W. L. and Selle, P. H. (1999), The influence of microbial phytase on the bioavailability of protein and energy in broiler chickens. In M. B Celho and E. T. Kornegay (eds), *Phytase in Animal Nutrition and Waste Management*, pp. 573–84. BASF Monograph Series, BASF Corporation, Parisippany, New Jersey, USA.

Ravindran, V., Selle, P., Ravindran, G., Morel, P. C. H., Kies, A. and Bryden, W. (2001), Microbial phytase improves performance, apparent metabolizable energy, and ileal amino acid digestibility of broilers fed a lysine-deficient diet. *Poult. Sci.* 80: 338–44.

Rebolé, A., Rodgrígues, L., Ortiz, L. T., Alzueta, C., Centeno, C., Viveros, A. Brenes, A. and Arija, I. (2006), Effect of dietary high-oleic sunflower seed, palm oil and vitamin E supplementation on broiler performance, fatty acid composition and oxidation susceptibility of meat. *British Poult. Sci.* 47: 581–91.

Renewable Fuels Association (2008), *Feeding the Future: The Role of the US Ethanol Industry in Food and Feed Production*. Renewable Fuels Association, Washington DC, USA.

Robinson, D. and Singh D. N. (2001), Alternative protein sources for laying hen. A report for the rural industries research and development corporation, Queensland Poultry Research and Development Centre, March 2001, Publication No., DAQ-241A.

Rusoff, L. L., Blakeney Jr, E. W. and Culley Jr, D. D. (1980), Duckweeds (Lemnaceae family): a potential source of protein and amino acids. *J. Agric. Food Chem.* 28 (4): 848–50.

Ryan, J. R., Kratzer, F. H., Grace, C. R. and Vohra, P. (1986), Landless cotten seed meal for laying and breeding hens and broiler chicks. *Poultry Sci.* 65: 949–55.

Schumacher, H., Paulsen, H. M., Gau, A. E., Link, W., Jürgens, H. U., Sass, O. and Dieterich, R. (2011), Seed protein amino acid composition of important local grain legumes *Lupinus angustifolius* L., Lupinus luteus L., *Pisum sativum* L. and *Vicia faba* L. *Plant Breed.* 130 (2): 156–64.

Selle, P. H., Ravindran, V., Bryden, W. L. and Scott, T. (2006), Influence of dietary phytate and exogenous phytase on amino acid digestibility in poultry: A review. *J. Poult. Sci.* 43: 89–103.

Senkoylu, N. and Dale, N. (1999), Sunflower meal in poultry diets: a review. *World's Poult. Sci. J.* 55: 153–74.

Senkoylu, N. and Dale, N. (2006), Nutritional evaluation of a high-oil sunflower meal in broiler starter diets. *J. Applied Poult. Res.* 15: 40–7.

Smithard, R. (2002), Secondary plant metabolites in poultry nutrition. In J. M. McNab and K. N. Boorman (eds), *Poultry Feedstuffs: Supply, Composition and Nutritive Value*, pp. 237–78. CAB International, London, UK.

Soetan, K. O. (September 2009), The need for adequate processing to reduce the antinutritional factors in plants used as human foods and animal feeds: A review. *African J. Food Sci.* 3 (9): 223–32.

Spragg, J. C. and Mailer, R. J. (2007), Canola Meal Value Chain Quality Improvement. A Final Report Prepared for AOF and Pork CRC. Project Code: 1B-103–0506.

Stegeman, D., Janssen, A. M., Helsper, J. P. F. G., van der Meer, I. M. and van Kernebeek, H. R. J. (2010), Technologie en grondstoffen voor vleesvervangers en hoogwaardige eiwitten. Rapport BO-08-008-002 in opdracht van ministerie van LNV. ISBN nummer 987-90-8585-756-3.

Sterling, K. G., Costa, E. F., Henry, M. H., Pesti, G. M. and Bakalli, R. I. (2002), Responses of broiler chickens to cottonseed and soybean meal-based diets at several protein levels. *Poult. Sci.* 81: 217–26.

Stock, R. A., Lewis, J. M., Klopfenstein, T. J. and Milton, C. T. (1999), Review of new information on the use of wet and dry milling feed by-products in feedlot diets. *Proceedings of the American Society of Animal Science*. Am. Soc. Anim. Sc., pp. 1–12.

Sundu, B., Kumar, A. and Dingle, J. (2006), Palm kernel meal in broiler diets: effect on chicken performance and health. *World Poult. Sci. J.* 62 (2): 316–25.

Suresh, H. N., Mahalingam, C. A. and Pallavi, (2012), Amount of chitin, chitosan and chitosan based on chitin weight in pure races of multivoltine and bivoltine silkworm pupae Bombyx mori L. *Int. J. Sci. Nature* 3 (1): 214–16.

Swetman, A. A., Nicolaides, L., Wareing, P. W., New, J. H., Wood, J. F. and Hammond, L. (2002). Food processing and preservation. In P. Golob, G. Farell and J. E. Orchard (eds), *Crop Post-harvest: Science and Technology, Volume 1, Principles and Practice*. John Wiley & Sons, QLD, Australia, 360–422.

Swiatkiewicz, S. and Koreleski, J. (2008), The use of distillers dried grains with soluble (DDGS) in poultry nutrition. *World's Poult. Sci. J.* 64 (2): 257–65.

Swiatkiewicz, S., Arczewska-Wlosek, A. and Jozefiak, D. (2015), Application of microalgae biomass in poultry nutrition. *World's Poult. Sci. J.* 71: 663–72.

Thompson, L. U., Tenebaum, A. V. and Hui, H. (1986), Effect of lectins and mixing of proteins on rate of protein digestibility. *J. Food Sci.* 51: 150–2.

Tiroesele, B. and Moreki, J. C. (2012). Termites and earthworms as potential alternative sources of protein for poultry. *Int. J. Agro Vet. Med. Sci.* 6: 368–76.

Trivedy, K., Kumar, S. N., Mondal, M. and Kumar Bhat, C. A. (2008), Protein banding pattern and major amino acid component in de-oiled pupal powder of silkworm, Bombyx mori Linn. *J. Entomol.* 5 (1): 10–16.

Usub, T., Lertsatitthanakorn, C., Poomsa-ad, N., Wiset, L., Yang, L. and Siriamornpun, S. (2008), Experimental performance of a solar tunnel dryer for drying silkworm pupae. *Biosystems Engineering* 101: 209–16.

Vadivel, V. and Pugalenthi, M. (2008), Effect of various processing methods on the levels of antinutritional constituents and protein digestibility of *Mucuna pruriens* (L.) DC. var. utilis (Wall. ex Wight) Baker ex Burck (velvet bean) seeds. *J. Food Biochem.* 32: 795–812.

Vadivel, V. and Janardhanan, K. (2000), Nutritional and antinutritional composition of velvet bean: An under-utilized food legume in South India. *Int. J. Food Sci. Nutr.* 51: 279–87.

Van der Poel, A. F. B, Dellaert, L. M. W, Van Norel, A. and Helsper, P. F. G. (1992), The digestibility in piglets of faba bean (*Viciafaba* L.) as affected by breeding towards the absence of condensed tannins. *Br. J. Nutr.* 68: 793–800.

Vander Pol, M., Hristov, A. N., Zaman, S. and Delano, N. (2008), Peas can replace soybean meal and corn grain in dairy cow diets. *J. Dairy Sci.* 91 (2): 698–703.

Van der Poel, A. F. B. (2013), Unconventional protein sources for poultry feeding: opportunities and threats. *Proceedings 19th Symposium on Poultry Nutrition*, Potsdam, Germany, 26–29 August 2013, 14–24.

Vijayakumari, K., Siddhuraju, P. and Jonardhaman, K. (1996), Effect of soaking, cooking and autoclaving on phytic acid and oligosaccharide contents of the tribal pulse, Mucunamonosperma DC. *Wight. Food Chem.* 55: 173–7.

Vijayakumari, K., Smitha, K. B. and Janardhanan, K. (2002), Biochemical characterization of the tribal pulse, *Mucuna utilis* Wall ex. Wight Seeds. *J. Food Sci. Tech.* 39: 650–3.

Villamide, M. J. and San Juan, L. D. (1998), Effect of chemical composition of sunflower seed meal on its true metabolizable energy and amino acid digestibility. *Poult. Sci.* 77: 1884–92.

Wei ZhaoJun, Lia AiMei, Zhang HaiXiang, Liu Jian and Jiang ShaoTong (2009), Optimization of supercritical carbon dioxide extraction of silkworm pupal oil applying the response surface methodology. *Bioresour. Technol.* 100: 4214–19.

Whitehead, C. C. (1985), Assessment of Biotin Deficiency in Animals. *Ann. New York Acad. Sci.* 447: 86–96.

Yasmin, A., Zeb, A., Khalil, A. W., Paracha, G. M. and Khattak, A. B. (2008), Effect of processing on anti-nutritional factors of red kidney bean (*Phaseolusvulgaris*) grains. *Food and Bioprocess Technol.* 1: 415–19.

Yust, M., Pedroche, J.,. Megías, C., Girón-Calle, J., Alaiz, M., Millán, F. and Vioque, J. (2003), Improvement of protein extraction from sunflower meal by hydrolysis with alcalase. *Grasas y Aceites* 54 (4): 419–23.

Zhang, Y. and Parsons, C. M. (1996), Effect of over-processing on the nutritional quality of peanut meal. *Poult. Sci.* 75: 514–18.

Zheng, L., Oh, S. T., Jeon, J. Y., Moon, B. H., Kwon, H. S., Lim, S. U., An, B. K. and Kang, C. W. (2012), The dietary effects of fermented Chlorella vulgaris (CBT) on production performance, liver lipids and intestinal microflora in laying hens. *Asian-Australasian J. Anim. Sci.* 25: 261–6.

Chapter 14

Maintaining the safety of poultry feed

G. Raj Murugesan and Chasity M. Pender, BIOMIN America Inc., USA

1 Introduction

Feed safety is reliant on a multitude of factors and can be affected by both natural and artificial contaminants including plant and fungal toxins, chemicals and pathogens to name a few. These contaminants may result in deteriorated feed quality, reduced performance and increased incidence of disease in poultry, ultimately leading to a significant economic loss for producers. Moreover, livestock feed is considered to be the beginning of the human food chain. As poultry are raised to produce meat and eggs for human consumption, these contaminants may also represent a human health concern. This chapter focuses on those contaminants considered to pose the most significant risk in poultry and the human food supply: mycotoxins, dioxins and bacterial pathogens. Each section will discuss the negative effects of these factors as well as possible control measures that may be implemented to reduce feed contamination and secure feed and food safety.

2 Mycotoxins

The term 'mycotoxin' is derived from 'mykes' meaning fungi and 'toxicon' meaning poison. Mycotoxins are secondary metabolites of low molecular weight produced by a wide range of fungi, principally moulds. There are over 200 species of moulds that produce mycotoxins. Aflatoxins (AF), zearalenone (ZEN), ochratoxin A (OTA), fumonisins (FUM) and trichothecenes such as deoxynivalenol (DON) and T-2 toxin are some of the mycotoxins that can significantly affect the health and productivity of poultry species (Fig. 1).

The conditions under which fungi and mycotoxins are produced in agricultural commodities depend highly on environmental factors such as water availability and

http://dx.doi.org/10.19103/AS.2016.0011.14

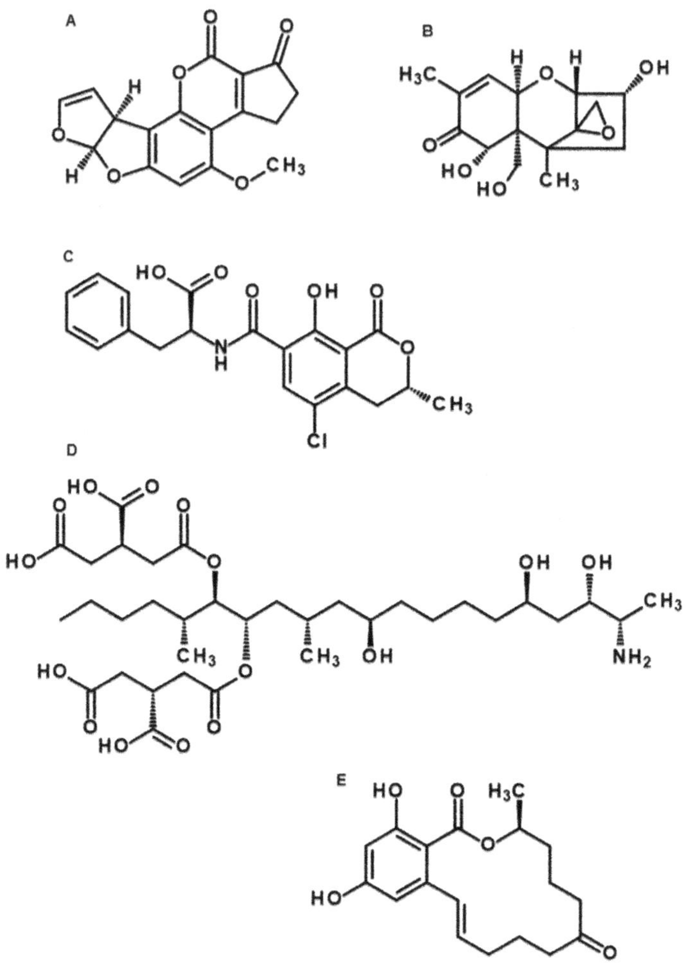

Figure 1 Chemical structures of most prevalent mycotoxins. A: Aflatoxin B1, B: Deoxynivalenol, C: Ochratoxin A, D: Fumonisin B1, E: Zearalenone.

temperature, but also slightly elevated CO_2 concentration may stimulate the growth of mycotoxigenic fungi (Magan et al., 2011). Extreme weather situations, precipitation and drought lead to plant stress and hence they become more vulnerable to fungal infection (Wu et al., 2011). A primary challenge is that fungi, for example *Fusarium* species, are capable of producing different mycotoxins with diverse toxigenic potentials.

2.1 Mycotoxicoses

Mycotoxins produce a variety of diseases, collectively called 'mycotoxicoses', directly or in combination with other primary stressors such as pathogens (Raju and Devegowda, 2000). These diseases are exhibited by symptoms and lesions, which can be used to clinically diagnose the presence of mycotoxins. However, these symptoms are not straightforward as co-occurrence is common and one mycotoxin may alter the effects of another. For example, when AF and OTA are co-contaminants of poultry feed, they interact in a synergistic manner (Huff and Doerr, 1981). During dual exposure of these toxins, OTA prevents the major effects of AF (i.e. fatty, yellow, enlarged and friable liver), thus reducing the ability to diagnose aflatoxicosis in the field.

Acute cases caused by ingestion of high levels of mycotoxins may result in mortality and a marked decline in productivity characterized by obvious clinical signs and post-mortem lesions. However, in most cases, mycotoxicoses is chronic and caused by low-level ingestion of fungal metabolites, resulting in a measurable decline in performance and the occurrence of non-specific changes, including subcutaneous haemorrhage in broilers and immunosuppression (D'mello et al., 1999). Under field conditions, suboptimal performance, in the absence of an obvious infectious, environmental or management factor, or a nutritional deficiency, suggests the possibility of mycotoxicoses.

2.2 Aflatoxins

AF are a class of mycotoxins produced by fungal species *Aspergillus flavus* and *Aspergillus parasiticus*. Most prevalent forms of AF include B_1, B_2, G_1 and G_2, with AF B_1 (AFB_1) being the most common and biologically active component (Busby and Wogan, 1981). AFs cause a multitude of effects in poultry, including decreased weight gain, poor feed efficiency, reduced egg production and egg weight, increased liver fat, changes in organ weights, reduction in serum protein levels, carcass bruising, poor pigmentation, liver damage (pale and enlarged), decreased activities of pancreatic enzymes and immunosuppression (Edds and Bortell, 1983; Leeson et al., 1995; Devegowda and Murthy, 2005). Immunosuppression caused by AF results in many disease outbreaks, vaccination failures and poor antibody titres (Devegowda and Murthy, 2005).

2.3 Trichothecenes

Trichothecenes are fungal metabolites with the same basic structure produced by *Fusarium, Myrothecium, Trichoderma, Verticimonosporium, Cephalosporium, Trichothecium* and *Stachybotrys*. The trichothecenes are classified into two groups: Type A and Type B. Type A trichothecenes include T-2 toxin, HT-2 toxin, neosolaniol, 8-acetoxyneosolaniol, 4-deacetylneosolaniol, monoacetoxyscirpenol (MAS) and diacetoxyscirpenol (DAS). Type B trichothecenes include deoxynivalenol (DON; also known as vomitoxin), nivalenol, 3-acetyldeoxynivalenol, 15-acetyldeoxynivalenol, 4-acetoxynivalenol (Fusarenon-X) and DON-3-Glucoside (Leeson et al., 1995). Trichothecenes are the most potent inhibitors of protein synthesis; the main toxic effect at the cellular level appears to be primary inhibition of protein synthesis followed by a secondary disruption of DNA and RNA synthesis (Leeson et al., 1995).

Toxic effects of trichothecenes include oral lesions, growth retardation, abnormal feathering, decreased egg production and eggshell quality, regression of the bursa of Fabricius, peroxidative changes in liver, abnormal blood coagulation, leucopoenia and proteinaemia and immunosuppression (Leeson et al., 1995; Dänicke, 2002). Concentrations of T-2 that cause oral lesions are lower (0.4 mg/kg) than concentrations reported to decrease chick performance (3–4 mg/kg; Leeson et al., 1995). In a comprehensive review, Dänicke (2002) concluded that broiler performance is affected at dietary concentrations of

3–4 mg/kg of T-2 toxin, whereas ducks were affected when the dietary concentration was as low as 0.4 mg/kg.

Deoxynivalenol is less toxic than T-2 toxin, and the level of DON that affects chick performance is still debated, with some researchers (Huff et al.,1986; Kubena et al., 1988, 1989) reporting toxic effects at 16 mg/kg diet, whereas others (Moran et al., 1982) report no toxic effect until dietary concentrations exceeded 116 mg/kg of DON. Dänicke et al. (2001) summarized results of 49 studies with DON and concluded that a dietary concentration of 5 mg/kg had no negative effects on performance. Deoxynivalenol has also been reported to have both immunosuppressive and immunomodulating effects in poultry (Dänicke, 2002). Recent studies indicate that DON at concentrations ranging from 1 to 7 mg/kg diet significantly alters several key functions of the intestinal tract including decreasing villus surface area available for absorption, altering the permeability of the intestinal tract, and increased susceptibility to clostridial and coccidial pathogens (Awad et al., 2011; Osselaere et al., 2013; Antonissen et al., 2014; Grenier et al., 2015a).

2.4 Fumonisins

The FUM are a group of mycotoxins produced by moulds of the genus *Fusarium* while some species of *Alternaria* have also been found to produce FB_1 (Chen et al., 1992). Six different FUM (A_1, A_2, B_1, B_2, B_3, B_4) have been identified and their structures defined so far (Cawood et al., 1991), and FUM B_1 (FB_1) has been the predominant form (Norred, 1993). While prior work with FUM suggested that chicks and turkeys were relatively resistant to their toxic effects along with low levels of absorption in the bird, recent evidence suggests that FUM can alter the control of inflammatory processes (Grenier and Applegate, 2013), especially in the lower small intestine, thus increasing their susceptibility to some enteric pathogens (Antonissen et al., 2014; Grenier et al., 2015b). Additionally, reductions in body weight gain and liver pathology have also been observed in chicks, ducks and turkeys (Broomhead et al., 2002; Tran et al., 2005).

The mechanism by which FUM cause toxicity in animals appears to be due to the disruption of sphingolipid metabolism (Wang et al., 1991). The FUM are specific inhibitors of ceramide synthase (sphinganine/sphingosine *N*-acyltransferase), a key enzyme required for the synthesis of ceramide and more complex sphingolipids. Inhibition of this enzyme system leads to an increase in tissue concentrations of the sphingolipids, sphingosine (SO) and sphinganine (SA), and a change in the SA:SO ratio. An increase in the SA:SO ratio has been observed in tissues of broilers, turkeys and ducklings fed FB_1 (Broomhead et al., 2002; Tran et al., 2005).

2.5 Ochratoxins

Ochratoxins are a group of structurally related metabolites that are produced by fungi belonging to the genera *Aspergillus* and *Penicillium*, and OTA is the most prevalent mycotoxin of this group. Signs of OTA toxicity in poultry include weakness, anaemia decreased feed consumption, reduced growth rate and egg production, poor feathering and excessive mortality at high dietary concentrations (Gibson et al., 1989). Increases in relative weights of liver, spleen, pancreas, proventriculus, gizzard and testes have also been reported in poultry fed OTA (Gibson et al., 1989).

OTA consists of an isocoumarin moiety linked through the 7-carboxy group to the amino acid L-(-phenylalanine. At the cellular level, OTA interferes with DNA, RNA and protein synthesis by inhibiting the enzyme phenylalanine-tRNA synthetase (Marquardt and Frohlich, 1992). OTA also affects renal carbohydrate metabolism through a reduction of renal mRNA coding for phosphoenolpyruvate carboxykinase (PEPCK), a key enzyme in gluconeogenesis (Leeson et al., 1995). The effects of OTA on DNA, RNA and protein synthesis are thought to be due to the phenylalanine moiety of the toxin competing with phenylalanine in the enzyme catalysed reaction (Marquardt and Frohlich, 1992). OTA also causes hypercarotenaemia in broilers (Huff and Hamilton, 1975) that is more severe than that caused by AF (Osborne et al., 1982; Schaeffer et al., 1987).

2.6 Interactions among mycotoxins

In general, contaminated feeds usually contain more than one mycotoxin and those co-occurring mycotoxins often show a synergistic or additive interaction on animal performance. The combination of multiple mycotoxins in feed can cause more adverse effects than a single mycotoxin due to additive or even synergistic interaction (CAST 2003). Moreover, the combination of mycotoxins, at concentrations that individually should not cause negative effects, may negatively affect animals (Grenier and Oswald, 2011). Another challenge is the global trade of agricultural raw materials used as feed ingredients, which can result in the distribution of mycotoxins across the world (Bryden, 2012). Hence, as part of a proper mycotoxin risk management, surveying the mycotoxin occurrence is very important to allow feed and animal producers to assess the risk of using certain feed ingredients or feeds from different regions.

2.7 Mycotoxin counteracting strategies

It is commonly known that mycotoxins vary in their chemical structures, resulting in vast differences in their chemical, physical and biochemical properties. While the biochemical properties define the toxicity of mycotoxins, chemical and physical properties determine the methods that can be used to detoxify them. Considering the great variety of mycotoxin structures, it is obvious that there is no single method which can be used to deactivate mycotoxins in feed. Therefore, different strategies have to be combined in order to specifically target individual mycotoxins without impacting the quality of feed.

The best-known method for mycotoxin deactivation is 'binding' with the use of binding agents, which are referred to as mycotoxin binders, adsorbents, or enterosorbents. They can be of organic (microbial) or inorganic (mainly clay minerals) in nature. The use of clay-based materials for toxin binding is not new. For centuries, humans and animals have been reported to eat clay minerals for various purposes and, in most cases, it is considered to be beneficial to health (Carretero, 2002). Thus, the inclusion of non-nutritive clay minerals in the diet of animals has been widely adopted for reducing toxin bioavailability and exposure from contaminated feeds.

Due to low feed inclusion requirements and easy management of enterosorbents, the widespread acceptance of these products by the farm animal industry has led to the introduction of a variety of diverse materials and/or complex mixtures for binding. These materials (and/or mixtures) are reported to contain smectite clays, zeolites, kaolinite, mica, silica, charcoal and various biological constituents including chlorophyllins, yeast products, lactic acid bacteria, plant extracts and algae. Some contain smectite or zeolite minerals that have been amended with natural or synthetic surfactants resulting in hydrophobic organoclays or organozeolites (Miller et al., 2014). There is considerable evidence to indicate that smectite clays are the most effective enterosorbents. The adsorption efficacy of binding agents or enterosorbents is limited to only a few mycotoxins, such as AF, ergot alkaloids and some other fungal toxins, while they have been shown to be ineffective for trichothecenes, FUMs, and OTA (Murugesan et al., 2015).

The approach to use microorganisms and their enzymes to detoxify specific mycotoxins not only works for non-adsorbable mycotoxins, but for all other toxins for which respective microbes can be isolated from nature. This approach has been known for a long time, even longer than the binder concept. Within a few years after the discovery of AF, the first report of a bacterium capable of detoxifying AF by catabolization was published (Ciegler, 1966). Since then, many microorganisms have been isolated from different habitats such as the gastrointestinal tract of animals, soil, mycotoxin contaminated materials (e.g. grains) and insects feeding on such materials; however, only a few of these organisms were useful or further investigated for practical applications in animal nutrition.

One of the microorganisms that have been further developed into practical application is *Trichosporon mycotoxinivorans*, a yeast strain capable of detoxifying OTA and ZEN (Molnar et al., 2004). Application of this yeast in poultry diets has been proven to detoxify OTA (Politis et al., 2005). Another organism has been an anaerobic rumen bacterium BBSH 797 (Genus novus of family Coriobacteriaceae, formerly *Eubacterium*), which was isolated and developed as a trichothecene detoxifying feed additive (Fuchs et al., 2002; Schatzmayr et al., 2006b). The BBSH 797 detoxifies trichothecenes by cleavage of the 12, 13 epoxide ring resulting in de-epoxy trichothecenes. Several microorganisms, mainly aerobic bacteria, but also yeasts, with FUM degradation properties were also explored and isolated in order to detoxify FUM; however, for various reasons, none of these microorganisms were useful as a mycotoxin-deactivating feed additive (Schatzmayr et al., 2006a). Therefore, the catabolic pathway of FUM degradation was investigated and the gene coding for the key enzyme of FUM detoxification – a carboxylesterase (FUM*zyme*®, BIOMIN Holding GmbH, Austria) – was identified, cloned and expressed in a yeast strain (Heinl et al., 2010; Hartinger and Moll, 2011). This carboxylesterase detoxifies FUM by cleaving its tricarballylic side chains and resulting in hydrolysed FUM, the non-toxic metabolite (HFB_1; Grenier et al., 2012, 2013).

3 Dioxins

3.1 Sources and risks of dioxin contamination

The generic term 'dioxins' refers to a large group of structurally and chemically related compounds including polychlorinated dibenzo-*p*-dioxins (PCDDs) and polychlorinated dibenzofurans (PCDFs). Dioxin-like polychlorinated biphenyls (PCBs) are also included under this term due to their similar toxic properties (Fig. 2; Fries, 1995). Approximately 419 types of dioxins have been identified, of which about 30 are considered to be highly toxic. Of those, 2, 3, 7, 8-tetrachlorodibenzo-*para*-dioxin (TCDD) is the most toxic congener.

Dioxins are predominantly the by-products of industrial processes (smelting, heat and power generation, waste incineration, etc.), but can also result from natural sources such as volcanic eruptions and forest fires. Due to their long half-life, dioxins are persistent and ubiquitous in the environment with soil being a typical site of accumulation. Since dioxins are lipophilic, they tend to accumulate in the fatty tissues; thus, they are commonly found

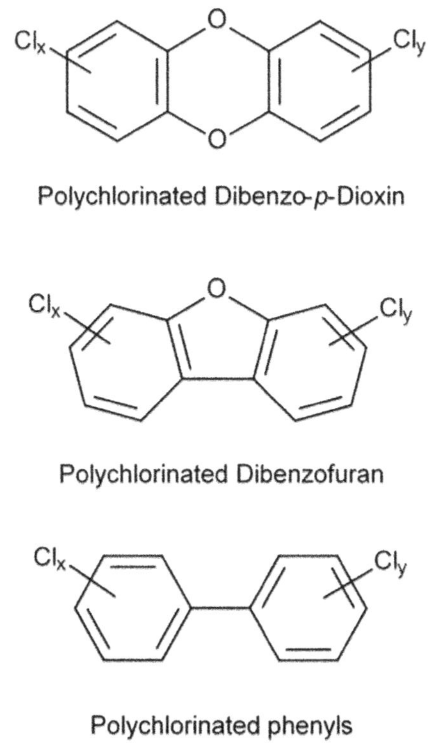

Figure 2 Chemical structures of most prevalent dioxins and dioxin-like compounds.

in fat containing foods such as meat, dairy products, eggs, fish and shellfish (De Vos et al., 2003; WHO, 2014).

In many instances, dioxin presence in the human food supply is a result of contaminated animal feed. In humans, short-term, high-level exposure of dioxins may result in skin lesions and altered liver function, while chronic exposure may lead to impaired immunity, nervous system development and endocrine and reproductive functions. The dioxin TCDD is also classified by the International Agency for Research on Cancer as a known human carcinogen (WHO, 2014). The major source of dioxin exposure in humans is the food supply with food of animal origin being the predominant source.

The first known incidents of dioxin-related toxicity in chickens were reported in 1957, when millions of broilers died due to ingestion of toxic fatty acids (Firestone, 1973). Dioxins were originally known as 'chicken oedema factor' due to their association with chicken oedema disease. It was not until 1966 that researchers identified dioxins as the causative agent (Cantrell et al., 1969).

Like mammals, the liver and adipose tissue are the major accumulation and storage sites of dioxins in poultry, although avian species appear to be more susceptible to their toxic effects. (Bursian et al., 2012; Fulton, 2008). Dioxins cause damage to vascular endothelium leading to vascular leakage and subsequent oedema, hydropericardium and ascites (Fulton, 2008; Pang et al., 1980). Damage to the epithelium of some parenchymal organs and degeneration of heart and skeletal muscle may also be observed. Depending on the level of dioxin contamination, some flocks will show signs of stunting, respiratory distress, weakness, ataxia and high mortality (Fulton, 2008). Dioxin toxicity has also been shown to result in inhibition of lymphoid development and atrophy of the thymus and bursa of Fabricius leading to immunosuppression (Andersoon et al., 1991; Fox and Grasman, 1999).

Several carry-over studies performed in poultry have concluded that once dioxins have accumulated in the body, concentrations tend to persist for an extended period of time in tissues and eggs after withdrawal (Hoogenboom et al., 2004, 2006; Iben et al., 2003). Additionally, it has been shown that PCDD/F and PCB accumulation in poultry is higher than in beef or pork (Huwe et al., 2009).

3.2 Dioxin counteracting strategies

Since food of animal origin is the predominant route of human exposure to dioxins, it is imperative to avoid contamination of animal feed as well as be cognizant of the many other potential sources including litter, soil, plants, worms and insects (Cardo et al., 2014). In terms of feed manufacturing, good manufacturing practices and Hazard Analysis and Critical Control Point principles should be implemented to control dioxin contamination throughout the process. Feed monitoring systems should also be enforced. Any material used in the production of animal feed should be guaranteed for quality and safety to ensure that it is not a source of contamination. Additionally, efforts should be made to reduce contamination of feed once on the farm (European Commission, 2000; Food and Agriculture Organization, 2006). Since dioxins retain high chemical stability and are not destroyed by microbial, photochemical, chemical or thermal degradation, measures directed at identifying and eradicating dioxin sources are essential for contamination reduction.

4 Bacterial contamination

Animal feed has been identified as a notable carrier for numerous bacteria that can cause health issues not only for the animal ingesting the contaminated feed, but also humans who may come in contact with products from those infected animals. Feed materials can become contaminated by various sources including soil, wind, water, insects and faeces (Maciorowski et al., 2007).

4.1 *Clostridium perfringens*

Animal feed is often manufactured with animal by-products such as feathers, bone and blood, which can significantly reduce feed costs; however, these ingredients provide an ideal environment for bacterial growth, most notably *C. perfringens*. Because *C. perfringens* spores are widespread in the environment, in soil, water and faeces, it is commonly found in a variety of other feed ingredients including wheat bran, barley, soybeans, corn and sunflowers, though it has been reported that *C. perfringens* is more prevalent in wheat- and barley-based diets than corn based diets (Annett et al., 2002).

C. perfringens is a Gram positive, anaerobic, spore-forming bacterium capable of producing a number of toxins. *C. perfringens* has been identified as the causative agent of necrotic enteritis in poultry. Necrotic enteritis is one of the world's most common and financially crippling poultry diseases, which, when triggered, can cause mortality rates of up to 50%. *C. perfringens* has also been linked to gangrenous dermatitis and cellulitis in broilers and turkeys, which may result in mortality rates of up to 1–2% per week (Thachil et al., 2012). Moreover, according to the Center for Disease Control and Prevention, *C. perfringens* is one of the most common causes of foodborne illness in the United States, with nearly 1 million cases reported every year (CDC, 2015).

Though it is considered to be a normal inhabitant of the intestines of humans and animals, some variants of *C. perfringens* could be considered as opportunistic pathogens that may produce toxins associated with disease when the intestinal microbiota become disrupted. Isolates producing alpha and netB toxins are most commonly associated with necrotic enteritis (Kulkarni et al., 2007; Keyburn et al., 2008).

Several factors have been implicated playing a role in necrotic enteritis outbreaks (Fig. 3). In general, diets rich in indigestible dietary protein, such as that found in animal proteins like meat and bone meal or fishmeal, can predispose birds to necrotic enteritis by altering the gut environment in such a way that it creates a favourable environment *for C. perfringens* growth. Since indigestible dietary protein cannot be digested and absorbed in the upper part of the intestinal tract, it can lead to higher concentrations of protein in the lower portion of the intestinal tract, which can then act as a substrate for the gut microbiota. The fermentation of protein produces unfavourable by-products such as amines and ammonia, and increases intestinal pH, which encourages the proliferation of pathogenic bacteria such as *C. perfringens*. Coccidiosis is another well-known predisposing factor for outbreaks of necrotic enteritis. As *Eimeria* oocysts invade, replicate and cycle

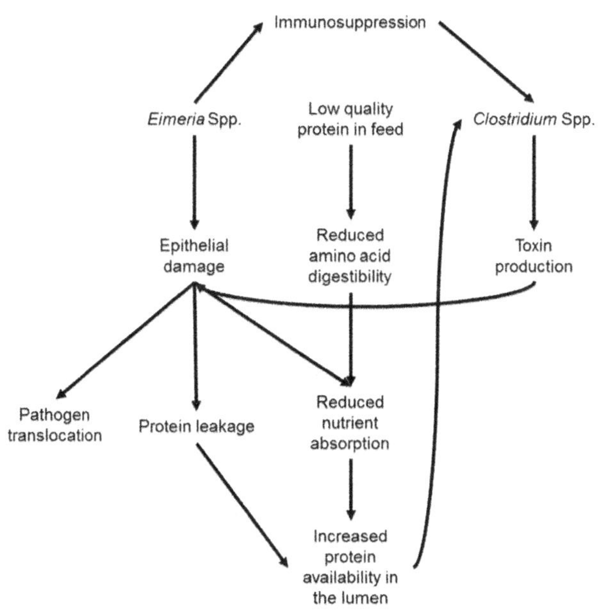

Figure 3 Factors influencing the development of necrotic enteritis.

through the intestinal epithelial cells, significant damage occurs. This damage allows the leakage of plasma proteins into the intestinal lumen, which is a rich-nutrient substrate that can be utilized by *C. perfringens* for proliferation and toxin production (Annett et al., 2002; Wu et al., 2015).

4.2 *Salmonella*

Bacterial contamination of poultry feed is an important consideration when aiming to reduce the spread of *Salmonella* in poultry production and potential food safety hazards. Numerous surveys of feed ingredients have been conducted and show that animal feed is frequently contaminated with *Salmonella* (Crump et al., 2002). As such, several international regulatory agencies, including United States Food and Drug Administration and the European Food Safety Authority have implemented guidelines to reduce feed contamination and abate food safety hazards (Jones and Richardson, 2004).

Salmonellosis is one of the most reported health concerns for humans, causing significant morbidity and mortality. According to the World Health Organization, it is estimated that there are tens of millions of humans cases of salmonellosis worldwide resulting in more than one hundred thousand deaths annually (WHO, 2013).

The genus *Salmonella* is made up of more than 2 500 different serovars. The most common *Salmonella* serotypes found in poultry are *S. Typhimurium*, *S. enteritidis*, *S. Kentucky* and *S. Heidelberg*. Although they are frequently found in poultry populations, these non-host adapted strains (collectively referred to as paratyphoid salmonellae) rarely induce disease in poultry, except in vulnerable young birds, and most birds remain asymptomatic carriers. These serotypes, however, can cause foodborne disease in humans (Foley et al., 2011).

Protein-rich vegetable and animal-derived protein sources, such as meat and fishmeal, have higher risks of becoming contaminated with microbes such as *Salmonella*, as the high levels of protein within these ingredients can act as a substrate (Veldman et al., 1995). Once ingested, these bacteria can multiply within the intestine and subsequently contaminate the environment when excreted in the faeces. *Salmonella* are also capable of translocating across the intestinal epithelia, multiplying in macrophages and being transported via the bloodstream or lymphatic system to infect vital organs such as lungs, liver and spleen as well as reproductive organs such as ovary and oviducts.

4.3 *Escherichia coli*

E. coli is a Gram-negative, non-spore-forming, facultative anaerobe. It is a normal inhabitant of animal and human intestinal microbiota, and thus are commonly utilized as an indicator of faecal contamination of feed. Since *E. coli* is commonly found in animal feed, it can be surmised that contamination with wildlife faeces or fertilization of crops could possibly be a major transmission source (Maciorowski et al., 2007). In a study conducted by da Costa et al. (2007), *E. coli* was present in 50% of commercial broiler feed and 32% of raw feed ingredients.

Colibacillosis, caused by certain *E. coli* strains designated as avian pathogenic *E. coli* (APEC), is considered to be the most common infectious bacterial disease in poultry of all ages. It is known to result in high incidence of morbidity and mortality and is associated with significant economic losses. Colibacillosis can manifest itself in a wide variety of disease syndromes, depending on the *E. coli* strain responsible, such as colisepticaemia, swollen-head syndrome, coliform cellulitis, airsacculitis and salpingitis. Secondary infections of

E. coli are also common after bouts of immunosuppression (Barnes et al., 2008; Maciorowski et al., 2007). Currently, there are an estimated 150 to 200 serotypes of *E. coli* with the most frequently isolated serotypes being O78:K80, O1:K1 and O2:K1 (Kabir, 2010).

4.4 Biogenic amines

It is common practice to include animal by-product meals, including meat and bone meal, blood meal, fishmeal, feather meal and poultry by-product meal, as a major component of poultry feed to meet protein requirements and reduce feed costs. Unfortunately, due to their high protein content, the use of these animal by-product meals provides an opportunity for the production of biogenic amines during putrefaction before rendering and/or when those feed components become spoiled (Tamim and Doerr, 2003).

Biogenic amines (histamine, tyramine, putrescine, cadaverine, gizzerosine, agmatine, spermine and spermidine) are the result of bacterial decomposition or putrefaction of animal proteins. They are low-molecular-weight compounds, which are formed predominantly by decarboxylation of particular amino acids into their corresponding amines through the action of bacterial enzymes during storage (Friday et al., 1999; Barnes et al., 2001). The presence of biogenic amines has been implicated in causing several detrimental effects in poultry including malabsorption syndrome, decreased body weight gain and feed efficiency, the passage of undigested feed, sloughing of intestinal epithelia and lesions in the proventriculus, gizzard and intestines (Bermudez and Firman, 1998; Barnes et al., 2001). Intestinal irritation caused by biogenic amines can also lead to secondary bacterial infections. Moreover, increased carcass contamination may be observed at processing due to intestinal rupture (Barnes et al., 2001).

Toxicity of these compounds vary depending on the particular amine in question. For example, Barnes and colleagues (2001) found that dietary inclusion of 0.1% histidine resulted in reduced body weight and feed conversion in 21-day-old broiler chicks. The toxic effects of spermine have been observed in 14-day-old chicks at levels as low as 0.2% (Sousadias and Smith, 1995). Gizzerosine, on the other hand, is considered to be 10 times as toxic as histamine in terms of its ability to promote hydrochloric acid production in the proventriculus and 300 times more toxic in its ability to cause gizzard erosion (Masumura et al., 1985).

4.5 Solutions to bacterial contamination

Animal feed is considered to be a primary source of bacterial contamination. These pathogens can be very difficult to control and every possible tool to mediate them needs to be utilized in order to establish an effective control programme. In regard to eliminating the presence of pathogenic bacteria in animal feeds, actions need to be taken to prevent contamination of feed and/or feed ingredients, prevent the growth of bacteria, and kill pathogens that are present.

Measures should be taken to prevent bacterial infections by obtaining pathogen-free, high-quality feed ingredients, particularly animal-derived by-product meals. In regard to specifically reducing biogenic amine production, a quality control programme must be implemented to ensure that raw products are stored properly to maintain ideal temperature (below 5°C) and prevent bacterial contamination after rendering as these are the most likely sources of biogenic amines (Barnes et al., 2001).

One of the most common and simplest means of feed decontamination is thermal treatment during pelleting. For example, pelleting has been shown to reduce *Salmonella* isolation rates from 50% to 93% (Jones and Richardson, 2004; Veldman et al., 1995). Despite efforts to reduce bacterial populations using heat treatments, there is always a risk of recontamination. Bactericidal chemical treatments, such as organic acids or formaldehyde, may also be used to kill microbes in feed and reduce recontamination. The antimicrobial effects of short-chain organic acids, such as propionic and formic acids, are derived from the ability of their undissociated forms to penetrate the outer membrane of Gram-negative bacteria and, due to the higher relative pH inside the cell, dissociate within the cell into anion and protons. The increased quantity of protons acidifies the intracellular environment and disrupts many cellular functions while the anion may inhibit DNA synthesis. In an effort to maintain a neutral cytoplasmic pH, the cell will actively export excess protons, thus consuming cellular adenosine triphosphate (ATP) and ultimately resulting in energy depletion and cell death (Ricke, 2003).

If pathogenic bacteria are ingested, there are several feed additives that could be utilized to protect intestinal health. A common means of lessening pathogenic threat in the intestinal tract is dietary supplementation of antibiotics. Antibiotic supplementation at sub-therapeutic levels has been a widely used practice for over 50 years to prevent disease and promote growth; however, concerns about antibiotic resistance has led many regulatory agencies and producers to re-evaluate this practice and limit antibiotic use.

Other feed additives are quickly gaining popularity to control pathogenic bacteria. Prebiotics and probiotics, for example, can be supplemented in the feed to reduce or prevent enteric infections and subsequent contamination of poultry products. Prebiotics are generally defined as non-digestible food/feed ingredients that beneficially affect the host by selectively stimulating the growth and/or activity of one or a limited number of bacteria in the intestine (Gibson and Roberfroid, 1995). In other words, prebiotics provide the environment and/or the nutrients for beneficial bacteria to better establish themselves in the intestine. A probiotic is defined as 'a live microbial feed supplement, which beneficially affects the host animal by improving intestinal balance' (Fuller, 1989). Probiotics are used to help maintain an ideal microbial balance within the intestine and promote gut integrity. This is achieved through three fundamental mechanisms: competitive exclusion, bacterial antagonism and stimulation of the immune system (Ohimain and Ofongo, 2012). It is important to note, however, that probiotic products are very diverse in their makeup; thus, they have varying modes of action depending on the strain(s) being utilized and doses used (Cox and Dalloul, 2015).

Phytogenic feed additives, or plant-derived compounds, may also alleviate microbial challenge and biogenic amine formation in the intestine. Herbs and essential oils have been used in humans and animals for centuries as flavour enhancers, food preservatives and medicines. Furthermore, several plant extracts are known to have antimicrobial, antiviral, anti-coccidial, fungicidal and/or antioxidant properties, making them ideal as feed additives (Applegate et al., 2010). For example, carvacrol and thymol, the major biologically active constituents found in oregano and thyme, have been shown to have antioxidant and antimicrobial properties in poultry (Hoffman-Pennesi and Wu, 2010). Moreover, they have been shown to control pathogenic bacteria, promote beneficial bacteria and augment nutrient utilization in the intestine (Mitsch et al., 2004; Murugesan et al., 2015). Overall, phytogenic feed additives, depending on their composition, may help improve performance by sustaining an ideal intestinal environment.

Finally, efforts to keep feed clean should include maintaining good feed practices and storage facilities and ensuring proper biosecurity to prevent access of natural carriers, such as rodents and wild birds, out of feed facilities.

5 Summary and future trends

Safety of animal feed plays an important role in not only maintaining the health of production animals, but also the health of humans consuming products from those animals. Ultimately, ensuring feed safety requires a multi-factorial approach as it is reliant on a multitude of factors and can be affected by a number of contaminants including plant and fungal toxins, chemicals and pathogens.

Mycotoxins, secondary fungal metabolites responsible for mycotoxicoses livestock and poultry, must be addressed before, during and after harvest to minimize risks. The most commonly used method of counteracting mycotoxins is the use of binding agents; however, due to the heterogeneity of mycotoxins in terms of their chemical, physical and biochemical properties, not all mycotoxins can be effectively bound and other strategies must be utilized. A more novel approach is the use of microorganisms and their enzymes to detoxify specific non-absorbable mycotoxins, such as DON and FUM.

Dioxins and PCBs are harmful environmental contaminants that can enter feedstuffs through air, soil or sediments. Not only do dioxins cause significant consequences in terms of animal health, but they are also commonly found in the human food supply as a result of contaminated animal feed. To avoid dioxin contamination, it is imperative to be cognizant of their many potential sources. Good manufacturing practices and Hazard Analysis and Critical Control Point principles should be implemented to control dioxin contamination during the feed manufacturing process and feed monitoring systems should also be enforced.

Animal feed is a significant source of numerous bacteria that can cause health issues for animals and humans alike. Bacterial contamination of feedstuffs can be mitigated through a thermal treatment during pelleting or application of bactericidal chemical treatments such as organic acids or formaldehyde. In instances where pathogenic bacteria are ingested, they can be controlled through the use of antibiotics or natural feed additives such as prebiotics, probiotics and phytogenic feed additives.

Safety of poultry feed, as a whole, will continue to improve as more sophisticated measures of identification and control are developed. As poultry production continues to evolve, more natural alternatives for eliminating mycotoxins and pathogens will begin to take position in the forefront. Furthermore, as biotechnology advances, so will our surveillance and control methods. Genomics, transcriptomics and proteomics are quickly transforming our approaches to the detection, prevention and treatment of biological feed contaminants. These advances are also allowing the development of control measures and treatments that are more specific in terms of their targets. Feedstuffs play an important role in maintaining the health of production animals as well as humans; therefore, it is essential that producers develop procedures and practices that ensure a supply of safe feed for animal production.

6 Where to look for further information

For an overview of mycotoxins, feed sampling, and mycotoxin analysis, please refer to Romer Lab's *Guide to Mycotoxins* (edited by E. M. Binder and R. Krska; http://www.foodriskmanagement.nl/wp-content/uploads/2013/03/Romer-Labs-Guide-to-Mycotoxin-Book_Original_41686.pdf). An in-depth read regarding masked mycotoxins can be found in *Masked Mycotoxins in Food: Formation, Occurrence and Toxicological Relevance* (edited by Chiara Dall'Asta and Franz Berthiller). The *Mycotoxin Factbook: Food & Feed Topics* (edited by D. Barug) is a great resource if you are interested in information on major mycotoxins and emerging problems in the food chain, the impact of mycotoxins in the feed chain, developments in mycotoxin prevention, trends in mycotoxin analysis, and regulatory issues related to mycotoxins. The journal article 'Prevalence and effects of mycotoxins on poultry health and performance, and recent development in mycotoxin counteracting strategies' (G. R. Murugesan et al., 2015; doi: 10.3382/ps/pev075), is also a good source of valuable information on the effects of mycotoxins in poultry and various counteracting strategies. The Food and Agriculture Organization of the United Nations provides a great amount of information covering basic facts about dioxins as well as prevention and control measures (http://www.fao.org/ag/againfo/home/en/news_archive/2009_IN_dioxin.html). A good source of information pertaining to *Salmonella* control can be found in this educational document produced by the American Feed Industry Association (http://ucfoodsafety.ucdavis.edu/files/172958.pdf) as well as in this article, F. T. Jones. 2011. A review of practical *Salmonella* control measures in animal feed. *J. Appl. Poult. Res.* 20 (1): 102–13. doi: 10.3382/japr.2010-00281. For a comprehensive review of animal feed contaminants, their negative effects on both animal and human health, as well as analysis and control measures please refer to *Animal Feed Contamination: Effects on Livestock and Food Safety* (edited by J. Fink-Gremmels).

7 References

Andersson, L., E. Nikolaidis, B. Brunstrom, A. Bergman, and L. Dencker. 1991. Effects of polychlorinated biphenyls with Ah receptor affinity on lymphoid development in the thymus and the bursa of Fabricius of chick embryos in ovo and in mouse thymus anlagen in vitro. *Toxicol. Appl. Pharmacol.* 107: 183–8.

Annett, C. B., J. R. Vistea, M. Chirino-Trejob, H. L. Classenc, D. M. Middletona, and E. Simkoa. 2002. Necrotic enteritis: Effect of barley, wheat and corn diets on proliferation of *Clostridium perfringens* type A. *Avian Pathol.* 31 (6): 598–601.

Antonissen, G., F. V. Immerseel, F. Pasmans, R. Ducatelle, F. Haesebrouck, L. Timbermont, M. Verlinden, G. P. J. Janssens, V. Eeckhaut, M. Eeckhout, S. De Saeger, S. Hessenberger, A. Martel, and S. Croubels. 2014. The Mycotoxin Deoxynivalenol predisposes for the development of *Clostridium perfringens*-induced Necrotic Enteritis in Broiler Chickens. *PLoS ONE* 9 (9): e108775.

Applegate, T. J., V. Klose, T. Steiner, A. Ganner, and G. Schatzmayr. 2010. Probiotics and phytogenics for poultry: Myth or reality? *J. Appl. Poult. Res.* 19 (2): 194–210.

Awad, W. A., M. Hess, M. Twaruzek, J. Grajewski, R. Kosicki, J. Bohm, and J. Zentek. 2011. The impact of the *Fusarium* Mycotoxin deoxynivalenol on the health and performance of broiler chicks. *Intl. J. Mol. Sci.* 12: 7996–8012.

Barnes, D. M., Y. K. Kirby, and K. G. Oliver. 2001. Effects of biogenic amines on growth and the incidence of proventricular lesions in Broiler Chickens. *Poult. Sci.* 80: 906–11.

Barnes, J. H., L. K. Nolan, and J. Vaillancourt. 2008. Colibacillosis. In Y. M. Saif (ed.), *Diseases of Poultry*, 691–732. Ames, IA: Iowa State Press.

Bermudez, A. J., and J. D. Firman. 1998. Effects of biogenic amines in broiler chickens. *Avian Dis.* 42: 199–203.

Broomhead, J. N., D. R. Ledoux, A. J. Bermudez, and G. E. Rottinghaus. 2002. Chronic effects of fumonisin B_1 in broilers and turkeys fed dietary treatments to market age. *Poult. Sci.* 81: 56–61.

Bryden, W. L. 2012. Mycotoxin contamination of the feed supply chain: Implications for animal productivity and feed security. *Anim. Feed Sci. Technol.* 173 (1–2): 134–58.

Bursian, S., J. Newsted, and M. Zwiernik. 2012. Polychlorinated biphenyls, polybrominated biphenys, polychlorinated debenzo-p-dioxins, and polychlorinated dibenzofurans. In R. Gupta (ed.), *Veterinary Toxicology: Basic and Clinical Principles*, 776–9. Waltham, MA: Academic Press.

Busby, W. F., Jr., and G. N. Wogan. 1981. Aflatoxins. In R. C. Shank (ed.), *Mycotoxins and N-Nitrosocompounds, Environmental Risks*, Vol. 2, 3–27. Boca Raton, FL: CRC Press Inc.

Cantrell, J. S., N. C. Webb, and A. J. Mabis. 1969. The identification and crystal structure of a hydropericardium-producing factor: 1,2,3,7,8,9-hexachlorodibenzo-p-dioxin. *Acta Cryst.* B25: 150–6.

Cardo, M. , M. Castel-Branco, V. Andreozzi, and F. Bernardo. 2014. Dioxins in the food chain: contamination fingerprint analysis in breeding hens, hatching eggs and broilers. *J. Environ. Prot.* 5: 1323–30.

Carretero, M. 2002. Clay minerals and their beneficial effects upon human health. *Appl. Clay Sci.* 21: 155–63.

CAST. 2003. Mycotoxins: Risks in plant, animal, and human systems, Task Force Report No. 139. Council for Agricultural Science and Technology, Ames, IA, 13–85.

Cawood, M. E., W. C. A. Gelderblom, R. Vleggaar, Y. Behrend, P. G. Thiel, and W. F. O. Marasas. 1991. Isolation of the fumonisin mycotoxins: A quantitative approach. *J. Agric. Food Chem.* 39: 1958–62.

Centers for Disease Control and Prevention. 2015. *Clostridium perfringens.* http://www.cdc.gov/foodsafety/diseases/clostridium-perfringens.html.

Chen, J., C. J. Mirocha, W. Xie, L. Hogge, and D. Olson. 1992. Production of the mycotoxin fumonisin B_1 by *Alternaria alternata f. sp. Lycopersici. Appl. Environ. Microbiol.* 8: 3928–31.

Ciegler, A., Lillehoj, B., Peterson, E., and H. H. Hall. 1966. Microbial detoxification of Aflatoxin. *Appl. Microbiol.* 14: 934–9.

Cox, C. M., and R. A. Dalloul. 2015. Immunomodulatory role of probiotics in poultry and potential in ovo application. *Benef. Microbes.* 6 (1): 45–52.

Crump, J. A., P. M. Griffen, and F. J. Angulo. 2002. Bacterial contamination of animal feed and its relationship to human foodborne illness. *Clin. Infect. Dis.* 35 (7): 859–65.

da Costa, P. M., M. Olivera, A. Bica, P. Vaz-Pires, and F. Bernardo. 2007. Antimicrobial resistance in *Enterococcus* spp. and *Escherichia coli* isolated from poultry feed and feed ingredients. *Vet. Microbiol.* 120 (1–2): 122–31.

Dänicke, S., M. Gareis, and J. Bauer, 2001. Orientation values for critical concentrations of deoxynivalenol and zearalenone in diets for pigs, ruminants and gallinaceous poultry. *Proc. Soc. Nutr. Physiol.* 10: 171–4.

Dänicke, S. 2002. Prevention and control of mycotoxins in the poultry production chain: A European view. *World. Poult. Sci. J.* 58: 451–74.

Devegowda, G., and T. N. K. Murthy. 2005. Mycotoxins: Their effects in poultry and some practical solutions. In D. E. Diaz (ed.), *The Mycotoxin Blue Book*, 25–56. Nottingham: Nottingham University Press.

De Vos, S., J. Maervoetb, P. Schepensb, and R. De Schrijver. 2003. Polychlorinated biphenyls in broiler diets: Their digestibility and incorporation in body tissues. *Chemosphere* 15 (1): 7–11.

D'mello, J., C. Placinta, and A. Macdonald. 1999. Fusarium mycotoxins: A review of global implications for animal health, welfare and productivity. *Anim. Feed Sci. Technol.* 80: 183–205.

Edds, G. T., and R. R. Bortell. 1983. Biological effects of aflatoxin in poultry. In U. L. Diener, R. L. Asquith and J. W. Dickens (eds.), Aflatoxin and *Aspergillus flavus* in corn, 55–61. Southern Cooperative Services Bulletin. 279. Auburn, AL: Alabama Agricultural Experiment Station, Auburn University.

European Commission. 2000. Opinion of the Scientific Committee on Animal Nutrition on the dioxin contamination of feedingstuffs and their contribution to the contamination of food of animal origin. http://ec.europa.eu/food/committees/scientific/out55_en.pdf.

Firestone, D. 1973. Etiology of chick edema disease. *Environ. Health Perspect.* 5: 59–66.

Foley, S. L., R. Nayak, I. B. Hanning, T. J. Johnson, J. Han, and S. C. Ricke. 2011. Population dynamics of *Salmonella enterica* serotypes in commercial egg and poultry production. *Appl. Environ. Microbiol.* 77 (13): 4273–9.

Food and Agriculture Organization. 2006. Code of practice for the prevention and reduction of dioxin and dioxin-like PCB contamination in foods and feeds.

Fox, L. L., and K. A. Grasman. 1999. Effects of PCB 126 on primary immune organ development in chicken embryos. *J. Toxicol. Environ. Health A* 58: 233–44.

Friday, M. L., J. D. Firman, and A. J. Bermudez. 1999. Effects of biogenic amines on broiler performance. *J. Appl. Poul. Res.* 8: 408–13.

Fries, G. F. 1995. A review of the significance of animal food products as potential pathways of human exposures to dioxins. *J. Anim. Sci.* 73: 1639–50.

Fuchs, E., E. M. Binder, D. Heidler, and R. Krska. 2002. Structural characterization of metabolites after the microbial degradation of type A trichothecenes by the bacterial strain BBSH 797. *Food Addit. Contam.* 19: 379–86.

Fuller, R. 1989. Probiotics in man and animals. *J. Appl. Bacteriol.* 66: 365–78.

Fulton, M. 2008. Other toxins and poisons. Y. M. Saif (ed.), *Diseases of Poultry*, 1231–58. Ames, IA: Iowa State Press.

Gibson, G. R., and M. B. Roberfroid. 1995. Dietary modulation of the human colonic microbiota: Introducing the concept of prebiotics. *J. Nutr.* 125: 1401–12.

Gibson, R. M., C. A. Bailey, L. F. Kubena, W. E. Huff, and R. B. Harvey. 1989. Ochratoxin A and dietary protein. 1. Effects on body weight, feed conversion, relative organ weight, and mortality in three-week-old broilers. *Poult. Sci.* 68: 1658–63.

Grenier, B., and I. P. Oswald. 2011. Mycotoxin co-contamination of food and feed: Meta-analysis of publications describing toxicological interactions. *World Mycotoxin J.* 4 (3): 285–313.

Grenier, B., A. P. F. L. Bracarense, H. E. Schwartz, C. Trumel, A. M. Cossalter, G. Schatzmayr, M. Kolf-Clauw, W. D. Moll, and I. P. Oswald. 2012. The low intestinal and hepatic toxicity of hydrolyzed fumonisin B 1 correlates with its inability to alter the metabolism of sphingolipids. *Biochem. Pharmacol.* 83: 1465–73.

Grenier, B., and T. J. Applegate. 2013. Invited review -modulation of intestinal functions upon mycotoxin ingestion: Meta-analysis of published experiments in animals. *Toxins.* 5: 396–430.

Grenier, B., A. P. F. Bracarense, H. E. Schwartz, J. Lucioli, A.-M. Cossalter, W.-D. Moll, G. Schatzmayr, and I. P. Oswald. 2013. Biotransformation approaches to alleviate the effects induced by Fusarium Mycotoxins in Swine. *J. Agric. Food Chem.* 61: 6711–19.

Grenier, B., I. Dohnal, R. Shanmugasundaram, S. D. Eicher, R. K. Selvaraj, G. Schatzmayr, and T. J. Applegate. 2015a. Susceptibility of broiler chickens to coccidiosis when fed subclinical doses of deoxynivalenol and fumonisins – special emphasis on the immunological response and the mycotoxin interaction. *Poult. Sci.* 94 (E-Suppl. 1): 83.

Grenier, B., H. E. Schwartz-Zimmerman, S. Caha, W. D. Moll, G. Schatzmayr, and T. J. Applegate. 2015b. Dose-dependent effects on sphingoid bases and cytokines in chickens fed diets prepared with Fusarium verticilloides culture material containing fumonisins. *Toxins* 7: 1253–72.

Hartinger, D., and W. D. Moll. 2011. Fumonisin elimination and prospects for detoxification by enzymatic transformation. *World Mycotoxin J.* 4: 271–83.

Heinl, S., D. Hartinger, M. Thamhesl, E. Vekiru, R. Krska, G. Schatzmayr, W. D. Moll, and R. Grabherr. 2010. Degradation of fumonisin B1 by the consecutive action of two bacterial enzymes. *J. Biotechnol.* 145: 120–9.

Hoffman-Pennesi, D. and C. Wu. 2010. The effect of thymol and thyme oil feed supplementation on growth performance, serum antioxidant levels, and cecal *Salmonella* population in broilers. *J. Appl. Poult. Res.* 19: 432–43.

Hoogenboom, L. A., C. A. Kan, T. F. Bovee, G. van der Weg, C. Onstenk and W. A. Traag. 2004. Residues of dioxins and PCBs in fat of growing pigs and broilers fed contaminated feed. *Chemosphere* 57 (1): 35–42.

Hoogenboom, L. A., C. A. Kan, M. J. Zeilmaker, J. Van Eijkeren, and W. A. Traag. 2006. Carry-over of dioxins and PCBs from feed and soil to eggs at low contamination levels – influence of mycotoxin binders on the carry-over from feed to eggs. *Food. Addit. Contam.* 23 (5): 518–27.
Huff, W. E., and P. B. Hamilton. 1975. Decreased plasma carotenoids during ochratoxicosis. *Poult. Sci.* 54: 1308–10.
Huff, W. E., and J. Doerr. 1981. Synergism between aflatoxin and ochratoxin A in broiler chickens. *Poult. Sci.* 60: 550–5.
Huff, W. E., L. F. Kubena, R. B. Harvey, W. M. Hagler, Jr., S. P. Swanson, T. D. Phillips, and C. R. Creger, 1986. Individual and combined effects of aflatoxin and deoxynivalenol (DON, vomitoxin) in broiler chickens. *Poult. Sci.* 65: 1291–8.
Huwe, J., D. Pagan-Rodriguez, N. Abdelmajid, N. Clinch, D. Gordon, J. Holterman, E. Zaki, M. Lorentzsen, and K. Dearfield. 2009. Survey of polychlorinated dibenzo-p-dioxins, polychlorinated dibenzofurans, and non-ortho-polychlorinated biphenyls in U.S. meat and poultry, 2007–2008: Effect of new toxic equivalency factors on toxic equivalency levels, patterns, and temporal trends. *J. Agric. Food Chem.* 57: 11194–200.
Iben, C., J. Böhm, H. Tausch, J. Leibetseder, and W. Luf. 2003. Dioxin residues in the edible tissue of broiler chicken. *J. Anim. Physiol. Anim. Nutr.* 87: 142–8.
Jones, F. T., and K. E. Richardson. 2004. *Salmonella* in commercially manufactured feeds. *Poult. Sci.* 83 (3): 384–91.
Kabir, S. M. 2010. Avian Colibacillosis and Salmonellosis: A closer look at epidemiology, pathogenesis, diagnosis, control and public health concerns. *Int. J. Environ. Res. Public Health* 7 (1): 89–114.
Keyburn, A. L., J. D. Boyce, P. Vaz, T. L Bannam, M. E. Ford, and D. Parker. 2008. NetB, a new toxin that is associated with avian necrotic enteritis caused by *Clostridium perfringens*. *PLoS Pathog.* 4: e26.
Kubena, L. F., W. E. Huff, R. B. Harvey, D. E. Corrier, T. D. Phillips, and C. R. Creger, 1988. Influence of ochratoxin a and deoxynivalenol (DON) on growing broiler chicks. *Poult. Sci.* 67: 253–60.
Kubena, L. F., W. E. Huff, R. B. Harvey, T. D. Phillips, and G. E. Rottinghaus, 1989. Individual and combined toxicity of deoxynivalenol and T-2 toxin in broiler chicks. *Poult. Sci.* 68: 622–6.
Kulkarni, R. R., V. R. Parreira, S. Sharif, and J. F. Prescott. 2007. Immunization of broiler chickens against *Clostridium perfringens*-induced necrotic enteritis. *Clin. Vaccine Immunol.* 14: 1070–7.
Leeson, S., G. J. Diaz, and J. D. Summers. 1995. *Poultry Metabolic Disorders and Mycotoxins*. Guelph, ON, Canada: University Books.
Maciorowski, K. G., P. Herrera, F. T. Jones, S. D. Pillai, and S. C Ricke. 2007. Effects on poultry and livestock of feed contamination with bacteria and fungi. *Anim. Feed Sci. Technol.* 133: 109–36
Magan, N., A. Medina, and D. Aldred. 2011. Possible climate-change effects on mycotoxin contamination of food crops pre- and postharvest. *Plant Pathol.* 60: 150–63.
Marquardt, R. R., and A. A. Frohlich. 1992. A review of recent advances in understanding ochratoxicosis. *J. Anim. Sci.* 70: 3968–88.
Masumura, T., M. Sugahara, T. Noguchi, K. Mori, and H. Naito. 1985. The effect gizzerosine, a recently discovered compound in overheated fish meal, on the gastric acid secretion in chicken. *Poult. Sci.* 64: 356–36.
Miller, J. D., A. W. Schaafsma, D. Bhatnagar, G. Bondy, I. Carbone, L. J. Harris, G. Harrison, G. P. Munkvold, I. P. Oswald, J. J. Pestka, L. Sharpe, M. W. Sumarah, S. A. Tittlemier, and T. Zhou. 2014. Mycotoxins that affect the North American agri-food sector: State of the art and directions for the future. *World Mycotoxin J.* 7: 63–82.
Mitsch, P., K. Zitterl-Eglseer, B. Kohler, C. Gabler, R. Losa, and I. Zimpernik. 2004. The effect of two different blends of essential oil components on the proliferation of *Clostridium perfringens* in the intestines of broiler chickens. *Poult. Sci.* 83: 669–75.
Molnar, O., G. Schatzmayr, E. Fuchs, and H. Prillinger. 2004. Trichosporon mycotoxinivorans sp. nov., a new yeast species useful in biological Detoxification of various Mycotoxins. *Syst. Appl. Microbiol.* 27: 661–71.

Moran, E. T., Jr., B. Hunter, P. Ferket, L. G. Young, and L. G. McGirr, 1982. High tolerance of broilers to vomitoxin from corn infected with *Fusarium graminearum*. *Poult. Sci.* 61: 1828–31.

Murugesan, G. R., B. Syed, S. Haldar, and C. M. Pender. 2015. Phytogenic feed additives as an alternative to antibiotic growth promotors in broiler chickens. *Front. Vet. Sci.* 2: 21.

Norred, W. P. 1993. Fumonisins – mycotoxins produced by *Fusarium moniliforme*. *J. Toxicol. Environ. Health* 38: 309–28.

Ohimain, E. I., and R. T. S. Ofongo. 2012. The effect of probiotic and prebiotic feed supplementation on chicken health and gut microflora: A review. *Int. J. Anim. Veter. Adv.* 4: 135–43.

Osborne, D. J., W. E. Huff, P. B. Hamilton, and H. R. Burmeister. 1982. Comparison of ochratoxin, aflatoxin, and T-2 toxin for their effects on elected parameters related to digestion and evidence for specific metabolism of carotenoids in chickens. *Poult. Sci.* 61: 1646–52.

Osselaere, A., R. Santos, V. Hautekiet, P. De Backer, K. Chiers, R. Ducatelle, and S. Croubels. 2013. Deoxynivalenol impairs hepatic and intestinal gene expression of selected oxidative stress, tight junction and inflammation proteins in broiler chickens, but addition of an adsorbing agent shifts the effects to the distal parts of the small intestine. *PLoS ONE*. 8 (7): e69014. doi:10.1371/journal.pone.0069014.

Pang, C. Y., G. D. Phillips, and L. D. Campbell. 1980. Chick edema disease syndrome in young turkey poults. *Can. Vet. J.* 21 (1): 12–16.

Politis, I., K. Fegeros, S. Nitsch, G. Schatzmayr, and D. Kantas. 2005. Use of Trichosporon mycotoxinivoransto suppress the effects of ochratoxicosis on the immune system of broiler chicks. *Brit. Poult. Sci.* 46: 58–65.

Raju, M., and G. Devegowda. 2000. Influence of esterified-glucomannan on performance and organ morphology, serum biochemistry and haematology in broilers exposed to individual and combined mycotoxicosis (aflatoxin, ochratoxin and T-2 toxin). *Brit. Poult. Sci.* 41: 640–50.

Ricke, S. C. 2003. Perspectives on the use of organic acids and short chain fatty acids as antimicrobials. *Poult. Sci.* 82: 632–9.

Schaeffer, J. L., J. J. Tyczkowski, and P. B. Hamilton. 1987. Alterations in carotenoid metabolism during ochratoxicosis in young broiler chickens. *Poult. Sci.* 66: 318–24.

Schatzmayr, G., M. Täubel, E. Vekiru, D. Moll, D. Schatzmayr, E. M. Binder, R. Krska, and A. P. Loibner. 2006a. Detoxification of mycotoxins by biotransformation. In D. Barug, D. Bhatnagar, H. P. van Egmond, J. W. van der Kamp, W. A. van Osenbruggen and A. Visconti (eds.), *The Mycotoxin Factbook, Food and Feed Topics*, 363–75. Wageningen, The Netherlands: Wageningen Academic Publishers.

Schatzmayr, G., F. Zehner, M. Täubel, D. Schatzmayr, A. Klimitsch, A. P. Loibner, and E. M. Binder. 2006b. Microbiologicals for deactivating mycotoxins. *Mol. Nutr. Food Res.* 50: 543–51.

Sousadias, M. G., and T. K. Smith. 1995. Toxicity and growth-promoting potential of spermine when fed to chicks. *J. Anim. Sci.* 73: 2375–81.

Tamim, N. M., and J. A. Doerr. 2003. Effect of putrefaction of poultry carcasses prior to rendering on biogenic amine production. *J. Appl. Poult. Res.* 12: 456–60.

Thachil, A. J., B. McComb, M. M. Early, C. Heeder, and K. V. Nagaraja. 2012. *Clostridium perfringens* and Clostridium septicum toxoid to control cellulitis in turkeys. *J. Appl. Poult. Res.* 21 (2): 358–66.

Tran, S. T., A. Auvergne, G. Benard, J. D. Bailly, D. Tardieu, R. Babile, and P. Guerre, 2005. Chronic effects of fumonisin B_1 on ducks. *Poult. Sci.* 84: 22–8.

Veldman, A., H. A. Vahl, G. J. Borggreve, and D. C. Fuller. 1995. A survey of the incidence of *Salmonella* species and *Enterobacteriaceae* in poultry feeds and feed components. *Vet Rec.* 136 (7): 169–72.

Wang, E., W. P. Norred, C. W. Bacon, R. T. Riley, and A. H. Merrill, Jr. 1991. Inhibition of sphingolipid biosynthesis by fumonisins: Implications for diseases associated with *Fusarium moniliforme*. *J. Biol. Chem.* 266: 14486–90.

World Health Organization. 2013. *Salmonella* (non-typhoidal). http://www.who.int/mediacentre/factsheets/fs139/en/.

World Health Organization. 2014. Dioxins and their effects on human health. http://www.who.int/mediacentre/factsheets/fs225/en/.

Wu, F., D. Bhatnagar, T. Bui-Klimke, I. Carbone, R. Hellmich, G. Munkvold, P. Paul, G. Payne, and E. Takle. 2011. Climate change impacts on mycotoxin risks in US maize. *World Mycotoxin J.* 4: 79–93.

Wu, S. B., D. Stanley, N. Rodgers, R. A. Swick, and R. J. Moore. 2015. Two necrotic enteritis predisposing factors, dietary fishmeal and *Eimeria* infection, induce large changes in the caecal microbiota of broiler chickens. *Vet. Microbiol.* 169 (3–4): 188–97.

Chapter 15

Thermal adaptation and tolerance of poultry

Shlomo Yahav, Institute of Animal Science, ARO, Israel

1 Introduction

Birds are endothermic species; they control and maintain body temperature (T_b) within a relatively narrow range, despite moderate to extreme changes in environmental conditions, and controlling T_b is a crucial physiological function in the dynamic steady state of endothermic birds. However, domestic fowls, especially the highly productive agricultural fowls (broilers, turkeys and laying hens), differ to some extent from other endothermic birds in their ability to maintain a sufficiently dynamic steady state despite severe changes in the environment (Yahav, 2014b).

Recent decades have seen significant developments in the genetic selection of the meat-type fowl, that is, broilers (Havenstein et al., 1994, 2003a; Zuidhof et al., 2014) and turkeys (Havenstein et al., 2007), which has led to rapid growth, accompanied by improved feed efficiency and increased metabolic rate (Janke et al., 2004), and which provides the poultry industry with heavy domestic fowls that have relatively short growth periods. The genetic improvements of Cobb, for example, over the last 30 years included a 10% increase in carcass yield, an 11% increase in relative breast meat weight

http://dx.doi.org/10.19103/AS.2016.0011.32

and a 0.60 reduction in feed conversion ratio, leading to 25% less feed per unit gain. In addition, there has been continuous improvement in broiler well-being, manifested in reductions in leg and skeletal health problems and metabolic disease. In turkeys, Havenstein et al. (2007) showed that the 2003 turkeys were approximately twice as heavy as the 1966 ones, with an improvement of approximately 20% in feed conversion rate (FCR).

Logically, such developments in the genetic selection for growth necessitate parallel increases in the size of the cardiovascular and respiratory systems, as well as enhancements in their functional efficiency. However, the inferior development of such major systems (Havenstein et al., 2003b) has led to relatively low capability to balance energy expenditure and maintain body water balance under extreme environmental conditions. Ignoring the necessity to select for improved physiological traits resulted in cardiopulmonary round heart disease in turkeys (Reed et al., 2007) and pulmonary hypertension disease, sudden death syndrome and endocrine (thyroid axis) inefficacy in broilers (Yahav, 2014a). Thus, the acute exposure of chickens to extreme conditions (hot spells or cold periods) has resulted in major economic losses.

Coincidentally, the global environment has changed and the average global surface temperature in 2050 has been predicted to increase by 0.6–2.5 °C (U.S. National Climatic Center, 2001). Global warming and climate change have contributed to desertification that has enlarged the extent of arid to semiarid land areas, year by year. Such lands have been identified in 110 countries and comprise 41.3% of the global terrestrial area. Approximately, 34.7% of the global population, which has been predicted to be nine billion by 2050, presently resides in these areas, which will have increasingly limited food production capacity. This reality has led the UN Secretary General to pronounce that 'the world farm production must be raised by 50% by 2030 to meet human demands for food'. Translating this announcement to numbers, based on the production levels of 2008 when this announcement was released, means that poultry meat production has to reach 137.6 million tons and egg production 99.2 million tons in 2030 (http://www.fao.org/docrep/015/am081m/PDF/am081m00b.pdf).

Thus, the poultry industry faces the dual challenge of improving production performance and quality, while also improving poultry thermotolerance. These challenges can be faced by means of the following: existing classical genetic selection, although only limited potential has been identified; elucidating the molecular pathways of genes involved in thermal adaptations, so that molecular approaches to genetic selection may be employed; using epigenetic approaches to improve performance and thermotolerance of domestic fowl; and using optimal environmental conditions (temperature, ventilation and relative humidity (RH)) while raising domestic fowl, taking into consideration their sensitivity to the environment.

One has to face various difficulties listed below in order to meet the challenge of increasing production:

- Global warming and its effect on production (thermotolerance);
- Physiological syndromes (ascites, skeletal disorders, sudden death syndrome, etc.) that might intensify as genetic selection continues;
- The poultry industry's effects on global warming – through emission of greenhouse gases;
- Welfare issues.

The aim of this chapter is to elucidate the crucial role played by the thermal status of domestic fowls in determining their performance capacity.

2 Body temperature control by endothermic birds

Body temperature is the main characteristic that reflects the thermal status of endothermic animals (birds and mammals). Endothermic birds control T_b, which is the most physiologically protected parameter of the body; therefore, the thermoregulatory system in these animals operates at a very high gain, in order to maintain T_b within a relatively narrow range, despite moderate to extreme changes in environmental conditions. The ability to maintain a stable T_b springs from the mechanisms that control heat production and heat loss, mechanisms that changed during the course of evolution, to enable endothermia to replace ectothermia. The evolutionary changes from ectothermia to endothermia were achieved because the developmental regulatory mechanisms maintained a balance between heat production and heat loss (Eq. 32.1). Both processes, especially heat production, are probably older than endothermy, but both are permanently activated and regulated by both neuronal and hormonal signals (Silva, 2006; Morrison et al., 2008; Richards and Proszkowiec-Weglarz, 2007).

2.1 The heat transfer model

Heat transfer modelling has been used to understand thermoregulation in endotherms. These models take into consideration a constant T_b for an animal not performing external work, based on the first law of thermodynamics:

$$S = M - E \pm R \pm C \pm K \tag{1}$$

where *S* is the bodily heat gain or loss that must be balanced by: *M*, metabolic heat production; *E*, evaporative heat loss; *R*, radiative heat gain or loss; *C*, convective heat loss or gain; and *K*, conductive heat loss or gain. Body temperature will remain unchanged when *S* is zero, that is, when heat gain matches heat loss. If more heat is produced and gained than lost, then *S* is positive and T_b will rise, and vice versa.

2.2 Metabolic heat production

Metabolic heat production (thermogenesis) is regulated by the thyroid hormones (McNabb and King, 1993; Silva, 2006) and can be divided into obligatory and facultative thermogenesis (Silva, 2006). Obligatory thermogenesis refers to the energy required to maintain T_b while the ambient temperature (T_a) lies in the thermoneutral zone – the range in which the body is in thermal equilibrium with the environment and produces energy at the resting metabolic rate (Gordon, 1993). Facultative thermogenesis refers to stimulated production of the energy required when T_a deviates below or, to some extent, above the thermoneutral zone. Facultative thermogenesis comprises short-term shivering thermogenesis (ST) and a long-term mechanism – non-shivering thermogenesis (NST).

In birds, NST occurs in the skeletal muscles (Duchamp and Barré, 1993; Dridi et al., 2008). The ambient temperature influences the rate of metabolic activity and, in turn, the amount of oxygen required by the bird (Freeman, 1964; Kühn et al., 1984; Buyse et al., 1999). The metabolic rate represents the free energy produced by the transformation of chemical energy during aerobic and anaerobic metabolic activities within the organism. This energy is obtained through the oxidation of feedstuffs; therefore, oxygen consumption can be used as a measure of energy metabolism.

2.3 Heat loss

In birds, heat is dissipated through respiratory–evaporative mechanisms (Richards, 1968, 1970, 1976; Seymour, 1972; Marder and Arad, 1989), a cutaneous evaporative mechanism (Webster and King, 1987; Ophir et al., 2002) and sensible heat loss (SHL) via radiation, convection (Yahav et al., 2005) and conduction (Wolfenson et al., 2001). Evaporative heat loss via panting is associated with loss of body water content, and excessive water loss will induce dehydration, followed by reduction of heat loss via this pathway. Increased SHL may enhance thermotolerance at high T_a. The difference between the surface and ambient temperatures is the main driving force for SHL.

Collectively, the ability to balance heat production and heat loss is the main challenge that domesticated fowl faces under varied environmental conditions, in the light of the fact that high performance contradicts thermotolerance.

3 Neuronal and endocrine T_b regulation

3.1 Neuronal regulation

The centre for T_b regulation (Fig. 1) is neuroanatomically located in the preoptic/anterior hypothalamus (PO/AH) (Boulant, 1996; DiMicco and Zaretsky, 2007). This area plays a dual role: it monitors local temperature changes and integrates temperature information from the periphery (Hellon and Taylor, 1982; Patapoutain et al., 2003; Wechselberger et al., 2006). Temperature is monitored by temperature-sensitive neurons that change their firing rate in accordance with the hypothalamic temperature (Griffin et al., 2001). These neurons form 40% of the PO/AH neurons; about 75% of them are warmth responsive – their firing rate increases or decreases in parallel with the hypothalamus temperature – and about 25% of them are cold responsive, and their firing rate changes in the opposite sense to the hypothalamus temperature (Kelso and Boulant, 1982; Dean and Boulant 1989; Tzschentke and Basta, 2002; Tzschentke et al., 2004). Changing the sensitivity of the warm- and/or cold-responsive neurons located in the PO/AH may change the threshold(s) for heat production and/or heat loss in the animal.

3.2 Endocrine regulation

The thermoregulatory response is mediated mainly by the level of metabolism induced or permitted by the hormonal axis. The metabolic rate is associated with the secretion of the hypothalamic–pituitary–thyroid gland axis. In birds, the hypothalamus produces thyrotropin-releasing hormone and corticotropin-releasing hormone (CRH), which have stimulatory effects, and somatostatin, which has an inhibitory effect on the production of thyroid stimulating hormone (TSH) in the anterior pituitary. TSH is the major controller of

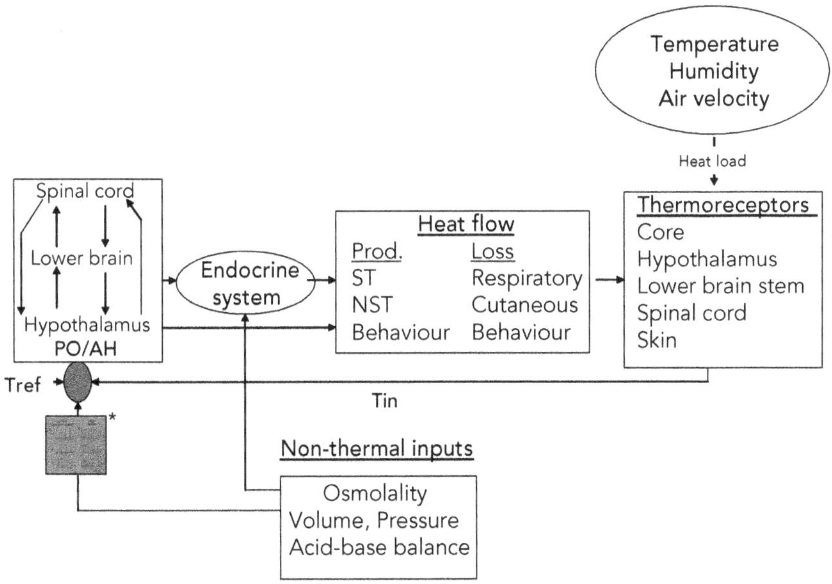

Figure 1 Schematic flow chart describing thermoreceptors' transformation of environmental heat load to thermal information and its transfer (Tin) to the PO/AH to elicit the animal's thermoregulatory responses (heat production and heat loss), mediated by the endocrine system. Non-thermal inputs from the blood system can modify the response. Tref = the set point temperature in the PO/AH; *Schematic figure of sensitive and non-sensitive neurons in the PO/AH (Boulant, 1996) (according to Uni and Yahav, 2008).

the production and release of thyroid hormones by the thyroid gland. Thyroxin (T_4), which is the main hormone secreted by the thyroid gland, is deiodinated to triiodothyronine (T_3) in all peripheral tissues. T_4 and T_3 exert a negative feedback on the pituitary and hypothalamus. T_3 is the main metabolism-stimulating hormone (McNabb and King, 1993; Gabarrou et al., 1997) that was found to be associated with temperature regulation (McNabb and King, 1993; Carew et al., 1998; Gonzales et al., 1999; Yahav, 2000a, 2014b).

The close association between energy and water balance – a non-thermal input – which is mediated by arginine vasotocin (AVT) also plays an important role in thermoregulation. The tight association between plasma osmolality and AVT was documented previously by Saito and Grossmann (1998), Yahav et al. (2004b) and Seth et al. (2004) and the association of dehydration – which induces AVT secretion – with hyperthermia was reported by Yahav (2014b).

4 Different strategies to cope with the environment

Domestic fowl, especially the highly productive species such as broiler chickens, turkeys and laying hens, are able to maintain their posthatch energy balance (represented by T_b) within a narrow range. Exposure to extreme environmental conditions, which dramatically affect the ability to control T_b within the normothermia range, may lead to a cascade of irreversible thermoregulatory events that could be lethal for the bird. These irreversible changes have been found particularly important for domestic fowls because of the improvement made in the last decade in the genetic selection of these birds. This caused difficulties in coping with environmental changes, especially with the acute one that results from their relatively low capability to balance energy expenditure and maintain body water balance.

Like other endotherms, domestic fowls can maintain thermal tolerance and avoid the deleterious consequences of thermal stresses by means of three direct responses: the rapid thermal stress response – RTSR (Parsell and Lindquist, 1994; Yahav et al., 1997a); acclimation/acclimatization (Yahav et al., 1997b) and embryonic or posthatch thermal manipulations (TMs), based on epigenetic adaptation during the perinatal period (Yahav and Hurwitz, 1996; Nichelmann et al., 1999; Tzschentke et al., 2001; Janke et al., 2002; Tzschentke and Basta, 2002; Piestun et al., 2008a,b).

The first two – RTSR and acclimation – depend on exposure duration, whereas TMs depend on timing. The RTSR is an immediate response to thermal stress exposure; it initiates the recruitment of bodily resources to maintain the animal in a dynamic steady state. This response cannot be maintained for a long period; therefore, under continuous exposure to severe environmental conditions, a cascade of events may lead to thermal stress (hyper- or hypothermia) followed by thermal stroke.

Acclimation/acclimatization is defined as physiological or behavioural changes occurring within the lifetime of an organism, which reduce the strain or enhance tolerance of the strain caused by experimentally induced stressful changes, in particular, climatic factors (IUPS Thermal Commission, 2001). It is initiated in response to persistent or repeated challenge of the thermoregulatory system by environmental temperatures and has a long-term effect.

Acclimation/acclimatization involves an array of autonomically controlled physiological mechanisms that work in concert to enhance heat or cold tolerance (Horowitz, 2002). Thermal acclimation/acclimatization can be characterized as an expansion of the dynamic thermoregulatory range of the bird. Basically, it involves the expansion of the body core temperature safety margins – alteration of heat loss and heat production – which is achieved through concerted procedures at all levels of the body. It induces a shift in the level of homeostasis so as to efficiently improve thermotolerance in the new environmental conditions. Inducing shifts in the level of homeostasis is based on the plasticity of the thermoregulatory control elements located in the PO/AH (Boulant, 1996; DiMicco and Zaretsky, 2007).

In domestic fowls, acclimation/acclimatization will take 4–7 days to efficiently control T_b under changed environmental conditions (see Yahav et al., 1997b; Yahav, 1999; and Yahav et al., 2000 for discussion of broilers, turkeys and laying hens, respectively). However, changes in the cardiovascular system can be detected only after approximately 2–3 weeks in broilers (Yahav et al., 1997b) and turkeys (Yahav, 1999).

On the other hand, the TM response is dependent on the critical time-segment of embryogenesis or posthatch, in which changes in the environmental conditions during 'critical developmental phases' occur (Dörner, 1974). It can induce a long-lasting physiological memory based on a strong influence on the determination of the 'set-point' for physiological control systems. Epigenetic temperature adaptation affects gene expression (Nichelmann and Tzschentke, 2002; Tzschentke and Basta, 2002; Tzschentke et al., 2004; Tzschentke and Plagemann, 2006) and seems to offer a suitable means of reaching the goal of improved acquisition of thermotolerance.

5 Physiological and cellular responses to changes in the environment

5.1 Physiological responses

The thermoregulatory system in homeotherms operates at a very high gain, in order to control T_b within a relatively narrow range. During heat exposure, the thermoregulatory system relies on the cardiovascular system to redistribute the blood flow in such a way that tissues and organs important for heat dissipation will receive an augmented supply. In birds, heat is dissipated through respiratory–evaporative mechanisms (Richards, 1968, 1970, 1976; Seymour, 1972; Marder and Arad, 1989), an evaporative cutaneous mechanism (Webster and King, 1987; Ophir et al., 2002), and SHL via radiation, convection (Yahav et al., 2005) and conduction (Wolfenson et al., 2001).

Evaporative heat loss via panting is associated with loss of body water content; therefore, dehydration will reduce heat loss via this pathway, as well as through the extensive passive cutaneous evaporation. Panting is also associated with respiratory alkalosis, which may affect the evaporation (Yahav et al., 1995). Both dehydration and respiratory alkalosis, which are characterized by elevated arterial blood *p*H and reduced arterial blood pCO_2, contribute to significant decline in poultry production. Therefore, an increase in SHL may reduce the intensity of evaporative heat loss, which will reduce the development of respiratory alkalosis and thereby contribute to better acquisition of thermotolerance at high T_a.

One of the immediate responses to heat stress and, to a lesser extent, to chronic heat is the development of hyperthermia (Etches et al., 1995), which initiates redistribution of the blood flow so as to increase the flow to the skin, especially to the non-feathered areas of chickens (Wolfenson et al., 1981; Wolfenson, 1986; Yahav et al., 1998b, 2004b, 2005) and to the upper respiratory passageways (Wolfenson et al., 1981), in order to transport heat from the viscera to the periphery. Subsequently, the elevation in skin temperature enables more efficient heat loss by radiation, convection and conduction as a result of enhancement of the main driving force for this kind of heat loss, that is, the gradient between surface temperature and T_a.

During acute heat exposure, chickens suffer from dehydration as a result of the high rates of panting (Etches et al., 1995) and also of passive water loss from the cutaneous surfaces, which depends on the ventilation rate (VR) (Yahav et al., 2005). Such dehydration may dramatically affect the functioning of the blood system, leading to heat stroke. During the first stages of acute heat stress, that is, before the development of dehydration, there is no impairment of the blood supply to other important tissues, although the supply to the skin and the upper respiratory passageways is elevated. This is mainly because of the increase in cardiac output that results from the increased heart rate and stroke volume, and the redistribution of blood flow among tissues, that is, relatively away from non-vital ones.

The increase in heart rate and the resulting enhanced venous return flow support an overall increase in cardiac output, which refills the arterial pressure reservoir more rapidly, and thereby prevents further reduction in arterial pressure (Whittow et al., 1964; Sturkie, 1967; Darre and Harrison, 1987; Zhou, 2000). However, as dehydration sets in, depletion of the blood volume causes a reduction in venous pressure which, in turn, diminishes blood flow to the skin and the upper respiratory tract, thereby affecting the efficacy of the heat-dissipation routes. This will lead to a severe hyperthermia that may reach 46 °C in domestic fowl, prior to death (Yahav, unpublished data). Severe hyperthermia is accompanied by detrimental changes at the cellular and molecular levels, which lead to the fatal cascade of events characterized by decreased blood pressure, brain hypoxia, neuronal dysfunction, cell fatigue, etc. (Hales et al., 1996).

An additional combined mechanism involves the modulation of heat production – thermogenesis – which is regulated by the thyroid hormones and the neuronal sympathetic system (Silva, 2006), and which will decline to an extent related to the duration and the level of heat stress, in order to prevent excessive accumulation of heat in the body. The metabolic rate and, therefore, the amount of oxygen required by the bird are related to the ambient temperature (Freeman, 1964; Kühn et al., 1984; Buys et al., 1999). The metabolic rate is associated with the secretion of the hormone T_4, which is secreted by the thyroid gland and is deiodinated to T_3 in all tissues. Under acute heat stress, the main – and immediately activated – mechanism to reduce production of T_3 and, consequently, of heat involves diminution of deiodination of peripheral T_4 to T_3.

Collectively, an immediate increase in heat loss by evaporation and SHL, coupled with changes in the cardiovascular system and a significant reduction in heat production, enables the chicken to cope with exposure to heat.

On the other hand, the negative effect of acute cold exposure in the adult domestic fowl (but not in the young one, Shinder et al., 2007) is limited and is likely to cause a reduction in T_b that would be limited to the lower range of normothermia. The factor that limits changes in T_b is the massive insulation provided by skin fat deposits and feather coverage, together with the small surface-to-volume ratio. This response is completely different from that in juveniles, in which lack of fat deposits and high surface-to-volume ratio predominate over the highly effective insulation provided by plumage coverage, leading to major difficulties in controlling T_b under acute cold exposure.

To withstand cold exposure, the bird minimizes heat loss by restricting blood flow to the skin by means of vasoconstriction that is initiated via the sympathetic nervous system. This occurs in parallel with redistribution of the blood flow to the viscera, in order to enhance heat production. In order to compensate for heat loss which, under extremely cold conditions may predominate over heat production, both ST (Hillman et al., 2005) and NST are employed (Duchamp and Barre, 1993). The latter is based on the involvement of avUCP in the skeletal muscle of chickens (Duchamp and Barre, 1993) and ducklings (Raimbault et al., 2001).

Collectively, an immediate decline in heat loss, coupled with changes in the cardiovascular system and a significant increase in heat production, enables poultry to cope with acute exposure to cold.

5.2 Cellular responses

The heat shock response at the cellular level is a coordinated genetic response to a wide range of physiological and environmental stressors; it culminates in the induction of genes that maintain cellular homeostasis and prevent damage associated with the stress.

The cellular response to heat stress basically can be divided into two phases. The first phase protects the cells in the short term, for up to approximately 1 h, and is characterized by the activation of pathways that involve the adenosine receptor, mitochondrial $_{K}$ATP-dependent channels and various kinases (Bogin et al., 1996, 1997; Hausenloy and Yellon, 2006).

The second phase is delayed; it is characterized by a long-lasting response that involves expression of the heat-shock genes and their encoded heat-shock proteins (HSPs) (Jaattela and Wissing, 1992; Parsell and Lindquist, 1994). Members of the superfamily of HSPs are among the most conserved proteins known in phylogeny, with respect to both function and structure (reviewed by Lindquist and Craig, 1988; Jaattela and Wissing, 1992; Welch, 1993). These proteins act as molecular chaperons by binding to other cellular proteins, assisting intracellular transport and folding, preventing protein denaturation or helping the cell to cope with denaturated proteins, preventing aggregation of these denatured proteins and facilitating their renaturation (Feige and Polla, 1994). The various families of HSPs (HSP 110, 90, 70, 60, 47 and small HSPs ranging from 16 to 40 kDa) include both proteins that are present prior to various stress treatments but whose synthesis is enhanced by them (constitutive HSPs) and proteins whose synthesis becomes detectable only after stress (inducible HSPs). The various families exhibit differing functions, and their responses are initiated by the activation and/or de-repression of the transcription factor HSF that binds to the heat-shock element in most of the genes expressing the HSPs (Pirkkala et al., 2001). In avian cells, the stress-induced transcription of heat shock is mediated by HSF1 and HSF3 (Nakai, 1999; Shabtay and Arad, 2006). The thermal stimulus that triggers the HSP response seems to be the perception of thermal sensation at the membrane level.

The HSP70 family is among the most prominent of heat-induced proteins (Parsell and Lindquist, 1994), and HSP70 was recognized for its prominent cytoprotective function (Volloch and Rits, 1999). It was further recognized as one of the major chaperones of the cell; it plays an important role in guiding the conformational status of the proteins during folding and translocation (Arya et al., 2007). Heat stress at the cellular level, which coincided with overexpression of HSP70 by as much as eightfold (Taylor et al., 1999), may be a factor responsible for the survival of the cells (Calderwood and Ciocca, 2008; Chakraborty et al., 2008). It is well established that there is a close correlation between the induction of these proteins and increased thermotolerance (Li and Laszlo, 1985; Wang and Edens, 1998; Yahav et al., 1997a).

Yu et al. (2008) exposed male broiler chickens to acute heat stress of 37 °C and monitored the expression of HSP70 and other HSPs. A significant increase in mRNA of HSP70 and HSP90 was measured in the chicken's heart after 2 h of heat exposure, emphasizing the protective response to stress-induced myocardial injury. Furthermore, HSP70 was located in the blood vessel walls and vascular endothelial cells of the broiler's heart, indicating the importance of the blood vessels and endothelial cells in the response to myocardial cell damage, and suggesting an association between HSP70 concentration in the blood vessels and the constriction functions that are activated after acute heat stress.

The activities of the HSP90 family are also considered to be of high importance. This family is among the most abundant cytosolic proteins, accounting for approximately 1–2% of all soluble proteins (Welch and Feramisco, 1982; Lanneau et al., 2007). HSP90 can bind to steroid receptors to form a complex that protects steroid hormone binding activity against inactivation (Joab et al., 1984; Jibard et al., 1999; Sceibel and Buchner, 1997). Thus, HSP90 may modulate the cellular response to steroid hormones. Like HSP70,

HSP90 is strongly expressed in endothelial cells of the cardiovascular system; it appears to regulate endothelial nitric oxide synthase (eNOS) by binding directly to it (Harris et al., 2000; Brouet et al., 2001; Fontana et al., 2002). In the cardiovascular system, eNOS is the primary source of nitric oxide, and is, therefore, a key regulator of systemic blood pressure, blood vessel proliferation and vascular lesion formation. It was found that prior heat shock resulted in increased expression of HSP90 in vascular endothelial with some variations that depended on the vascular bed, passage and degree of heat shock (Ketis et al., 1988).

A study in broiler chickens exposed to high temperature (Lei et al., 2009) found the induction of HSP90 mRNA and of its protein at an early stressing stage, indicating a fast cell-protection response. This was further expressed in endothelial cells and blood vessel walls, suggesting a relationship between vasomotor response and HSP90 localization. As for the heart itself, Richard et al. (1995) showed that myocytes that constitutively expressed high levels of HSP90 had significantly higher survival capability under imposed heat stress.

In a study to improve broilers' thermotolerance, Yahav and Hurwitz (1996) showed that thermal conditioning at the age of three or five days improved acquisition of thermotolerance in broilers exposed to heat stress at six weeks of age. The improved thermotolerance was characterized by the development of relatively mild hyperthermia, which suggested better ability to maintain T_b than that of the control chickens, and was coupled with lower HSP70, HSP90 and HSP27 production in the heart and lungs of the mildly hyperthermic birds (Yahav et al., 1997a). The association between T_b and *in ovo* production of HSPs in broilers demonstrated significantly greater production of HSP70, HSP90 and HSP27 in the lungs of severely hyperthermic broilers (T_b = 46.4 °C) than in those of mildly hyperthermic ones (T_b = 43.1 °C). Furthermore, Friedman-Einat et al. (1996) found a new member of the small-HSP family, with an apparent molecular weight of 29 kDa, in six-week-old heat-stressed broilers. This protein appeared after 3 h of *in vivo* heat stress, suggesting that it has a specific role as a 'second-stage defense protein'.

Together, the duration and the severity of heat stress determine the HSP response, and the outcome for the cell can be either survival or death.

It can be concluded that the physiological and cellular array of body responses to acute temperature stress reduces the deleterious consequences of stress for a short period. Inability to cope with temperature stress, mainly with heat stress, will result in heat stroke.

6 Ambient temperature, ventilation and RH: the effects on thermal status and performance

The main environmental parameters affecting poultry performance are ambient temperature (T_a), RH and VR. Whereas each of these parameters has been studied separately, reports on the effect of the three in combination on performance and thermoregulation are scarce. The use of such combinations must take into consideration the following facts:

- The development of genetic selection reduced the range of the thermoneutral zone and shifted it towards lower ambient temperatures.

- A wrong combination of the environmental conditions could induce a chilling effect, followed by hypothermia or heat stress, followed, in turn, by hyperthermia.
- The age of the bird, which affects its volume-to-surface ratio and the relative coverage of plumage or feather.
- The use of appropriate and accurate environmental control facilities in farm houses is of high importance.

Of the three parameters, T_a has the most prominent effects on domestic fowl performance (Charles, 1986; Yahav et al., 1996; May and Lott, 2001; Dozier et al., 2006, 2007; Syafwan et al., 2012). This relates to the effect of T_a on feed intake and on the energy required for maintenance; optimal T_a is required to obtain the best FCR. A large volume of research has addressed the determination of optimal T_a during brooding, but it shifted in association with the selection for growth performance. Nevertheless, the effect of a combination of all three environmental parameters on the performance of broilers, but especially of turkeys and laying hens, is not adequate.

Only a few studies have examined the effect of RH on broilers, and most of them have focused on its effect on performance (Prince et al., 1965; Winn and Godfrey 1967; Adams and Rogler 1968; Canton et al. 1983; Yahav et al., 1995; Yahav, 2000b). Yahav et al. (1995) found that the effects of acclimation to high T_a at various RH levels led to maximal growth rate and feed intake in male and female broiler chickens at RH 60–65%, and a similar pattern was observed in broilers exposed to T_a of 30 or 28 °C at the same RH values. In turkeys, the response of performance to RH had been found to be age dependent (Yahav et al., 1995, 1998a; Yahav, 2000a). At T_a of 35 °C and RH ranging from 40 to 75%, maximal growth rate and feed intake of male turkeys aged 4–8 weeks were obtained at RH of 40–45%, but at 30 °C no effect of RH on performance was observed. Turkeys aged 10–19 weeks, exposed to 35 °C showed a bell-shaped pattern of response to RH, with maximum body weight and feed intake at RH 70–75%. Laying hens, however, did not exhibit any significant response to changes in RH combined with high ambient temperature (Yahav et al., 2000).

Ventilation is a powerful determinant for the major component of SHL, that is, convection. It has been assumed that SHL does not play an important role in domestic fowls when T_a is above the upper limit of the thermoneutral zone (for review, see Hillman et al., 1985) and, therefore, that its influence on T_b could be neglected. This assumption was based on (a) the small differences between the body surface temperature (T_s) and T_a and (b) the fact that in fully feathered birds only limited areas, that is, legs, head, wattle and comb, are unfeathered, so that convection at high T_a could be neglected (Tzschentke et al., 1996). However, SHL has been reported to be important in desert mammals (Mohler and Heath, 1988; Klir et al., 1990; Klir and Heath, 1992) and in domestic fowls; in fact, several studies demonstrated the effect of ventilation on performance (Drury, 1966; Wathes and Clark, 1981; Mitchell, 1985; Lacy and Czarick, 1992; Timmons and Hillman, 1993; Phillips and Sanborn, 1994; Tzschentke et al., 1996; Simmons et al., 1997, 2003; Lott et al., 1998; Yahav et al., 2004, 2005; May et al., 2000; Czarick et al., 2000; Tzschentke and Nichelmann, 2000).

As already mentioned, studies on the effects of combinations of environmental conditions on thermal status and performance of domestic fowls are scarce. Findings of such studies on broilers, turkeys and laying hens are summarized in Tables 1–3. In broilers and turkeys at 3–6 weeks of age, the exposure was for 2 weeks, after 1 week of acclimation, whereas in laying hens during the period of peak production – 34–38 weeks of age – it was for 3 weeks, after 1 week of acclimation to environmental conditions.

6.1 Broilers' thermal status and performance under combinations of environmental conditions

The effects of exposure of broilers to combinations of environmental conditions are summarized in Table 1 and Fig. 1. It seems that by changing T_a, varying the VR and keeping RH constant, thermoregulatory and performance responses can be altered. Maximal performance was recorded at VRs of 1.5–2.0 m/s with T_a held at 35 °C. This performance was accompanied by elevated feed intake but with no significant differences in FCR. The best performance was accompanied by the lowest T_b. Although the broilers developed hyperthermia at this T_a, it was relatively mild compared to that observed at the lowest and highest VRs. It must be noted that FCR was relatively high, which means that under this high T_a a significant amount of energy was transferred towards maintenance.

Table 1 Effects of ventilation rate at different ambient temperatures and optimal RH on T_b, BW, feed intake and FCR of six-week-old broilers

	Ventilation rate (m/s)			
	0.5	1.5	2.0	2.5
	Broiler exposed to 35 °C			
T_b (°C)	43.8 ± 0.08[a]	42.7 ± 0.06[c]	42.8 ± 0.04[c]	43.3 ± 0.10[b]
BW (g/42 days)	1952 ± 12[c]	2071 ± 12[a]	2082 ± 19[a]	2047 ± 16[b]
Feed intake (g/28–42 days)	1546 ± 20[b]	1637 ± 33[a]	1670 ± 27[a]	1609 ± 23[ab]
FCR (g/g)	2.29 ± 0.37	2.06 ± 0.36	2.07 ± 0.71	2.09 ± 0.69
	Broiler exposed to 30 °C			
T_b (°C)	40.5 ± 0.10[b]	40.4 ± 0.06[b]	40.9 ± 0.09[a]	41.0 ± 0.09[a]
BW (g/42 days)	2145 ± 19[b]	2189 ± 17[ab]	2177 ± 21[ab]	2249 ± 24[a]
Feed intake (g/28–42 days)	1750 ± 17[b]	1803 ± 22[ab]	1817 ± 11[ab]	1891 ± 20[a]
FCR (g/g)	1.28 ± 0.10	1.27 ± 0.08	1.30 ± 0.05	1.29 ± 0.04
	Broiler exposed to 25 °C			
T_b (°C)	40.7 ± 0.06[b]	41.1 ± 0.07[a]	41.1 ± 0.06[a]	41.0 ± 0.05[a]
BW (g/42 days)	2507 ± 29[a]	2407 ± 33[b]	2489 ± 31[ab]	2417 ± 28[b]
Feed intake (g/28–42 days)	1302 ± 23	1306 ± 33	1321 ± 20	1260 ± 18
FCR (g/g)	1.24 ± 0.06	1.29 ± 0.03	1.21 ± 0.04	1.29 ± 0.07

[a, b, c]Within rows, values designated by different letters differ significantly ($P \le 0.05$).

Figure 2 Correlation between body weight and body temperature of six-week-old broiler chickens, under various environmental conditions (ambient temperatures, ventilation rates and RH). Each point represents 60 broilers.

At T_a of 30 °C, the maximal performance was obtained at a VR of 2.5 m/s, accompanied again by the highest feed intake, but with no effect whatsoever on FCR. The thermal status of all birds was excellent, with their T_b in the normothermia range. This could account for the significantly lower FCR. However, at 25 °C the best performance was obtained with ventilation at 0.8 m/s (Table 1), with no significant differences among treatments, in feed intake or FCR. The shift of best performance to the lowest VR can be attributed to the need to eliminate effects of chilling. Differences in feed intake with changes in T_a can be attributed to differences in the amount of energy needed for maintenance (mainly for thermoregulation; Yahav, 2014b). Whereas 30 °C is above the thermoneutral zone, meaning energy for maintenance is elevated, 25 °C is within this zone and therefore much less energy is required for maintenance, which leads to a significant drop in feed intake but nevertheless allows more energy for growth.

Under harsh environmental conditions (35 °C), there was a significant and high correlation (R^2 = 0.93) between T_b and body weight, but the main environmental factor was VR. Reduction of T_a to 30 or 25 °C weakened this correlation (Fig. 2), but further reduction in ambient temperature would cause a chilling effect.

6.2 Turkeys' thermal status and performance under combinations of environmental conditions

Turkeys are known to be less sensitive than broilers to changes in environmental conditions (Yahav, 2000). Indeed, in all experiments in which turkeys were exposed to varied T_a levels (20–35 °C) and VRs (0.8–2.5 m/sec), the birds exhibited T_b at the bottom to the middle (40.6–41.5 °C; Table 2) of the normothermia range known for domestic fowls (Yahav, 2014b). The finding that turkeys' T_b was not affected by the highest T_a (35 °C) is related to their having better thermoregulation capability than broiler chickens, which, under the

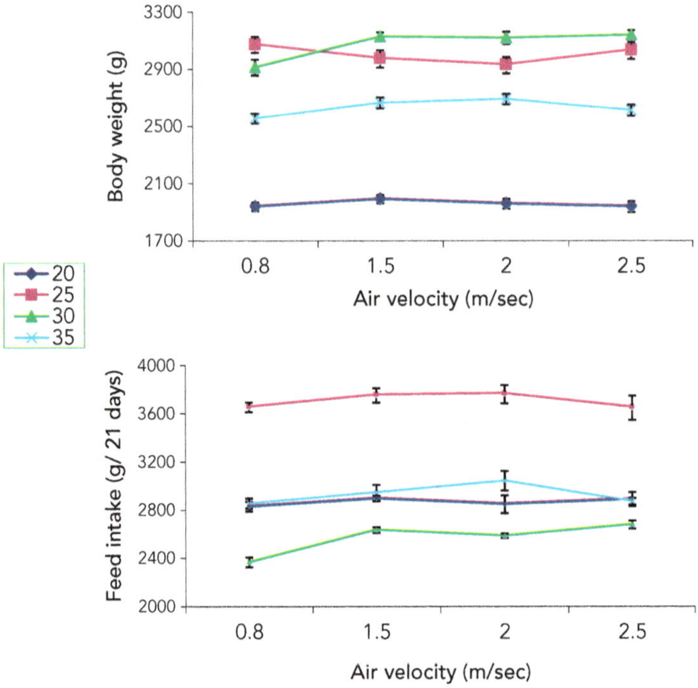

Figure 3 Effects of exposing turkeys aged from three through six weeks to different constant ambient temperatures (20, 25, 30 and 35 °C) coupled with various ventilation rates. Upper graph shows effects on body weight at six weeks of age; lower graph shows effects on cumulative feed intake from three through six weeks.

same environmental conditions, exhibited significantly higher T_b (42.8–43.9 °C) (Yahav et al., 2004). However, there is no doubt that the ability to maintain T_b was significantly related to feed intake. Comparing body weight (BW) and feed intake at different T_a levels demonstrated the strong effect of this parameter on turkeys: at 35 °C, BW was significantly lower than at 30 °C, despite the fact that feed intake was significantly higher (Table 2; Fig. 3).

This finding is closely related to the high energy demand for maintenance at 35 °C, which is manifested in the relatively inefficient FCR. At 25 °C, young turkeys (six weeks old) begin to exhibit difficulties in coping with such a temperature, irrespective of VR. This is indicated by the dramatic increase in feed intake, accompanied by a drop in BW and increase in FCR compared with those at 30 °C. A further decline in T_a elicited a significant decrease in BW as a result of a dramatic reduction in feed intake followed by a dramatic increase in FCR. This may be related to a physical limitation to increases in energy intake as T_a declines (Yahav, 2002).

Thus, overall, despite the fact that at 35 and 30 °C VR had an effect on turkeys' performance, T_a was the major factor affecting their performance.

Table 2 Effects of ventilation rate at different ambient temperatures and optimal RH on T_b, BW, feed intake and FCR of six-week-old turkeys

	Ventilation rate (m/s)			
	0.8	1.5	2.0	2.5
	Turkeys exposed to 35°C			
T_b (°C)	41.5 ± 0.06[a]	41.2 ± 0.01[ab]	40.6 ± 0.09[a]	41.2 ± 0.07[ab]
BW (g/42 days)	2556 ± 35[b]	2666 ± 38[a]	2692 ± 36[a]	2612 ± 40[ab]
Feed intake (g/21–42 days)	2856 ± 41[b]	2948 ± 61[ab]	3043 ± 81[a]	2873 ± 27[ab]
FCR (g/g)	1.61 ± 0.08	1.58 ± 0.10	1.60 ± 0.06	1.58 ± 0.09
	Turkeys exposed to 30°C			
T_b (°C)	40.9 ± 0.12[b]	41.1 ± 0.08[ab]	41.4 ± 0.07[a]	41.3 ± 0.08[a]
BW (g/42 days)	2913 ± 55[b]	3128 ± 31[a]	3119 ± 43[a]	3139 ± 33[a]
Feed intake (g/21–42 days)	2366 ± 40[b]	2634 ± 24[a]	2585 ± 19[a]	2678 ± 34[a]
FCR (g/g)	1.11 ± 0.05	1.13 ± 0.09	1.11 ± 0.06	1.14 ± 0.07
	Turkeys exposed to 25°C			
T_b (°C)	40.6 ± 0.06	40.8 ± 0.04	40.7 ± 0.09	40.7 ± 0.13
BW (g/42 days)	3071 ± 56	2973 ± 60	2928 ± 57	3030 ± 59
Feed intake (g/21–42 days)	3652 ± 40	3752 ± 61	3761 ± 77	3647 ± 101
FCR (g/g)	1.63 ± 0.08[b]	1.77 ± 0.06[a]	1.79 ± 0.10[a]	1.65 ± 0.04[ab]
	Turkeys exposed to 20°C			
T_b (°C)	41.4 ± 0.06	41.0 ± 0.09	41.3 ± 0.10	41.2 ± 0.09
BW (g/42 days)	1939 ± 30	1991 ± 29	1958 ± 34	1938 ± 38
Feed intake (g/21–42 days)	2830 ± 39	2897 ± 22	2849 ± 74	2891 ± 58
FCR (g/g)	2.59 ± 0.12	2.53 ± 0.09	2.55 ± 0.14	2.62 ± 0.10

[a, b]Within rows, values designated by different letters differ significantly ($P \leq 0.05$).

6.3 Laying hens' thermal status and performance under combinations of environmental conditions

For laying hens, VRs ranged between 0.5 and 3.0 m/s but, despite the extended range of VRs, the birds always held T_b within the normothermic range (Table 3). With regard to egg weight: at 35 °C the optimal VR was 2 m/s, but this was associated with a significantly increased feed intake that negatively affected FCR. Therefore, it seems that, although egg weight was significantly lower with ventilation at 1.5 m/s, this could be the optimal VR, in the light of the lower feed intake and the best FCR. At 30 °C, the

Table 3 Effects of ventilation rate at different ambient temperatures and optimal RH on T_b, BW, feed intake and FCR of laying hens at the peak of production

	Ventilation rate (m/s)			
	0.5	1.5	2.0	3.0
	Laying hens exposed to 35 °C			
T_b (°C)	41.6 ± 0.06	41.5 ± 0.09	41.6 ± 0.07	41.3 ± 0.14
Egg weight (g)	55.3 ± 0.59 ^{c}	58.9 ± 0.56^{b}	60.5 ± 0.71^{a}	59.9 ± 0.66^{ab}
Feed intake (g/week)	89.9 ± 3.29^{bc}	87.5 ± 4.07^{c}	105.1 ± 4.22^{a}	96.7 ± 2.66^{ab}
FCR (g/g)	1.62 ± 0.04^{ab}	1.48 ± 0.06^{c}	1.73 ± 0.08^{a}	1.52 ± 0.04^{b}
	Laying hens exposed to 30 °C			
T_b (°C)	41.6 ± 0.14	41.6 ± 0.10	41.6 ± 0.05	41.5 ± 0.11
Egg weight (g)	62.8 ± 0.61	63.41 ± 0.55	63.93 ± 0.59	62.3 ± 0.71
Feed intake (g/week)	104.3 ± 0.71	108.2 ± 3.34	111.6 ± 1.44	108.6 ± 0.96
FCR (g/g)	1.66 ± 0.07^{b}	1.71 ± 0.05^{ab}	1.79 ± 0.10^{a}	1.74 ± 0.08^{ab}
	Laying hens exposed to 25 °C			
T_b (°C)	41.5 ± 0.06	41.4 ± 0.03	41.4 ± 0.11	41.5 ± 0.08
Egg weight (g)	64.0 ± 0.59^{b}	64.6 ± 0.62^{ab}	64.3 ± 0.55^{b}	65.9 ± 0.70^{a}
Feed intake (g/week)	118.9 ± 5.88	114.1 ± 2.03	116.6 ± 1.87	112.6 ± 4.64
FCR (g/g)	1.86 ± 0.08^{a}	1.77 ± 0.04^{b}	1.81 ± 0.09^{ab}	1.71 ± 0.06^{c}
	Laying hens exposed to 20 °C			
T_b (°C)	41.5 ± 0.04	41.5 ± 0.10	41.7 ± 0.08	41.6 ± 0.07
Egg weight (g)	81.7 ± 0.87^{b}	81.9 ± 0.66^{ab}	82.8 ± 0.80^{ab}	83.6 ± 0.72^{a}
Feed intake (g/week)	112.6 ± 4.27	116.0 ± 1.06	112.1 ± 2.18	108.8 ± 2.17
FCR (g/g)	1.38 ± 0.06^{ab}	1.42 ± 0.09^{a}	1.35 ± 0.11^{b}	1.30 ± 0.05^{c}

$^{a, b, c}$Within rows, values designated by different letters differ significantly ($P \leq 0.05$).

same trend continued, but at 25 and 20 °C, the combination of highest egg weight, lowest feed intake and significantly best FCR was obtained at the highest VR of 3.0 m/s.

These results are contrary to the findings for broilers and young turkeys, despite the significantly heavier BW of those two, a finding that reflects surface-to-volume ratios. It would have been expected that at 25 and 20 °C, broilers and young turkeys would prefer higher VRs than laying hens would, but this was not the case. Furthermore, optimal ventilation could have been associated with higher egg production intensity, but no significant differences among treatments were obtained. The only factor that might account for broilers' and turkeys' preference for the lower VR – although laying hens preferred the higher one – is the differences in ages, but the mechanism through which age might affect the response is unknown.

Collectively, it is most important to challenge poultry with combinations of environmental parameters in order to elucidate their effects on performance. Furthermore, thermal status alone will not be an indicator for performance, at least for the strains that are less sensitive to changes in the environment, that is, turkeys and laying hens. However, it is a well-defined indicator for broilers, which are very sensitive to environmental changes.

7 Thermal manipulations during incubation – an epigenetic approach to improving thermotolerance and performance

7.1 The concept

Thermal manipulation (TM) during incubation offers a means to alter environmental conditions during a specific and critical period of development. Although it contrasts with the maintenance of uniform temperature used in commercial incubation, it is in keeping with conditions in nature, where incubation conditions are non-uniform because of the birds' need to search for food, to evade predators and to tolerate non-uniform nest insulation (Webb, 1987). This may be one of the reasons why birds in the wild are quite capable of coping with extreme environmental conditions. However, it must be noted that changes in environmental conditions during incubation cannot and must not be allowed to affect the performance and the quality of the chickens.

The hypotheses underlying TM during embryogenesis are (a) during embryogenesis it is possible to induce long-lasting physiological memory, based on epigenetic adaptation; (b) long-lasting memory can be defined, most probably, as alteration in the hypothalamic threshold response to changes in the environment and (c) TM that involves specific levels and durations of temperature exposure during sensitive periods within embryogenesis will improve thermotolerance during the bird's entire lifespan.

Three critical parameters have to be considered in the application of TM during chick embryogenesis: (1) the timing of the critical phase, (2) the temperature level and (3) the duration of exposure.

Identification of the critical phase of embryogenesis that would enable successful application of TM to improve the acquisition of thermotolerance was based on the hypothesis that the 'set point' or 'response threshold' of controlling systems could be altered most efficiently during the development/maturation of the hypothalamus–hypophyseal–thyroid axis, via thermoregulation, and/or the hypothalamus–hypophyseal–adrenal axis, via application of stress. The necessity to adopt both axes was based on the finding that during embryogenesis the two axes interact: CRH stimulates the secretion of TSH, and corticosterone inhibits the degradation of T_3 to diiodothyronine.

7.2 The effects of TM on thermotolerance

In the light of the development and maturation periods of the thyroid and adrenal axes, the TM treatment adopted embryonic days E7 through E16 as the critical period. During this period, several experiments were done in order to determine the best duration, within a range from 3 to 24 h, and to examine various incubation temperatures, ranging from 37.5 to 41 °C. These experiments indicated the optimal treatment to comprise TM of 39.5 °C at 65% RH, for 12 h per day during days E7 through E16. This was applied to broilers (Piestun et al., 2008a,b), turkeys (Piestun et al., 2015b) and laying hens (Yahav, unpublished data).

Application of TM under the above conditions caused a significant decline in the metabolic rate during the endothermic phase of embryogenesis (from E17 to E18 onwards). This decline was manifested in significant reductions in eggshell temperature, heart rate, oxygen consumption and plasma concentrations of T_4, T_3 and corticosterone (Piestun et al., 2009a). A similar trend was obtained in turkey embryos (Piestun et al., 2015b) and laying hens (Yahav, unpublished data). In broilers, the effect was studied posthatch and up to the marketing age. Thermotolerance acquisition was manifested in a significantly reduced T_b, accompanied by significantly reduced plasma T_4 and T_3 concentrations (Piestun et al., 2008a). Although T_3 is the most potent hormone with regard to the chicks' metabolic rate, the fact that plasma T_4 concentration was also significantly reduced in the TM chicks at hatch and subsequently suggests that the activity of the thyroid gland was reduced. Heat challenge of 35 °C for several hours at marketing age emphasized the improved thermotolerance of the TM chickens.

This was manifested in significantly lower T_b; however, despite the fact that thermal challenge induced hyperthermia in all chickens, the effect on the TM ones was more moderate. Furthermore, the plasma thyroid hormone concentrations were significantly lower in the TM chickens (Piestun et al., 2011). The effects on these two parameters (T_b and hormone concentrations) highlight the reduced metabolic rate. The reduced metabolic rate was associated with significantly reduced plasma corticosterone concentration, suggesting that the chickens were less stressed. It is obvious that if a chicken can maintain a reduced metabolic rate despite heat challenge, it will also exhibit a less pronounced stress response. On the other hand, the TM chickens achieved better thermotolerance not only through reduction of their metabolic rate, but also through enhanced SHL. The improved capacity for sensible heat dissipation – mainly through convection and radiation – at the end of the heat challenge highlights improved vasodilatation capacity (Druyan et al., 2012).

This accumulated evidence shows that the epigenetic adaptation approach and its association with changes in the incubation environment, with emphasis on fine-tuning the level and duration of stress to coincide with the 'critical phase', can elicit efficient epigenetic temperature adaptation in broiler chickens.

7.3 Effects on performance

Improvement of thermotolerance is crucial, especially for domesticated fowls undergoing intensive genetic selection for growth and FCR. Nevertheless, any thermotolerance improvement that coincides with negative effects on performance parameters will not be adopted because of economic losses.

The effect of 12 h of TM from E7 through E16 did not affect hatchability. However, it negatively affected the completion of internal piping, by 3% on average and also induced unhealed navel that was 1.3% higher in the treated chicks (Piestun et al., 2008a). Body weight on hatch was similar in the control and the treated chicks. This similarity in BW continued up to marketing age (Table 4; based on different experiments and environmental conditions: regular conditions up to 35 days of age, hot conditions up to 35 days of age; regular conditions up to 70 days of age). TM chickens could achieve similar body weights to those of the control despite the significant reduction in feed intake that resulted from their lower metabolic rate; or, in other words, as a result of requiring less energy for maintenance and therefore being able to invest similar energy in performance.

In all three different experiments (Table 4), the TM chickens, both males and females, exhibited significantly improved FCR, significantly heavier breast muscle weight (as percentage of BW) and lower to significantly lower abdominal fat pad weight (as percentage of BW), with the exception of females raised to 70 days of age.

Although no significant differences in BW were observed between TM and control chickens of 35 or 70 days of age, under either regular or hot conditions, heavier breast muscle was recorded in the TM birds, suggesting that TM during E7 through E16 promoted breast muscle growth. This was indicated by higher muscle hypertrophy from two weeks posthatch through marketing day (Piestun et al., 2009b). Heavier breast muscle can be related to increased myofibre diameter and/or increased myofibre number, which requires an increased number of muscle nuclei in the tissue (Cheek and Hill, 1970; Allen et al., 1979); therefore, it is reasonable to consider that TM caused increased muscle cell proliferation in the embryo or posthatch (Piestun et al., 2015a).

Table 4 Effects of thermal manipulations during broiler embryogenesis, on BW, FCR and breast muscle and abdominal fat pad relative weights

Treatment and species	Body weight (g)	FCR	Breast muscle (% BW)	Abdominal fat (% BW)
Broiler males (control; 35 d; RC*)	2008	1.72[a]	18.5[b]	1.54[a]
Broiler males (12H; 35 d; RC*)	2006	1.62[b]	19.2[a]	1.17[b]
Broiler females (control; 35 d; RC*)	1756	1.63[a]	17.7[b]	1.99
Broiler females (12H; 35 d; RC*)	1732	1.55[b]	18.9[a]	1.91
Broiler males (control; 35 d; HC**)	1849	1.92[a]	18.0[b]	1.27
Broiler males (12H; 35 d; HC**)	1842	1.83[b]	19.0[a]	1.11
Broiler females (control; 35 d; HC**)	1643	1.74[a]	17.5[b]	1.68[a]
Broiler females (12H; 35 d; HC**)	1629	1.68[b]	18.2[a]	1.36[b]
Broiler males (control; 70 d; RC*)	5812	2.05[a]	22.7[b]	2.11[a]
Broiler males (12H; 70 d; RC*)	5874	1.97[b]	23.4[a]	1.88[b]
Broiler females (control; 70 d; RC*)	4499	2.32[a]	23.3[b]	3.80
Broiler females (12H; 70 d; RC*)	4372	2.22[b]	24.3[a]	3.94

[a, b]Within each column, sex and experiment data with different superscripts differ significantly ($P \leq 0.05$). *RC – regular environmental conditions; **HC – hot conditions: 32/25 °C, 12/12 h, from 21 to 35 days of age (according to Yahav, 2014a).

The negative effect of TM on abdominal fat pad accumulation can be related to increased embryonic movement and associated energy expenditure that resulted from increased incubation temperature. This would reduce adipocyte diameter and size of the fat pad in the embryo (Hammond et al., 2007).

In sum, the accumulating evidence shows that the epigenetic adaptation approach, and its association with changes in the incubation environment, can elicit efficient improvement of several performance parameters of broiler chickens during their lifespan.

8 Conclusions

- The fact that heavy production conflicts with thermotolerance accounts for the difficulties of domestic fowls in withstanding environmental changes. It further highlights the essential association between the environment and the performance and thermal status of domestic fowls.
- Turkeys and laying hens are less sensitive than broilers to changes in environmental conditions.
- For further research, it is most important to challenge poultry with combinations of environmental parameters, in order to elucidate the effects of environmental changes on performance.
- Thermal status alone cannot serve as an indicator for performance, at least not for strains that are less sensitive to changes in the environment, for example, turkeys and laying hens.
- The epigenetic adaptation approach and its association with changes in the incubation environment, with emphasis on fine-tuning the level, timing and duration of stress to coincide with the 'critical phase', can elicit efficient epigenetic temperature adaptation in broiler chickens.
- Accumulating evidence shows that the epigenetic temperature adaptation approach can efficiently elicit improvement of several performance parameters of broiler chickens throughout their lifespan. However, significant research is still required to understand the epigenetic changes in the PO/AH.

9 Where to look for further information

For further information: *Sturkie's Avian Physiology*, Elsevier Publications, London, New-York, Tokyo (Last edition 2014).

10 References

Adams, R. L. and Rogler, J. C. (1968), 'The effects of dietary aspirin and humidity on the performance of light and heavy breed chicks', *Poult. Sci.*, 47, 1344–8.

Allen, R. E., Merkel, R. B. and Young, R. B. (1979), 'Cellular aspects of muscle growth: Myogenic cell proliferation', *J. Anim. Sci.*, 49, 115–27.

Arya, R., Mallik, M. and Lakhotia, S. C. (2007), 'Heat shock gene – integrating cell survival and death', *J. Cell Sci.*, 32, 595–610.

Bogin, E., Peh, H. C., Avidar, Y., Israeli, B. A. and Cahaner, A. (1996), 'The effects of long term high environmental temperature on cellular enzyme activities from different organs', *Euro. J. Clinic. Chem. Clinic. Biochem.*, 34, 625–9.

Bogin, E., Peh, H. C., Avidar, Y., Israeli, B. A., Kevkhaye, E., Lombardi, P. and Cahaner, A. (1997), 'Sex and genotype dependence on the effects of long term high environmental temperatures on cellular enzymes activities from chickens organs', *Avian Pathol.*, 26, 511–24.

Boulant, J. A. (1996), *Environmental Physiology: Hypothalamic Neurons Regulating Body Temperature, Handbook of Physiology: Section 4*, New York: APS, Oxford University Press.

Brouet, A., Sonveaux, P., Dessy, C., Balligand, J. L. and Feron, O. (2001), 'HSP90 ensures the transition from the early Ca^{2+}-dependent to the late phosphorelation-dependent activation of the endothelial nitric oxide synthase (eNOS) in VEGF exposed endothelial cells', *J. Biol. Chem.*, 276, 32663–9.

Buys, J., Scheele, C. W., Kwakernaak, C. and Decuypere, E. (1999), 'Performance and physiological variables in broiler chicken lines differing in susceptibility to the ascites syndrome: 2. Effect of ambient temperature on partial efficiencies of protein and fat retention and plasma hormone concentrations', *Br. Poult. Sci.*, 40, 140–4.

Calderwood, S. K. and Ciocca, D. R. (2008), 'Heat shock proteins: Stress proteins with Janus-like properties in cancer', *Inter. J. Hypertherm.*, 24, 31–9.

Canton, G. H., Buffington, D. E. and Mather, F. B. (1983), 'Evaporative cooling effects on mature, male broiler breeders', *Am. Soci. Agri. Eng.*, 1794.

Carew, L. B., Evarts, K. G. and Alster, F. A. (1998), 'Growth, feed intake, and plasma thyroid hormone levels in chicks fed dietary excesses of essential amino acids', *Poult. Sci.*, 77, 295–8.

Chakraborty, P. K., Mustafi, S. B., Ganguly, S., Chatterjee, M. and Raha, S. (2008), 'Resveratrol induces apoptosis in K562 (chronic myelogenous leukemia) cells by targeting a key survival protein, heat shock protein 70', *Cancer Sci.*, 99, 1109–16.

Charles, D. R. (1986), 'Temperature for broilers', *J. World's Poult. Sci.* 42, 249–58.

Cheek, D. B. and Hill, D. E. (1970), 'Muscle and liver cell growth: Role of hormones and nutritional factors', *Fed. Proc.*, 29, 1503–9.

Czarick, M., Lott, B. and Lacy, M. (2000), 'Poultry housing tips: Is an air speed of 600 ft/min in a tunnel house harmful?', vol. 12, no. 6. The University of Georgia, Cooperative Extension Service.

Darre, M. J. and Harrison, P. C. (1987), 'Heart rate, blood pressure, cardiac output and total peripheral resistance of single comb white leghorn hens during an acute exposure to 35C ambient temperature', *Poult. Sci.*, 66, 541–7.

Dean, J. B. and Boulant, J. A. (1989), 'Effects of synaptic blockade on thermosensitive neurons in rat diecephalon in vitro', *Am. J. Physiol.*, 257, R65–73.

DiMicco, J. A. and Zaretsky, D. V. (2007), 'The dorsomedial hypothalamus: A new player in thermoregulation', *Am. J. Physiol. – Regul. Integr. Comp. Physiol.*, 292, R47–63.

Dörner, G. (1974), 'Environment-dependent brain differentiation and fundamental processes of life', *Acta Biol. Med. Germanica*, 33, 129–48.

Dozier, W. A., Price, C. J., Kidd, M. T., Corzo, A., Anderson, J. and Branton, S. L. (2006), 'Growth performance, meat yield and economic responses of Ross X Ross308 broilers provided diets varying in metabolizable energy from 30 to 59 days of age during low and moderate temperatures', *J. Appl. Poult. Res.*, 15, 367–82.

Dozier, W. A., Purswell, J. L,. Kidd, M. T., Corzo, A. and Branton, S. L. (2007), 'Apparent metabolizable energy needs of broilers from two to four kilograms as influenced by ambient temperature', *J. Appl. Poult. Res.* 16, 206–18.

Dridi, S., Temim, S., Derouet, M., Tesseraud, S. and Taouis, M. (2008), 'Acute cold- and chronic heat-exposure upregulate hepatic leptin and muscle uncoupling protein (UCP) gene expression in broiler chickens', *J. Exp. Zool. – A*, 309, 381–8.

Drury, L. N. (1966), 'Air velocity and broiler growth in diurnally cycled hot environment', *ASAE*, 9, 329–32.

Druyan, S., Piestun, Y. and Yahav, S. (2012), *Heat Stress in Domestic Fowl – Genetic and Physiological Aspects*, New York: NovaScience Publishers Inc.

Duchamp, C. and Barré, H. (1993), 'Skeletal muscle as the major site of non-shivering thermogenesis in cold adapted ducklings', *Am. J. Physiol., Regul. Integr. Comp. Physiol.*, 265, R5469–79.

Etches, R. J., John, T. M. and Verrinder-Gibbins, G. A. M. (1995), *Behavioral, Physiological, Neuroendocrine and Molecular Responses to Heat Stress, Poultry Production in Hot Climate*, Cambridge, UK: CAB International.

Feige, U. and Polla, B. S. (1994), 'HSP70: multi gene, multi structure, multifunction family with potential clinical applications', *Experientia*, 50, 979–86.

Fontana, J., Fulton, D., Chen, Y., Fairchild, T. A., McCab, T. J., Fujita, N., Tsuruo, T. and Sessa, W. C. (2002), 'Domain mapping studies reveal that the M domain of HSP90 serves as molecular scaffold to regulate Akt-depended phosphorelation of endothelial nitric oxide synthase and NO release', *Circ. Res.*, 90, 866–73.

Freeman, B. M. (1964), 'The emergence of the homeothermic-metabolic response in the fowl (*Gallus domesticus*)', *Comp. Biochem. Physiol.*, 13, 413–22.

Friedman-Einat, M., Haberfeld, A., Shamai, A., Horev, G., Hurwitz, S. and Yahav, S. (1996), 'A novel 29-kDa chicken heat shock protein', *Poult. Sci.*, 75, 1528–30.

Gabarrou, J. F., Duchamp, C., Williams, J. and Géraert, P. A. (1997), 'A role for thyroid hormones in the regulation of diet-induced thermogenesis in birds', *Br. J. Nutr.*, 78, 963–73.

Gonzales, E., Buyse, J., Sartori, J. R., Loddi, M. and Decuypere, E. (1999), 'Metabolic disturbances in male broilers of different strains. 2. Relationship between the thyroid and somatotropic axes with growth and mortality', *Poult. Sci.*, 78, 516–21.

Gordon, C. J. (1993), *Temperature Regulation in Laboratory Rodents*, New York: Cambridge University Press.

Griffin, J. D., Saper, C. B. and Boulant, J. A. (2001), 'Synaptic and morphological characteristic of temperature-sensitive and -insensitive rat hypothalamic neurons', *J. Physiol.*, 537, 521–35.

Hales, J. R. S., Hubbard, R. W. and Gaffin, S. L. (1996), *Limitation of Heat Tolerance, Handbook of Physiology, Section 4, Environmental Physiology*, Oxford: Oxford University Press.

Hammond, C. L., Simbi, B. H. and Stickland, N. C. (2007), 'In ovo temperature manipulation influences embryonic motility and growth of limb tissues in the chick (*Gallus gallus*)', *J. Exp. Biol.*, 210, 2667–75.

Harris, M. B., Ju, H., Venema, W. J., Blackstone, M. and Venema, R. C. (2000), 'Role of heat shock 90 in bradykinin-stimulated endothelial nitric oxide release', *Gen. Pharmacol.*, 35, 165–70.

Havenstein, G. B., Ferket, P. R., Scheideler, S. E. and Larson, B. T. (1994), 'Growth, livability, and feed conversion of 1991 versus 1957 broilers when fed "typical" 1957 and 1991 broiler diets', *Poult. Sci.*, 73, 1785–94.

Havenstein, G. B., Ferket, P. R. and Qureshi, M. A. (2003a), 'Growth, livability, and feed conversion of 1957 versus 2001 broilers when fed representative 1957 and 2001 broiler diets', *Poult. Sci.*, 82, 1500–8.

Havenstein, G. B., Ferket, P. R. and Qureshi, M. A. (2003b), 'Carcass composition and yield of 1957 versus 2001 broilers when fed representative 1957 and 2001 broiler diets', *Poult. Sci.*, 82, 1509–18.

Havenstein, G. B., Ferket, P. R., Grimes, J. L., Qureshi, M. A. and Nestor, K. E. (2007), 'Comparison of the performance of 1966- versus 2003-type turkeys when fed representative 1966 and 2003 turkey diet: Growth rate, livability, and feed conversion', *Poult. Sci.*, 86, 232–40.

Hausenloy, D. J. and Yellon, D. M. (2006), 'Survival kinases in ischemic preconditioning and postconditioning', *Cardiovasc. Res.*, 70, 240–53.

Hellon, R. F. and Taylor, D. C. M. (1982), 'An analysis of a thermal afferent pathway in the rat', *J. Physiol.*, 326, 319–28.

Hillman, P. E., Scott, N. R. and Van Tienhoven, A. (1985), *Physiological Responses and Adaptations to Hot and Cold Environments, Stress Physiology in Livestock (vol. 3, Poultry)*, Boca Raton, FL: CRC Press.

Horowitz, M. (2002), 'From molecular and cellular to integrative heat defense during exposure to chronic heat', *Comp. Physiol. Biochem.*, 131, 475–83.

IUPS Thermal Commission (2001), 'Glossary of terms for thermal physiology', *Jap. J. Physiol.*, 51, 245–80.

Jaattela, M. and Wissing, D. (1992), 'Emerging role of heat shock proteins in biology and medicine', *Ann. Med.*, 24, 249–58.

Janke, O., Tzschentke, B., Höchel, J. and Nichelmann, M. (2002), 'Metabolic responses of chicken and muscovy duck embryos to high incubation temperatures', *Comp. Biochem. Physiol.*, 131(A), 741–50.

Janke, O., Tzschentke, B. and Boerjan, B. (2004), 'Comparative investigations of heat production and body temperature in modern chicken breeds', *Avian Poult. Biol. Rev.*, 15, 191–6.

Jibard, N., Meng, X., Leclerc, P., Rajkowski, K., Fortin, D., Schweizer-Groyer, G., Catelli, M. G., Baulieu, E. E. and Cadepond, F. (1999), 'Delimitation of two regions in the 90-kDa heat shock protein (Hsp90) able to interact with the glucocorticosteroid receptor (GR)', *Exp. Cell Res.*, 247, 461–74.

Joab, I., Radanyi, C., Renoir, M., Buchou, T., Catelli, M. G., Binart, N., Mester, J. and Baulieu, E. E. (1984), 'Common non-hormone binding component in non-transformed chick oviduct receptors of four steroid hormones', *Nature*, 308, 850–3.

Kelso, S. R. and Boulant, J. A. (1982), 'Effect of synaptic blockade on thermosensitive neurons in hypothalamic tissue slices', *Am. J. Physiol.*, 243, R480–90.

Klir, J. J. and Heath, J. E. (1992), 'An infrared thermographic study of surface temperature in relation to thermal stress in three species of foxes: The red fox (*Vulpes vulpes*), arctic fox (*Alopex lagopus*), and kit fox (*Vulpes macrotis*)', *Physiol. Zool.*, 65, 1011–21.

Klir, J. J., Heath, J. E. and Benanni, N. (1990), 'An infrared thermographic study of surface temperature in relation to thermal stress in the Mongolian gerbil, *Meriones unguiculatus*', *Comp. Biochem. Physiol.*, 96A, 141–6.

Ketis, N. V., Hoover, R. L. and Karnovsky, M. J. (1988), 'Effects of hyperthermia on cell survival and patterns of protein synthesis in endothelial cells from different origins', *Cancer Res.*, 48, 2101–6.

Kühn, E. R., Decuypere, E. and Rudas, P. (1984), 'Hormonal and environmental interactions on thyroid function in the chick embryo and post-hatching chickens', *J. Exp. Zool.*, 232, 653–8.

Lacy, M. P. and Czarick, M. (1992), 'Tunnel ventilated broiler houses: Broiler performance and operating cost', *J. Appl. Poult. Res.*, 1, 104–9.

Lanneau, D., de Thonel, A., Maurel, S., Didelot, C. and Garrido, C. (2007), 'Apoptosis versus cell differentiation: Role of heat shock proteins HSP90, HSP70 and HSP27', *Prion*, 1, 53–60.

Lei, L., Yu, J. and Bao, E. (2009), 'Expression of heat shock protein 90 (Hsp90) and transcription of its corresponding mRNA in broilers exposed to high temperature', *Br. Poult. Sci.*, 50, 504–11.

Li, G. C. and Laszlo, A. (1985), *Thermotolerance in Mammalian Cells: A Possible Role for Heat Shock Proteins, Changes in Eukaryotic Gene Expression in Response to Environmental Stress*, Orlando, FL, USA: Academic Press Inc.

Lindquist, S. and Craig, E. A. (1988), 'The heat-shock proteins', *Annu. Rev. Genet.*, 22, 631–77.

Lott, B. D., Simmons, J. D. and May, J. D. (1998), 'Air velocity and high temperature effects on broiler performance', *Poult. Sci.*, 77, 391–3.

Marder, J. and Arad, Z. (1989), 'Panting and acid-base regulation in heat stressed birds', *Comp. Biochem. Physiol.*, 94A, 395–400.

May, J. D. and Lott, B. D. (2001), 'Related weight gain and feed: Gain of male and female broilers to rearing temperature', *Poult. Sci.*, 80, 581–4.

May, J. D., Lott, B. D. and Simmons, J. D. (2000), 'The effect of air velocity on broiler performance and feed and water consumption', *Poult. Sci.*, 179, 1396–1400.

McNabb, F. and King, D. B. (1993), *Thyroid Hormones Effects on Growth Development and Metabolism: The Endocrinology of Growth Development and Metabolism in Vertebrates*, New York: Academic Press.

Mitchell, M. A. (1985), 'Effects of air velocity on convective and radiant heat transfer from domestic fowls at environmental temperatures of 20 and 30°C', *Br. Poult. Sci.*, 26, 413–23.

Mohler, F. S. and Heath, J. E. (1988), 'Comparison of IR thermography and thermocouple measurement of heat loss from rabbit pinna', *Am. J. Physiol.*, 254, 389–95.

Morrison, S. F., Nakamura, K. and Madden, C. J. (2008), 'Central control of thermogenesis in mammals', *J. Exp. Physiol.*, 93, 773–97.
Nakai, A. (1999), 'New aspects in the vertebrate heat shock factor system: Hsf3 and Hsf4', *Cell Stress Chaperones*, 4, 86–93.
Nichelmann, M. and Tzschentke, B. (2002), 'Ontogeny of thermoregulation in precocial birds', *Comp. Biochem. Physiol.*, 131(A), 751–63.
Nichelmann, M., Höchel, J. and Tzschentke, B. (1999), 'Biological rhythms in birds – development, insights and perspectives', *Comp. Biochem. Physiol.*, 124A, 437–9.
Ophir, E., Arieli, Y., Marder, J. and Horowitz, M. (2002), 'Cutaneous blood flow in the pigeon *Columba livia*: Its possible relevance to cutaneous water evaporation', *J. Exp. Biol.*, 205, 2627–36.
Parsell, D. A. and Lindquist, S. (1994), *Heat Shock Proteins and Stress Tolerance, Biology of Heat Shock Proteins and Molecular Chaperones*, Cold Spring Harbor, NY: Cold Spring Harbor Laboratory Press.
Patapoutain, A., Peier, A. M., Story, G. M. and Viswanath, V. (2003), 'Thermo TRP channels and beyond: mechanisms of temperature sensation', *Nat. Rev. Neurosci.*, 4, 529–39.
Phillips, P. K. and Sanborn, A. F. (1994), 'An infrared, thermographic study of surface temperature in three ratites: Ostrich, emu and double-wattled cassowary', *J. Therm. Biol.*, 19, 423–30.
Piestun, Y., Shinder, D., Ruzal, M., Halevy, O., Brake, J. and Yahav, S. (2008a), 'Thermal manipulations during broiler's embryogenesis – its effect on the improvement of thermotolerance acquisition', *Poult. Sci.*, 87, 1516–25.
Piestun, Y., Shinder, D., Halevy, O. and Yahav, S. (2008b), 'The effect of thermal manipulations during development of the thyroid and adrenal axes on in-hatch and post-hatch thermoregulation', *J. Therm. Biol.*, 33, 413–18.
Piestun, Y., Halevy, O. and Yahav, S. (2009a), 'Thermal manipulations of broiler embryos – the effect on thermoregulation and development during embryogenesis', *Poult. Sci.*, 88, 2677–88.
Piestun, Y., Halevy, O. and Yahav, S. (2009b), 'Thermal manipulations of broiler embryos – the effect on thermoregulation and development during embryogenesis', *Poult. Sci.*, 88, 2677–88.
Piestun, Y., Halevy, O., Shinder, D., Ruzal, M., Druyan, S. and Yahav, S. (2011), 'Thermal manipulations during broiler embryogenesis improves post-hatch performance under hot conditions', *J. Therm. Biol.*, 36, 469–74.
Piestun, Y., Yahav, S. and Halevy, O. (2015a), 'Thermal manipulation during embryogenesis affects myoblast proliferation and skeletal muscle growth in meat-type chickens', *Poult. Sci.*, 94, 2528–36.
Piestun, Y., Zimmerman, I. and Yahav, S. (2015b), 'Thermal manipulations of turkey embryos – the effect on thermoregulation and development during embryogenesis', *Poult. Sci.*, 94, 273–80.
Pirkkala, L., Nykanen, P. and Sistonen, L. (2001), 'Roles of the heat shock transcription factors in regulation of the heat shock response and beyond', *Faseb J.*, 15, 1118–31.
Prince, R. R., Whitaker, J. H., Matterson, L. D. and Luginbuhl, R. E. (1965), 'Response of chickens to temperature and relative humidity environments', *Poult. Sci.*, 44, 73–7.
Raimbault, S., Dridi, S., Denjean, F., Lachuer, J., Couplan, E., Bouillaud, F., Bordas, A., Duchamp, C., Taouis, M. and Ricquier, D. (2001), 'An uncoupling protein homologue putatively involved in facultative muscle thermogenesis in birds', *Biochem. J.*, 353, 441–4.
Reed, K. M., Mendoza, K. M., Hu, G. R., Sullivan, L. R., Grace, M. W., Chaves, L. D. and Kooyman, D. L. (2007), 'Genomic analysis of genetic markers associated with inherited cardiomyopathy (round heart disease) in the turkey (Meleagris gallopavo)', *Anim. Genet.*, 23, 211–17.
Richard, J., Derek, M. and David, S. L. (1995), 'Differential cytoprotection against heat stress or hypoxia following expression of specific stress protein genes in myogenic cells', *J. Mol. Cell Cardiol.*, 27, 1669–78.
Richards, M. P. and Proszkowiec-Weglarz, M. (2007), 'Mechanisms regulating feed intake, energy expenditure, and body weight in poultry', *Poult. Sci.*, 86, 1478–90.
Richards, S. A. (1968), 'Vagal control of thermal panting in mammals and birds', *J. Physiol.*, 199, 89–101.

Richards, S. A. (1970), 'The biology and comparative physiology of thermal panting', *Biol. Rev.*, 45, 223–64.

Richards, S. A. (1976), 'Evaporative water loss in domestic fowls and its partition in relation to ambient temperature', *J. Agri. Sci.*, 87, 527–32.

Saito, N. and Grossmann, R. (1998), 'Effect of short-term dehydration on plasma osmolality, levels of arginine vasotocin and its hypothalamic gene expression in the laying hen', *Comp. Biochem. Physiol.*, 121A, 235–9.

Sceibel, T. and Buchner, J. (1997), *Protein Aggregation as a Cause for Disease, Guidebook to Molecular Chaperons and Protein-Folding Catalysts*, Oxford: Sambrook and Tooze.

Seth, R., Köhler, A., Grossmann, R. and Chaturvedi, C. M. (2004), 'Expression of hypothalamic arginine vasotocin in response to water deprivation and sex steroid administration in female Japanese quail', *J. Exp. Biol.*, 207, 3025–33.

Seymour, S. R. (1972), 'Convective heat transfer in the respiratory systems of panting animals', *Comp. Biochem. Physiol.*, 35, 119–27.

Shabtay, A. and Arad, Z. (2006), 'Reciprocal activation of HSF1 and HSF3 in brain and blood tissues: Is redundancy developmentally related?', *Am. J. Physiol. Regul. Integr. Comp. Physiol.*, 291, R566–72.

Shinder, D., Rusal, M., Tanny, Y., Druyan, D. and Yahav, S. (2007), 'Thermoregulatory Response of Chicks *(Gallus domesticus)* to Low Ambient Temperatures at an Early Age', *Poult. Sci.*, 86, 2000–9.

Silva, J. E. (2006), 'Thermogenic mechanisms and their hormonal regulation', *Physiol. Rev.*, 86, 435–64.

Simmons, J. D., Lott, B. D. and May, J. D. (1997), 'Heat loss from broiler chickens subjected to various wind speeds and ambient temperatures', *Appl. Eng. Agric.*, 13, 665–9.

Simmons, J. D., Lott, B. D. and Miles, D. M. (2003), 'The effects of high air velocity on broiler performance', *Poult. Sci.*, 82, 232–4.

Sturkie, P. D. (1967), 'Cardiovascular effects of acclimatization to heat and cold in chickens', *J. Appl. Physiol.*, 22, 13–15.

Syafwan, S., Wermink, G. J., Kwakkel, R. P. and Verstegen, N. W. (2012), 'Dietary self- selection by broilers at normal and high temperature changes feed intake, behavior, nutrient intake and performance', *Poult. Sci.*, 91, 537–49.

Taylor, R. P., Harris, M. B. and Starnes, J. W. (1999), 'Acute exercise can improve cardioprotection without increasing heat shock protein', *Am. J. Physiol. Heart Circ. Physiol.*, 276, H1098–102.

Timmons, M. B. and Hillman, P. E. (1993), 'Partitional heat losses in heat-stressed poultry as affected by wind speed', *4th Int. Livestock Environ. Symp.*, ASAE Special Publication, St. Joseph, MI.

Tzschentke, B. and Basta, D. (2002), 'Early development of neuronal hypothalamic thermosensitivity in birds: Influence of epigenetic temperature adaptation', *Comp. Biochem. Physiol.*, 131, 825–32.

Tzschentke, B. and Nichelmann, M. (2000), 'Influence of age and wind speed on total effective ambient temperature in three poultry species (*Gallus domesticus, Cairina moschata, Meleagris gallapavo*)', *Arch. Geflügel*, 64, 1–8.

Tzschentke, B. and Plagemann, A. (2006), 'Imprinting and critical periods in early development', *J. World's Poult. Sci.*, 62, 626–37.

Tzschentke, B., Nichelmann, M. and Postel, T. (1996), 'Effects of ambient temperature, age and wind speed on the thermal balance of layer-strain fowl', *Br. Poult. Sci.*, 37, 501–20.

Tzschentke, B., Basta, D. and Nichelmann, M. (2001), 'Epigenetic temperature adaptation in birds: Peculiarities and similarities in comparison to acclimation', *News Biomed. Sci.*, 1, 26–31.

Tzschentke, B., Basta, D., Janke, O. and Maier, I. (2004), 'Characteristics of early development of body functions and epigenetic adaptation to the environment in poultry: Focused on development of central nervous mechanisms', *Avian Poult. Biol. Rev.*, 15, 107–18.

Uni, Z. and Yahav, S. (2008), *Managing Pre-natal Development of Broiler Chickens to Improve Productivity and Thermotolerance. Managing Prenatal Development to Enhance Livestock Productivity*, Dordrecht-Heidelberg, London, New York: Springer Press.

US National Climatic Center (2001), http://www.ncdc.noaa.gov/monitoring-references/faq/global-warming.php.
Volloch, V. and Rits, S. (1999), 'A natural extracellular factor that induces HSP72, inhibits apoptosis and restores stress resistance in aged human cells', *Exp. Cell Res.*, 253, 483–92.
Wang, S. and Edens, F. W. (1998), 'Heat conditioning induces heat shock proteins in broiler chickens and turkey poults', *Poult. Sci.*, 77, 1636–45.
Wathes, C. M. and Clark, J. A. (1981), 'Sensible heat transfer from the fowl: Radiative and convective heat losses from a flock of broiler chickens', *Br. Poult. Sci.*, 22, 185–96.
Webb, D. R. (1987), 'Thermal tolerance of avian embryos: A review', *Condor*, 89, 874–98.
Webster, M. D. and King, J. R. (1987), 'Temperature and humidity dynamics of cutaneous and respiratory evaporation in pigeons', *Columba livia', J. Comp. Physiol.*, 157, 253–60.
Wechselberger, M., Wright, C. L., Bishop, G. A. and Boulant, J. A. (2006), 'Ionic channels and conductance-based models for hypothalamic neuronal thermosensitivity', *Am. J. Physiol., Regul. Integr. Comp. Physiol.*, 291, R518–29.
Welch, W. J. (1993), 'How cells respond to stress', *Sci. Am.*, 268, 56–64.
Welch, W. J. and Feramisco, J. R. (1982), 'Purification of the major mammalian heat shock proteins', *J. Biol. Chem.*, 257, 14949–59.
Whittow, G. C., Sturkie, P. D. and Stein, G. J. (1964), 'Cardiovascular changes associated with thermal polypnea in the chicken', *Am. J. Physiol.*, 207, 1349–53.
Winn, P. N. and Godfrey, E. F. (1967), 'The effect of humidity on growth and feed conversion of broiler chickens', *Int. J. Biometeorol.*, 11, 39–50.
Wolfenson, D. (1986), 'The effect of acclimatization on blood flow and its distribution in normothermic and hyperthermic domestic fowl', *Comp. Biochem. Physiol. A Comp. Physiol.*, 85, 739–42.
Wolfenson, D., Bachrach, D., Maman, M., Graber, Y. and Rozenboim, I. (2001), 'Evaporative cooling of ventral regions of the skin in heat-stressed laying hens', *Poult. Sci.*, 80, 958–64.
Wolfenson, D., Frei, Y. F., Snapir, N. and Berman, A. (1981), 'Heat stress effects on capillary blood flow and its redistribution in the laying hen', *Pflugers Arch.*, 390, 86–93.
Yahav, S. (1999), 'The effect of constant and diurnal cyclic temperatures on performance and blood system of young turkeys', *J. Therm. Biol.*, 24, 71–8.
Yahav, S. (2000a), 'Domestic fowl – strategies to confront environmental conditions', *Poult. Avian Biol. Rev.*, 11, 81–95.
Yahav, S. (2000b), 'Relative humidity at moderate temperatures, its effect on male broiler chickens and turkeys', *Br. Poult. Sci.*, 41, 94–100.
Yahav, S. (2002), 'Limitation in energy consumption affects the ability of young turkeys to cope with low ambient temperature', *J. Therm. Biol.*, 27, 103–8.
Yahav, S. (2014a), 'Challenges in poultry production during the 21st century', *XIVth Eur. Conference*, Stavanger, Norway.
Yahav, S. (2014b), *Regulation of Body Temperature – Strategies and Mechanisms, Sturkie's Avian Physiology*, London, New York, Tokyo: Elsevier Publications.
Yahav, S. and Hurwitz, S. (1996), 'Induction of thermotolerance in male broiler chickens by temperature conditioning at an early age', *Poult. Sci.*, 75, 402–6.
Yahav, S., Goldfeld, S., Plavnik, I. and Hurwitz, S. (1995), 'Physiological responses of chickens and turkeys to relative humidity during exposure to high ambient temperature', *J. Therm. Biol.*, 20, 245–53.
Yahav, S., Straschnow, A., Plavnik, I. and Hurwitz, S. (1996), 'Effect of diurnal cyclic versus constant temperatures on chicken growth and food intake', *Br. Poult. Sci.*, 37, 43–54.
Yahav, S., Shamay, A., Horev, G., Bar-Ilan, D., Genina, O. and Friedman-Einat, M. (1997a), 'Effect of acquisition of improved thermotolerance on the induction of heat shock proteins in broiler chickens', *Poult. Sci.*, 76, 1428–34.
Yahav, S., Straschnow, A., Plavnik, I. and Hurwitz, S. (1997b), 'Blood system response of chickens to changes in environmental temperature', *Poult. Sci.*, 76, 627–33.

Yahav, S., Plavnik, I., Rusal, M. and Hurwitz, S. (1998a), 'The response of turkeys to relative humidity at high ambient temperature', *Br. Poult. Sci.*, 39, 340–5.

Yahav, S., Luger, D., Cahaner, A., Dotan, M., Ruzal, M. and Hurwitz, S. (1998b), 'Thermoregulation in naked-neck chickens subjected to different ambient temperatures', *Br. Poult. Sci.*, 39, 133–8.

Yahav, S., Shinder, D., Razpakovski, V., Rusal, M. and Bar, A. (2000), 'Lack of response of laying hens to relative humidity at high ambient temperature', *Br. Poult. Sci.*, 41, 660–3.

Yahav, S., Straschnow, A., Luger, D., Shinder, D., Tanny, J. and Cohen, S. (2004), 'Ventilation, sensible heat loss, broiler energy and water balance under harsh environmental conditions', *Poult. Sci.*, 83, 253–8.

Yahav, S., Shinder, D., Tanny, J. and Cohen, S. (2005), 'Sensible heat loss – the broiler's paradox', *J. World's Poult. Sci.*, 61, 419–35.

Yu, J., Bao, E., Yan, J. and Lei, L. (2008), 'Expression and localization of HSPs in the heart and blood vessel of heat-stressed broilers', *Cell Stress Chaperones*, 13, 327–35.

Zhou, W. (2000), 'Physiological significance of the change in blood viscosity of broiler chickens under high ambient temperature', *Jpn. Poult. Sci.*, 37, 201–11.

Zuidhof, M. J., Schneider, B. L., Carney, V. L., Korver, D. R. and Robinson, F. E. (2014), 'Growth, efficiency, and yield of commercial broilers from 1957, 1978, and 2005', *Poult. Sci.*, 93, 2970–82.

Index

Lightning Source UK Ltd.
Milton Keynes UK
UKOW06n0910140917
309119UK00001B/8/P